Lecture Notes on Data Engineering and Communications Technologies   **176**

Series Editor

Fatos Xhafa, *Technical University of Catalonia, Barcelona, Spain*

The aim of the book series is to present cutting edge engineering approaches to data technologies and communications. It will publish latest advances on the engineering task of building and deploying distributed, scalable and reliable data infrastructures and communication systems.

The series will have a prominent applied focus on data technologies and communications with aim to promote the bridging from fundamental research on data science and networking to data engineering and communications that lead to industry products, business knowledge and standardisation.

Indexed by SCOPUS, INSPEC, EI Compendex.

All books published in the series are submitted for consideration in Web of Science.

Leonard Barolli
Editor

Complex, Intelligent and Software Intensive Systems

Proceedings of the 17th International Conference on Complex, Intelligent and Software Intensive Systems (CISIS-2023)

 Springer

Editor
Leonard Barolli
Department of Information and Communication
Engineering
Fukuoka Institute of Technology
Fukuoka, Japan

ISSN 2367-4512   ISSN 2367-4520 (electronic)
Lecture Notes on Data Engineering and Communications Technologies
ISBN 978-3-031-35733-6   ISBN 978-3-031-35734-3 (eBook)
https://doi.org/10.1007/978-3-031-35734-3

This Springer imprint is published by the registered company Springer Nature Switzerland AG
The registered company address is: Gewerbestrasse 11, 6330 Cham, Switzerland

Welcome Message of CISIS-2023 International Conference Organizers

Welcome to the 17th International Conference on Complex, Intelligent and Software Intensive Systems (CISIS-2023), which will be held from July 5 to July 7, 2023, in conjunction with the 17th International Conference on Innovative Mobile and Internet Services in Ubiquitous Computing (IMIS 2023).

The aim of the conference is to deliver a platform of scientific interaction between the three interwoven challenging areas of research and development of future ICT-enabled applications: software intensive systems, complex systems, and intelligent systems.

Software intensive systems are systems, which heavily interact with other systems, sensors, actuators, devices, other software systems, and users. More and more domains are involved with software intensive systems, e.g., automotive, telecommunication systems, embedded systems in general, industrial automation systems, and business applications. Moreover, the outcome of Web services delivers a new platform for enabling software intensive systems. The conference is thus focused on tools, practically relevant, and theoretical foundations for engineering software intensive systems.

Complex systems research is focused on the overall understanding of systems rather than its components. Complex systems are very much characterized by the changing environments in which they act by their multiple internal and external interactions. They evolve and adapt through internal and external dynamic interactions.

The development of intelligent systems and agents, which is each time more characterized by the use of ontologies and their logical foundations build a fruitful impulse for both software intensive systems and complex systems. Recent research in the field of intelligent systems, robotics, neuroscience, artificial intelligence, and cognitive sciences are very important factors for the future development and innovation of software intensive and complex systems.

This conference is aiming at delivering a forum for in-depth scientific discussions among the three communities. The papers included in the proceedings cover all aspects of theory, design, and application of complex systems, intelligent systems, and software intensive systems.

We are very proud and honored to have two distinguished keynote talks by Dr. Salvatore Venticinque, University of Campania "Luigi Vanvitelli", Italy, and Prof. Sanjay Kumar Dhurandher, Netaji Subhas University of Technology, India, who will present their recent work and will give new insights and ideas to the conference participants.

The organization of an International Conference requires the support and help of many people. A lot of people have helped and worked hard to produce a successful technical program and conference proceedings. First, we would like to thank all authors for submitting their papers, the Program Committee Members, and the reviewers who carried out the most difficult work by carefully evaluating the submitted papers. We are grateful to Honorary Chair Prof. Makoto Takizawa, Hosei University, Japan, for his guidance and support.

Finally, we would like to thank Web Administrator Co-Chairs for their excellent and timely work.

We hope you will enjoy the conference proceedings.

CISIS-2023 Organizing Committee

Honorary Chair

Makoto Takizawa — Hosei University, Japan

General Co-chairs

Isaac Woungang — Toronto Metropolitan University, Canada
Tomoya Enokido — Rissho University, Japan

Program Committee Co-chairs

Marek Ogiela — AGH University of Technology, Poland
Naohiro Hayashibara — Kyoto Sangyo University, Japan
Sanjay Kumar Dhurandher — University of Delhi, India

International Advisory Board

David Taniar — Monash University, Australia
Minoru Uehara — Toyo University, Japan
Arjan Durresi — IUPUI, USA
Beniamino Di Martino — University of Campania "L. Vanvitelli", Italy

Award Co-chairs

Keita Matsuo — Fukuoka Institute of Technology, Japan
Kin Fun Li — University of Victoria, Canada
Olivier Terzo — LINKS Foundation, Italy

International Liaison Co-chairs

Wenny Rahayu — La Trobe University, Australia
Markus Aleksy — ABB AG Corporate Research Center, Germany

Flora Amato   University of Naples Federico II, Italy
Omar Hussain   University of New South Wales, Australia

Publicity Co-chairs

Takahiro Uchiya   Nagoya Institute of Technology, Japan
Antonio Esposito   University of Campania "Luigi Vanvitelli", Italy
Farookh Hussain   University of Technology Sydney, Australia

Finance Chair

Makoto Ikeda   Fukuoka Institute of Technology, Japan

Local Arrangement Co-chairs

Mehrdad Tirandazian   Toronto Metropolitan University, Canada
Glaucio Carvalho   Toronto Metropolitan University, Canada

Web Administrator Chairs

Phudit Ampririt   Fukuoka Institute of Technology, Japan
Ermioni Qafzezi   Fukuoka Institute of Technology, Japan

Steering Committee Chair

Leonard Barolli   Fukuoka Institute of Technology, Japan

Track Areas and PC Members

1. Database and Data Mining Applications

Track Co-chairs

Kin Fun Li   University of Victoria, Canada
Pavel Krömer   Technical University of Ostrava, Czech Republic

PC Members

Antonio Attanasio	LINKS Foundation, Italy
Tibebe Beshah	Addis Ababa University, Ethiopia
Jana Heckenbergerova	University of Pardubice, Czech Republic
Konrad Jackowski	Wroclaw University of Technology, Poland
Petr Musílek	University of Alberta, Canada
Aleš Zamuda	University of Maribor, Slovenia
Genoveva Vargas-Solar	French Council of Scientific Research, LIG-LAFMIA, France
Xiaolan Sha	Sky, UK
Kosuke Takano	Kanagawa Institute of Technology, Japan
Masahiro Ito	Toshiba Lab, Japan
Watheq ElKharashi	Ain Shams University, Egypt
Mohamed Elhaddad	University of Victoria, Canada
Wei Lu	Keene State College, USA

2. Artificial Intelligence and Bio-Inspired Computing

Track Co-chairs

Hai Dong	Royal Melbourne Institute of Technology, Australia
Salvatore Vitabile	University of Palermo, Italy
Urszula Ogiela	AGH University of Science and Technology, Poland

PC Members

Kit Yan Chan	Curtin University, Australia
Shang-Pin Ma	National Taiwan Ocean University, Taiwan
Pengcheng Zhang	Hohai University, China
Le Sun	Nanjing University of Information Science and Technology, China
Sajib Mistry	Curtin University, Australia
Carmelo Militello	Italian National Research Council, Italy
Klodiana Goga	LINKS Foundation, Italy
Vincenzo Conti	University of Enna Kore, Italy
Minoru Uehara	Toyo University, Japan
Philip Moore	Lanzhou University, China
Mauro Migliardi	University of Padua, Italy

Dario Bonino	CHILI, Italy
Andrea Tettamanzi	University of Nice, France
Cornelius Weber	Hamburg University, Germany
Tim Niesen	German Research Center for Artificial Intelligence (DFKI), Germany
Rocco Raso	German Research Center for Artificial Intelligence (DFKI), Germany
Fulvio Corno	Politecnico di Torino, Italy

3. Multimedia Systems and Virtual Reality

Track Co-chairs

Yoshinari Nomura	Okayama University, Japan
Francesco Orciuoli	University of Salerno, Italy
Shinji Sugawara	Chiba Institute of Technology, Japan

PC Members

Shunsuke Mihara	Lockon Inc., Japan
Shunsuke Oshima	Kumamoto National College of Technology, Japan
Yuuichi Teranishi	NICT, Japan
Kazunori Ueda	Kochi University of Technology, Japan
Hideaki Yanagisawa	National Institute of Technology, Tokuyama College, Japan
Kaoru Sugita	Fukuoka Institute of Technology, Japan
Keita Matsuo	Fukuoka Institute of Technology, Japan
Santi Caballé	Open University of Catalonia, Spain
Nobuo Funabiki	Okayama University, Japan
Yoshihiro Okada	Kyushu University, Japan
Tomoyuki Ishida	Fukuoka Institute of Technology, Japan
Nicola Capuano	University of Basilicata, Italy
Jordi Conesa	Universitat Oberta de Catalunya, Spain
Farzin Asadi	Kocaeli University, Kocaeli, Turkey
David Gañan	Universitat Oberta de Catalunya, Spain
Le Hoang Son	Vietnam National University, Vietnam
Jorge Miguel	Grupo San Valero, Spain
David Newell	Bournemouth University, UK

4. Next-Generation Wireless Networks

Track Co-chairs

Marek Bolanowski	Rzeszow University of Technology, Poland
Sriram Chellappan	Missouri University of Science and Technology, USA

PC Members

Yunfei Chen	University of Warwick, UK
Elis Kulla	Fukuoka Institute of Technology, Japan
Santi Caballé	Open University of Catalonia, Spain
Admir Barolli	Aleksander Moisiu University of Durres, Albania
Makoto Ikeda	Fukuoka Institute of Technology, Japan
Keita Matsuo	Fukuoka Institute of Technology, Japan
Shinji Sakamoto	Kanazawa Institute of Technology, Japan
Omer Wagar	University of Engineering & Technology, Poland
Zhibin Xie	Jiangsu University of Science and Technology, China
Jun Wang	Nanjing University of Post and Telecommunication, China
Vamsi Paruchuri	University of Central Arkansas, USA
Arjan Durresi	IUPUI, USA
Bhed Bista	Iwate Prefectural University, Japan
Tadeusz Czachórski	Polish Academy of Sciences, Poland
Andrzej Paszkiewicz	Rzeszow University of Technology, Poland

5. Semantic Web and Web Services

Track Co-chairs

Antonio Messina	Italian National Research Center (CNR), Italy
Aneta Poniszewska-Maranda	Lodz University of Technology, Poland
Salvatore Venticinque	University of Campania "Luigi Vanvitelli", Italy

PC Members

Alba Amato	Italian National Research Center (CNR), Italy
Nik Bessis	Edge Hill University, UK

Robert Bestak	Czech Technical University in Prague, Czech Republic
Ivan Demydov	Lviv Polytechnic National University, Ukraine
Marouane El Mabrouk	Abdelmalek Essaadi University, Morocco
Corinna Engelhardt-Nowitzki	University of Applied Sciences, Austria
Michal Gregus	Comenius University in Bratislava, Slovakia
Jozef Juhar	Technical University of Košice, Slovakia
Nikolay Kazantsev	National Research University Higher School of Economics, Russia
Manuele Kirsch Pinheiro	Université Paris 1 Panthéon Sorbonne, France
Cristian Lai	CRS4 Center for Advanced Studies, Research and Development in Sardinia, Italy
Michele Melchiori	Universita' degli Studi di Brescia, Italy
Giovanni Merlino	University of Messina, Italy
Kamal Bashah Nor Shahniza	Universiti Teknologi MARA, Malaysia
Eric Pardede	La Trobe University, Australia
Pethuru Raj	IBM Global Cloud Center of Excellence, India
Jose Luis Vazquez Avila	University of Quintana Roo, México
Anna Derezinska	Warsaw University of Technology, Poland

6. Security and Trusted Computing

Track Co-chairs

Hiroaki Kikuchi	Meiji University, Japan
Jindan Zhang	Xianyang Vocational Technical College, China
Lidia Fotia	University of Calabria, Italy

PC Members

Saqib Ali	Sultan Qaboos University, Oman
Zia Rehman	COMSATS University Islamabad, Pakistan
Morteza Saberi	UNSW Canberra, Australia
Sazia Parvin	UNSW Canberra, Australia
Farookh Hussain	University of Technology Sydney, Australia
Walayat Hussain	University of Technology Sydney, Australia
Sabu Thampi	Indian Institute of Information Technology and Management - Kerala (IIITM-K) Technopark Campus, India
Sun Jingtao	National Institute of Informatics, Japan
Anitta Patience Namanya	University of Bradford, UK

Smita Rai	Uttarakhand Board of Technical Education Roorkee, India
Abhishek Saxena	American Tower Corporation Limited, India
Ilias K. Savvas	University of Thessaly, Greece
Fabrizio Messina	University of Catania, Italy
Domenico Rosaci	University Mediterranea of Reggio Calabria

7. HPC & Cloud Computing Services and Orchestration Tools

Track Co-chairs

Olivier Terzo	LINKS Foundation, Italy
Jan Martinovič	IT4Innovations National Supercomputing Center, VSB Technical University of Ostrava, Czech Republic
Jose Luis Vazquez-Poletti	Universidad Complutense de Madrid, Spain

PC Members

Alberto Scionti	LINKS Foundation, Italy
Antonio Attanasio	LINKS Foundation, Italy
Jan Platos	VŠB-Technical University of Ostrava, Czech Republic
Rustem Dautov	Kazan Federal University, Russia
Giovanni Merlino	University of Messina, Italy
Francesco Longo	University of Messina, Italy
Dario Bruneo	University of Messina, Italy
Nik Bessis	Edge Hill University, UK
Ming Xue Wang	Ericsson, Ireland
Luciano Gaido	Istituto Nazionale di Fisica Nucleare (INFN), Italy
Giacinto Donvito	Istituto Nazionale di Fisica Nucleare (INFN), Italy
Andrea Tosatto	Open-Xchange, Germany
Mario Cannataro	University "Magna Græcia" of Catanzaro, Italy
Agustin C. Caminero	Universidad Nacional de Educación a Distancia, Spain
Dana Petcu	West University of Timisoara, Romania
Marcin Paprzycki	Systems Research Institute, Polish Academy of Sciences, Poland
Rafael Tolosana	Universidad de Zaragoza, Spain

8. Parallel, Distributed, and Multicore Computing

Track Co-chairs

Eduardo Alchieri	University of Brasilia, Brazil
Valentina Casola	University of Naples "Federico II", Italy
Lidia Ogiela	AGH University of Science and Technology, Poland

PC Members

Aldelir Luiz	Catarinense Federal Institute, Brazil
Edson Tavares	Federal University of Technology - Parana, Brazil
Fernando Dotti	Pontificia Universidade Catolica do Rio Grande do Sul, Brazil
Hylson Neto	Catarinense Federal Institute, Brazil
Jacir Bordim	University of Brasilia, Brazil
Lasaro Camargos	Federal University of Uberlandia, Brazil
Luiz Rodrigues	Western Parana State University, Brazil
Marcos Caetano	University of Brasilia, Brazil
Flora Amato	University of Naples "Federico II", Italy
Urszula Ogiela	AGH University of Science and Technology, Poland

9. Energy-Aware Computing and Systems

Track Co-chairs

Muzammil Behzad	University of Oulu, Finland
Zahoor Ali Khan	Higher Colleges of Technology, United Arab Emirates
Shigenari Nakamura	Tokyo Denki University, Japan

PC Members

Naveed Ilyas	Gwangju Institute of Science and Technology, South Korea
Muhammad Sharjeel Javaid	University of Hafr Al Batin, Saudi Arabia
Muhammad Talal Hassan	COMSATS University Islamabad, Pakistan
Waseem Raza	University of Lahore, Pakistan

Ayesha Hussain	COMSATS University Islamabad, Pakistan
Umar Qasim	University of Engineering and Technology, Pakistan
Nadeem Javaid	COMSATS University Islamabad, Pakistan
Yasir Javed	Higher Colleges of Technology, UAE
Kashif Saleem	King Saud University, Saudi Arabia
Hai Wang	Saint Mary's University, Canada

10. Multi-agent Systems, SLA Cloud, and Social Computing

Track Co-chairs

Giuseppe Sarnè	Mediterranean University of Reggio Calabria, Italy
Douglas Macedo	Federal University of Santa Catarina, Brazil
Takahiro Uchiya	Nagoya Institute of Technology, Japan

PC Members

Mario Dantas	Federal University of Juiz de Fora, Brazil
Luiz Bona	Federal University of Parana, Brazil
Márcio Castro	Federal University of Santa Catarina, Brazil
Fabrizio Messina	University of Catania, Italy
Hideyuki Takahashi	Tohoku University, Japan
Kazuto Sasai	Ibaraki University, Japan
Satoru Izumi	Tohoku University, Japan
Domenico Rosaci	Mediterranean University of Reggio Calabria, Italy
Lidia Fotia	Mediterranean University of Reggio Calabria, Italy

11. Internet of Everything and Machine Learning

Track Co-chairs

Omid Ameri Sianaki	Victoria University Sydney, Australia
Khandakar Ahmed	Victoria University, Australia
Inmaculada Medina Bulo	Universidad de Cádiz, Spain

PC Members

Farhad Daneshgar	Victoria University Sydney, Australia
M. Reza Hoseiny F.	University of Sydney, Australia
Kamanashis Biswas (KB)	Australian Catholic University, Australia
Khaled Kourouche	Victoria University Sydney, Australia
Huai Liu	Victoria University Sydney, Australia
Mark A. Gregory	RMIT University, Australia
Nazmus Nafi	Victoria Institute of Technology, Australia
Mashud Rana	CSIRO, Australia
Farshid Hajati	Victoria University Sydney, Australia
Ashkan Yousefi	Victoria University Sydney, Australia
Nedal Ababneh	Abu Dhabi Polytechnic, Abu Dhabi, UAE
Lorena Gutiérrez-Madroñal	University of Cádiz, Spain
Juan Boubeta-Puig	University of Cádiz, Spain
Guadalupe Ortiz	University of Cádiz, Spain
Alfonso García del Prado	University of Cádiz, Spain
Luis Llana	Complutense University of Madrid, Spain

CISIS-2023 Reviewers

Adhiatma Ardian
Ali Khan Zahoor
Amato Alba
Amato Flora
Barolli Admir
Barolli Leonard
Bista Bhed
Buhari Seyed
Chellappan Sriram
Chen Hsing-Chung
Cui Baojiang
Dantas Mario
Di Martino Beniamino
Dong Hai
Durresi Arjan
Enokido Tomoya
Esposito Antonio
Fachrunnisa Olivia
Ficco Massimo
Fotia Lidia

Fun Li Kin
Funabiki Nobuo
Gotoh Yusuke
Hussain Farookh
Hussain Omar
Javaid Nadeem
Ikeda Makoto
Ishida Tomoyuki
Kikuchi Hiroaki
Kushida Takayuki
KullaElis
Matsuo Keita
Mizera-Pietraszko Jolanta
Mostarda Leonardo
Nakashima Makoto
Oda Tetsuya
Ogiela Lidia
Ogiela Marek
Okada Yoshihiro
Palmieri Francesco

Park Hyunhee
Paruchuri Vamsi Krishna
Poniszewska-Maranda
    Aneta
Rahayu Wenny
Sakamoto Shinji
Scionti Alberto
Sianaki Omid Ameri
Spaho Evjola
Takizawa Makoto
Taniar David
Terzo Olivier
Uehara Minoru
Venticinque Salvatore
Woungang Isaac
Xhafa Fatos
Yim Kangbin
Yoshihisa Tomoki

CISIS-2023 Keynote Talks

Evolution of Intelligent Software Agents

Salvatore Venticinque

University of Campania "Luigi Vanvitelli", Caserta, Italy

Abstract. The talk will focus on the evolution of models, techniques, technologies, and applications of software agents in the last years. Rapidly evolving areas of software agents range from programming paradigms to artificial intelligence. Driven by different motivations, an heterogeneous body of research is carried out under this banner. In each research area, the acceptance of agents has always been at once critical or skeptical and enthusiastic for promising future opportunities. Nevertheless, the efforts have been continuously spent to advance the research in this field. One example is the semantic Web vision, whereby machine-readable Web data could be automatically actioned upon by intelligent software Web agents. Maybe it has yet to be realized; however, semantic enrichment of Web metadata of digital archives is constantly growing including links to domain vocabularies and ontologies by supporting more and more advanced reasoning.

Securing Mobile Wireless Networks

Sanjay Kumar Dhurandher

Netaji Subhas University of Technology, New Delhi, India

Abstract. The area of mobile computing aims toward providing connectivity to various mobile users. There is an increasing demand by users that the information be available to them at any place and at any time. This has led to more use of mobile devices and networks. Since the wireless networks such as WLAN and Wi-Fi require the use of the unlicensed ISM band for data communication, there are increased threats to users because the data may be modified/fabricated. Additionally, these types of networks are further prone to various other threats which may even result in cyber-attacks and cyber-crime. Thus, it is a need to protect the users/devices from such threats leading to loss of important financial data and in some cases leakage of important defense documents of certain targeted countries.

Contents

Quantum Algorithms for Trust-Based AI Applications

Davinder Kaur, Suleyman Uslu, and Arjan Durresi[✉]

Indiana University-Purdue University Indianapolis, Indianapolis, IN, USA
{davikaur,suslu}@iu.edu, adurresi@iupui.edu

Abstract. Quantum computing is a rapidly growing field of computing that leverages the principles of quantum mechanics to significantly speed up computations that are beyond the capabilities of classical computing. This type of computing can revolutionize the field of trustworthy artificial intelligence, where decision-making is data-driven, complex, and time-consuming. Different trust-based AI systems have been proposed for different AI applications. In this paper, we have reviewed different trust-based AI systems and summarized their alternative quantum algorithms. This review provides an overview of quantum algorithms for three trust-based AI applications: fake user detection in social networks, medical diagnostic system, and finding the shortest path used in social network trust aggregation.

1 Introduction

Quantum Computing is the field that uses quantum mechanical phenomena such as superposition and entanglement to perform operations on data much more efficiently than classical computing. It is the intersection of physics, mathematics, and computer science. Quantum computing can perform many computations simultaneously. This computing technology is based on qubits, which can exist in multiple states simultaneously. It provides several advantages over the classical computing methods because it drastically reduces the execution time and energy consumption [37].

Quantum computing is an innovative and life-changing technology. Recently, Google has been investing billions of dollars in building its quantum computer by 2029 [1]. IBM and Microsoft are also working on providing quantum computing benefits to customers [25]. This technology can solve complex problems that are difficult to solve using classical computing. Inspired by the help of quantum computing, in this paper, we have reviewed the quantum algorithms for some of the real-world trust-based artificial intelligence a pplications that are generally time-consuming using classical algorithms. The trust-based AI applications are widely used in our day-to-day lives. It is essential to make these applications safe, reliable, and trustworthy [16,20] by integrating AI applications with trust assessment techniques [31,33]. Different researchers have proposed trust-based methods to create various AI applications trustworthy. Some researchers

L. Barolli (Ed.): CISIS 2023, LNDECT 176, pp. 1–12, 2023.
https://doi.org/10.1007/978-3-031-35734-3_1

proposed a trust framework for AI applications in Food, Energy, and Water sectors [38–43], some proposed for fake user detection [19,30], and some suggested for medical diagnostics systems [17,18]. So, in this review, we have discussed the quantum alternatives of some widely used algorithms for trust-based AI systems.

This paper is organized as follows. Section 2 presents the background of quantum computing. Section 3 reviews the quantum algorithms for widely used trust-based AI applications. And in Sect. 4, we conclude the paper.

2   Background

Quantum Computing is a new paradigm that leverages the concept of quantum mechanics to process information differently from classical computing approaches. Quantum mechanics explains the behavior of the particles on the quantum level, i.e., sub-atomic and atomic levels [35]. This type of computing method is used to solve complex problems which are challenging to solve using traditional approaches. A quantum computer consists of various components, including:

- Qubit: Qubit is the building block of quantum computers. It is the fundamental information-carrying unit. It is a quantum analog of classical bits and can exist in multiple states simultaneously. More precisely, classical computers use binary bits: 0s and 1s, whereas quantum computer uses 0s, 1s, and both a 0 and 1 simultaneously. The capability of having multiple states at the same time gives quantum computers immense processing power.
- Superposition: Superposition is the ability of a qubit to exist in multiple states simultaneously. It refers to the linear combination of two quantum states. In quantum computing, a qubit can exist as a superposition of two states (0 and 1) and can perform multiple computations simultaneously.
- Entanglement: Entanglement is the phenomenon in which two or more qubits become correlated so that their states become dependent on each other. Quantum entanglement enables qubits separated by large distances to interact with each other instantaneously. This property type enables one to perform certain types of computations more efficiently.
- Quantum Gates: Quantum gates are basic building blocks for quantum circuits. They are similar to logic gates used in classical computing. These gates operate on qubits to manipulate their states to perform complex computations. These gates perform operations like superposition, entanglement, and measurements.
- Quantum Memory: Quantum memory stores quantum information, which is fragile and can be easily lost due to environmental disturbances. Several types of quantum memory exist, including superconducting qubits, trapped ions, and topological qubits.
- Quantum Algorithms: Quantum algorithms are a set of algorithms that take advantage of the unique properties of quantum computers to perform specific calculations more efficiently than classical computers [1].

3 Review

This section provides a review of three trust-based AI applications, their widely used classical algorithms, and the quantum alternative of those algorithms.

3.1 Fake User Detection in Social Networks

Social networks have become an integral part of people's daily lives. However, with the increased use of social networks also comes certain risks, such as the spread of fake news, malicious content, and viruses by creating fake accounts [15]. It is essential to detect these fake user accounts as soon as possible and take action to prevent the spread of harmful content. Detecting fake user clusters in social networks is challenging, as fake users can use various tactics to evade detection. However, several methods can be used to detect fake user clusters in social networks. Some researchers proposed analyzing user profiles and behavior to detect fake users [45]. Other researchers proposed methods that utilize the graphical properties of social networks to detect fake users [6]. Another set of researchers uses trust information between users to detect malicious and fake users [19,30].

Combining all these approaches for clustering social network graphs is an effective way to detect clusters of fake users in social networks. The traditional clustering algorithms on large social network graphs are computationally expensive and time-consuming. Quantum computers and quantum clustering algorithms provide exponential speed-ups to the conventional clustering approaches. Following subsections explain the traditional clustering algorithms and their quantum alternatives.

3.1.1 Traditional DBSCAN Clustering Approach

DBSCAN (Density-Based Spatial Clustering of Applications with Noise) is an unsupervised machine learning method to find arbitrary shape clusters and clusters with noise. This algorithm group together the points that are close to each other in terms of distance and density [11]. The main idea behind this clustering technique is that if a point is close to many points from a cluster, it belongs to that cluster. It takes two parameters as input: epsilon and minPts. Epsilon is a distance threshold that defines the radius of a neighborhood around each point, and minPts is the minimum number of points to define a cluster. The algorithm starts by selecting the random unvisited point, and its neighborhood is determined using epsilon. If the neighborhood has at least minPts, cluster formation starts. The algorithm expands recursively. In the next step, the algorithm chooses another point that has not been visited in the previous step, and the process continues until all the points have been visited [9].

The advantages of DBSCAN include its ability to find clusters of arbitrary shapes and sizes, its tolerance for noise points, and its ability to handle datasets with varying densities. However, the algorithm can be sensitive to the choice of epsilon and minPts, may not perform well on datasets with different local density clusters, and is time-consuming.

3.1.2   Quantum DBSCAN Clustering Algorithm

Inspired by the advantages of quantum computing, [46] proposed a quantum Mutual MinPts-nearest Neighbor Graph (MMNG) based DBSCAN algorithm. This algorithm performs better on datasets with different low-density clusters and dramatically increases speed compared to the traditional approach. The proposed algorithm comprises two sub-algorithms: a Quantum mutual MinPts-nearest neighbor graph algorithm and a quantum DBSCAN algorithm. The Quantum mutual MinPts-nearest neighbor graph algorithm divides the dataset into subsets. And on each of the generated subsets, the quantum DBSCAN algorithm is applied to obtain clusters and noise set. Different subsets have different epsilon for this algorithm. In the quantum DBSCAN algorithm, the distance calculation needed to determine the Epsilon neighborhood is done using quantum search. The steps of the Quantum MMNG DBSCAN algorithm are given below:

Algorithm 1. Quantum MMNG DBSCAN

Input: Dataset, minPts
Output: Cluster and noise set
Procedure:
Step 1: Divide the dataset into subsets using the quantum - MMNG algorithm [46].
Step 2: For every subset obtained in Step 1, calculate the epsilon and get clusters and noise using the quantum DBSCAN algorithm [46].
Step 3: Return all the clusters and noise set.

The complexity of the proposed algorithm is $O(N\sqrt{minPts * n})$, where n is the number of data points, and minPts is the minimum number of data points required to define the cluster.

3.1.3   Louvain Community Detection Algorithm

Louvain algorithm is an unsupervised community detection algorithm used to detect communities from large networks [7]. This algorithm does not require the user input of community size or the number of communities before execution. The algorithm comprises the modularity optimization phase and the community aggregation phase [3]. In the modularity optimization phase, each node is assigned to its community, and the algorithm iteratively evaluates the modularity gain resulting from merging nodes with its neighbors. Only the nodes that result in the highest modularity gain will be moved to the community. In the second phase, communities detected in the first phase are aggregated, and the first phase is repeated on this new network. This process is repeated until no further modularity gain can be achieved. Figure 1 shows different social network communities detected by the Louvain algorithm in detecting fake users [19].

Louvain algorithm has several advantages like speed, scalability, and flexibility, making it suitable for detecting communities in large social network graphs [3]. The quantum variant of this algorithm provides quantum speedups to the task of community detection in large complex networks [5]. The following section discusses the quantum variant of the Louvain community detection algorithm.

3.1.4 Quantum Variant of Louvain Algorithm

The Quantum variant of the Louvain algorithm is the EdgeQLouvain. This algorithm utilizes a single Grover search over an ample search space (the entire set of vertices) rather than searching over vertices and their neighbors [5] to find a good move. Given the input graph with directed edges (u,v), this algorithm searches for the edge that will increase the modularity if u is moved to the neighboring community. This version of the quantum Louvain algorithm has several advantages as it does not need nested Grover search, making it much simplified and faster than other quantum variants [5]. Given an edge set E of the graph that contains undirected edges, a directed graph is obtained by replacing every undirected edge {u,v} with (u,v) and (v,u), and on this directed graph, the algorithm as described below is applied.

This algorithm shows polynomial speedup as compared to the traditional Louvain algorithm. The query complexity of this algorithm for every step k is $O(1/\sqrt{h_k})$ where h_k is the fraction of the edges.

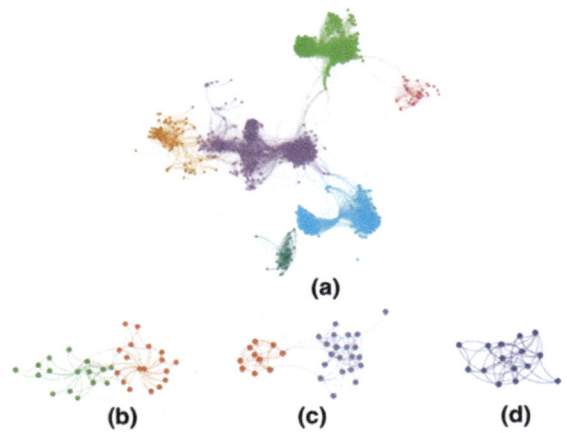

Fig. 1. Communities detected by the Louvain algorithm for different social networks. a) Facebook ego network. b) Karate friends network. c) Fake user network 1 and d) Fake User Network 2

Algorithm 2. Edge Quantum Louvain Algorithm

Input: Graph edge set
Output: Cluster set
Procedure:
Step 1: Initially, assign every vertex to its community.
Step 2: Utilize the quantum algorithm QSearch [4] to search all over the edges(u,v) to find one that yields a good move.
Step 3: Find the best neighboring community of u using the quantum maximum finding algorithm [8].
Step 4: Repeat steps 2 and 3 until there is no modularity increase.
Step 5: Aggregate communities to make a new graph and repeat steps 2-4 until there is no more change.

3.2   AI System for Medical Diagnostics

Artificial intelligence and machine learning systems have completely changed our lives. Many high stake applications like medical diagnostics are widely adopting these systems. With the vast amount of data and computing power available, these algorithms have become very good at predicting diagnostic results and saving time and money [36]. These algorithms are helping doctors with a cancer diagnosis by analyzing the image dataset of old cancer cases by detecting, measuring, and exploring the tumor cells [17,18].

The widely used machine learning algorithms in medical diagnostics are classification algorithms. The classification algorithms are the supervised algorithms that take training data as the input to predict the likelihood or probability of the new data to belong into the predetermined categories. To speed up the classical machine learning classification algorithms, quantum machine learning is introduced, integrating the quantum algorithms with the classification algorithms [2]. The following subsection discusses the traditional classification algorithms and their quantum alternatives.

3.2.1   Traditional Support Vector Machine

Support vector machine (SVM) is a supervised machine learning algorithm for classification and regression tasks. This algorithm is widely used in biological applications as it learns to assign labels by learning from examples [27]. SVM works by finding the best possible boundary or hyperplane that distinctly classifies the data points of different classes in the high dimensional space. This algorithm aims to find a hyperplane that maximizes the margin, that is, the distance between the hyperplane and the closest data points from each class. This algorithm is helpful in high-dimensional spaces as it uses kernel functions to transform the input data into high-dimensional space where it is easier to find a separating hyperplane.

This algorithm has several advantages, including its robustness to the outliers and effectiveness in cases where the number of samples is lesser than the number of features. However, this algorithm can be computationally intensive and not suitable for large data sets [44].

3.2.2 Quantum Support Vector Machine

The quantum support vector machine algorithm performs the least square-SVM using a quantum computer [29]. This quantum algorithm uses phase estimation and a quantum matrix inversion algorithm to maximize the algorithm's speed. Considering there are N data points (x_i, y_i): i = 1, 2.....N), where x_i is the feature vector, and y_i is the binary label of the data, the goal of the SVM is to find the hyperplane $w.x + b = 0$ that divides the data points into two categories. The quantum least square SVM algorithm calculates the kernel matrix using the quantum random access memory [10] and solves the linear equation using the quantum algorithm for solving linear equations and then performs classification using the trained qubits. Following are the steps of the quantum support vector machine algorithm.

Algorithm 3. Quantum Support Vector Machine

Input: Training Data and Test Data
Output: Classification: +1 or -1
Procedure:
Step 1: Calculate the kernel metrics using the quantum inner product algorithm [24].
Step 2: Solve the linear equation using the quantum algorithm for linear equations [10].
Step 3: Classify the test data using the training results, using a quantum algorithm [29].

The complexity of the quantum SVM algorithm is $O(log NM)$ as compared to the traditional SVM, which has the complexity of $O(M^2(M + N))$. Here N is the N-dimensional feature vector, and M is the number of data points.

3.2.3 Traditional Logistic Regression

Logistic regression is a supervised machine learning algorithm widely used for classification tasks. This algorithm classifies observations into a discrete set of classes. Logistic regression predicts the probability of an event occurring based on a set of independent variables or predictors. The algorithm models the relationship between the dependent variable and one or more independent variables using a logistic function, which produces an S-shaped curve [21]. This logistic function transforms the linear combination of input variables and coefficients into a probability value of 0 and 1. The logistic regression algorithm minimizes the difference between the predicted probabilities and the actual output in the training data by finding the optimal values for the coefficients.

Logistic regression is widely used in many fields, including healthcare, finance, and social sciences. It has many advantages: it can handle categorical and continuous variables as input features. However, this algorithm does not perform well if there is a non-linear relationship between the input and output variables and is time-consuming when applied to big datasets.

3.2.4 Quantum Logistic Regression

Logistic regression is an important algorithm used for classification tasks. But this algorithm can be slow for large data sets as it involves a gradient descent method at each iteration which is quite time-consuming. To overcome this, [23] proposed a quantum logistic regression algorithm that implements the critical task of the gradient descent at each iteration, making the algorithm exponentially faster than the classical logistic regression. The quantum algorithm is divided into two steps: the first is to generate the quantum state using the amplitude estimation [4], and the second step is using the swap test [34] to obtain a gradient in the classical form. To ensure the new data can be classified by this algorithm directly, it outputs the model parameters in classical form. Following are the steps for the quantum logistic regression algorithm.

The quantum logistic regression algorithm provides exponential speed compared to the traditional algorithm. The complexity of this algorithm is $O(polylogN)$ for every iteration, where N is the number of data points.

Algorithm 4. Quantum Logistic Regression

Input: Training Data and Test Data
Output: Classification: +1 or -1
Procedure:
Step 1: Initialize all parameters.
Step 2: Calculate the dependent variable.
Step 3: Calculate the cost function.
Step 4: Calculate the gradient of the cost function using a quantum algorithm that consists of amplitude estimation [4] and swap test [34].
Step 5: Update all parameters.
Step 6: Repeat steps 2–5.

3.3   Finding Shortest Path

Shortest path algorithms are used to find the shortest path between points. Given a graph, these algorithms find the shortest path from one point to another or from one point to all other points [26]. These algorithms have many applications. In traffic information systems, these algorithms are used to find the optimal path from one point to another point. In networking, these algorithms are used in routing protocols to find the optimal path to transmit data packets. These algorithms are also used in social network analysis and autonomous vehicle route planning. For example, the highest confidence path can be used to speed up calculations in stock market prediction using Twitter trust networks [32].

To make the process of finding the shortest path faster, quantum algorithms for finding the shortest path are proposed as they are capable of performing several operations simultaneously [22]. The following subsections discuss the widely used traditional shortest path algorithm and its quantum alternative.

3.3.1   Traditional Dijkstra Algorithm

Dijkstra algorithm is a popular algorithm for solving the single source shortest path search in the weighted graphs with non-negative weights [13]. This algorithm finds the shortest path from the source node to all other nodes. This algorithm is handy in the traffic information system to find the shortest path between the current location and the destination and also in modeling networks.

The algorithm keeps track of visited and non-visited nodes. Initially, it starts with the source node, whose distance is zero. Then for each non-visited neighbor of the current node, the algorithm calculates the distance from the source node to that neighbor by using the weights of the edges connecting them. If the distance is less than the current distance, it updates it. This process is repeated until all the nodes have been visited [12]. This way, the algorithm finds the shortest path between the source and any other node in the graph.

3.3.2 Quantum Dijkstra Algorithm

Researchers [28] have proposed the quantum version of the Dijkstra algorithm that utilizes the principles of quantum superposition and inference to find the shortest path in the graph. This algorithm proves to be better in terms of time complexity than the traditional Dijkstra algorithm. This algorithm utilizes the quantum search algorithm and phase estimation to speed up the search operation. The following steps explain the quantum Dijkstra algorithm:

Algorithm 5. Quantum Dijkstra Algorithm

Input: Set of nodes, source node.
Output: The shortest path from the source node to all the other nodes.
Procedure:
Step 1: Initialize the distance to the source node to zero, and all other nodes to infinity.
Step 2: Create a set of visited and unvisited nodes.
Step 3: While there are unvisited nodes:
Select the smallest distance unvisited neighbor and find the minimum distance path from the source node to the neighbor using Grover's algorithm [14] and quantum minimum searching algorithm [8].
Step 4: Output the best path from the source to all the nodes.

The complexity of this algorithm is $O(\sqrt{NM}log^2 N)$ as compared to the traditional Dijkstra algorithm, which has the complexity of $O(M + NlogN)$. Here N is the number of vertexes, and M is the number of edges.

4 Conclusion

Quantum computing provides several benefits as compared to classical computing. Different researchers have proposed different quantum algorithms that offer significant benefits. However, there was a lack of mapping between the quantum algorithms and the real-life applications. This review summarizes quantum algorithms for trust-based AI applications for fake user detection, medical diagnostics, and finding the shortest path in trust networks.

References

1. Adedoyin, A., et al.: Quantum algorithm implementations for beginners. arXiv preprint arXiv:1804.03719 (2018)
2. Biamonte, J., Wittek, P., Pancotti, N., Rebentrost, P., Wiebe, N., Lloyd, S.: Quantum machine learning. Nature **549**(7671), 195–202 (2017)
3. Blondel, V.D., Guillaume, J.L., Lambiotte, R., Lefebvre, E.: Fast unfolding of communities in large networks. J. Stat. Mech. Theory Exp. **2008**(10), P10,008 (2008)
4. Brassard, G., Hoyer, P., Mosca, M., Tapp, A.: Quantum amplitude amplification and estimation. Contemp. Math. **305**, 53–74 (2002)
5. Cade, C., Folkertsma, M., Niesen, I., Weggemans, J.: Quantum algorithms for community detection and their empirical run-times. arXiv preprint arXiv:2203.06208 (2022)
6. Cao, Q., Sirivianos, M., Yang, X., Pregueiro, T.: Aiding the detection of fake accounts in large scale social online services. In: Presented as part of the 9th {USENIX} Symposium on Networked Systems Design and Implementation ({NSDI} 12), pp. 197–210 (2012)
7. De Meo, P., Ferrara, E., Fiumara, G., Provetti, A.: Generalized louvain method for community detection in large networks. In: 2011 11th International Conference on Intelligent Systems Design and Applications, pp. 88–93. IEEE (2011)
8. Durr, C., Hoyer, P.: A quantum algorithm for finding the minimum. arXiv preprint quant-ph/9607014 (1996)
9. Ester, M., Kriegel, H.P., Sander, J., Xu, X., et al.: A density-based algorithm for discovering clusters in large spatial databases with noise. In: KDD, vol. 96, pp. 226–231 (1996)
10. Giovannetti, V., Lloyd, S., Maccone, L.: Quantum random access memory. Phys. Rev. Lett. **100**(16), 160501 (2008)
11. Hinneburg, A.: A density based algorithm for discovering clusters in large spatial databases with noise. In: KDD Conference, 1996 (1996)
12. Javaid, A.: Understanding dijkstra's algorithm. SSRN 2340905 (2013)
13. Johnson, D.B.: A note on dijkstra's shortest path algorithm. J. ACM (JACM) **20**(3), 385–388 (1973)
14. Jozsa, R.: Searching in grover's algorithm. arXiv preprint quant-ph/9901021 (1999)
15. Kaur, D., Uslu, S., Durresi, A.: Trust-based security mechanism for detecting clusters of fake users in social networks. In: Barolli, L., Takizawa, M., Xhafa, F., Enokido, T. (eds.) WAINA 2019. AISC, vol. 927, pp. 641–650. Springer, Cham (2019). https://doi.org/10.1007/978-3-030-15035-8_62
16. Kaur, D., Uslu, S., Durresi, A.: Requirements for trustworthy artificial intelligence – a review. In: Barolli, L., Li, K.F., Enokido, T., Takizawa, M. (eds.) NBiS 2020. AISC, vol. 1264, pp. 105–115. Springer, Cham (2021). https://doi.org/10.1007/978-3-030-57811-4_11
17. Kaur, D., Uslu, S., Durresi, A.: Trustworthy AI explanations as an interface in medical diagnostic systems. In: Advances in Network-Based Information Systems: The 25th International Conference on Network-Based Information Systems (NBiS-2022), pp. 119–130. Springer, Heidelberg (2022). https://doi.org/10.1007/978-3-031-14314-4_12

18. Kaur, D., Uslu, S., Durresi, A., Badve, S., Dundar, M.: Trustworthy explainability acceptance: a new metric to measure the trustworthiness of interpretable ai medical diagnostic systems. In: Barolli, L., Yim, K., Enokido, T. (eds.) CISIS 2021. LNNS, vol. 278, pp. 35–46. Springer, Cham (2021). https://doi.org/10.1007/978-3-030-79725-6_4

19. Kaur, D., Uslu, S., Durresi, M., Durresi, A.: A geo-location and trust-based framework with community detection algorithms to filter attackers in 5g social networks. Wirel. Netw. **2022**, 1–9 (2022)

20. Kaur, D., Uslu, S., Rittichier, K.J., Durresi, A.: Trustworthy artificial intelligence: a review. ACM Comput. Surv. (CSUR) **55**(2), 1–38 (2022)

21. Kleinbaum, D.G., Dietz, K., Gail, M., Klein, M., Klein, M.: Logistic Regression. Springer, Heidelberg (2002)

22. Krauss, T., McCollum, J.: Solving the network shortest path problem on a quantum annealer. IEEE Trans. Quant. Eng. **1**, 1–12 (2020)

23. Liu, H.L., et al.: Quantum algorithm for logistic regression. arXiv preprint arXiv:1906.03834 (2019)

24. Lloyd, S., Mohseni, M., Rebentrost, P.: Quantum algorithms for supervised and unsupervised machine learning. arXiv preprint arXiv:1307.0411 (2013)

25. MacQuarrie, E.R., Simon, C., Simmons, S., Maine, E.: The emerging commercial landscape of quantum computing. Nat. Rev. Phys. **2**(11), 596–598 (2020)

26. Magzhan, K., Jani, H.M.: A review and evaluations of shortest path algorithms. Int. J. Sci. Technol. Res. **2**(6), 99–104 (2013)

27. Noble, W.S.: What is a support vector machine? Nat. Biotechnol. **24**(12), 1565–1567 (2006)

28. Ray, P.: Quantum simulation of dijkstra' algorithm. Int. J. Adv. Res. Comput. Sci. Manag. Stud. **2**, 30–43 (2014)

29. Rebentrost, P., Mohseni, M., Lloyd, S.: Quantum support vector machine for big data classification. Phys. Rev. Lett. **113**(13), 130503 (2014)

30. Rittichier, K.J., Kaur, D., Uslu, S., Durresi, A.: A trust-based tool for detecting potentially damaging users in social networks. In: Barolli, L., Chen, H.-C., Enokido, T. (eds.) NBiS 2021. LNNS, vol. 313, pp. 94–104. Springer, Cham (2022). https://doi.org/10.1007/978-3-030-84913-9_9

31. Ruan, Y., Durresi, A.: A survey of trust management systems for online social communities-trust modeling, trust inference and attacks. Knowl.-Based Syst. **106**, 150–163 (2016)

32. Ruan, Y., Durresi, A., Alfantoukh, L.: Using twitter trust network for stock market analysis. Knowl.-Based Syst. **145**, 207–218 (2018)

33. Ruan, Y., Zhang, P., Alfantoukh, L., Durresi, A.: Measurement theory-based trust management framework for online social communities. ACM Trans. Internet Technol. (TOIT) **17**(2), 16 (2017)

34. Schuld, M., Sinayskiy, I., Petruccione, F.: Prediction by linear regression on a quantum computer. Phys. Rev. A **94**(2), 022342 (2016)

35. National Academies of Sciences, E., Medicine, et al.: Quantum computing: progress and prospects (2019)

36. Sidey-Gibbons, J.A., Sidey-Gibbons, C.J.: Machine learning in medicine: a practical introduction. BMC Med. Res. Methodol. **19**, 1–18 (2019)

37. Steane, A.: Quantum computing. Rep. Progr. Phys. **61**(2), 117 (1998)

38. Uslu, S., Kaur, D., Rivera, S.J., Durresi, A., Babbar-Sebens, M.: Decision support system using trust planning among food-energy-water actors. In: Barolli, L., Takizawa, M., Xhafa, F., Enokido, T. (eds.) AINA 2019. AISC, vol. 926, pp. 1169–1180. Springer, Cham (2020). https://doi.org/10.1007/978-3-030-15032-7_98

39. Uslu, S., Kaur, D., Rivera, S.J., Durresi, A., Babbar-Sebens, M.: Trust-based game-theoretical decision making for food-energy-water management. In: Barolli, L., Hellinckx, P., Enokido, T. (eds.) BWCCA 2019. LNNS, vol. 97, pp. 125–136. Springer, Cham (2020). https://doi.org/10.1007/978-3-030-33506-9_12

40. Uslu, S., Kaur, D., Rivera, S.J., Durresi, A., Babbar-Sebens, M.: Trust-based decision making for food-energy-water actors. In: Barolli, L., Amato, F., Moscato, F., Enokido, T., Takizawa, M. (eds.) AINA 2020. AISC, vol. 1151, pp. 591–602. Springer, Cham (2020). https://doi.org/10.1007/978-3-030-44041-1_53

41. Uslu, S., Kaur, D., Rivera, S.J., Durresi, A., Babbar-Sebens, M., Tilt, J.H.: Control theoretical modeling of trust-based decision making in food-energy-water management. In: Barolli, L., Poniszewska-Maranda, A., Enokido, T. (eds.) CISIS 2020. AISC, vol. 1194, pp. 97–107. Springer, Cham (2021). https://doi.org/10.1007/978-3-030-50454-0_10

42. Uslu, S., Kaur, D., Rivera, S.J., Durresi, A., Babbar-Sebens, M., Tilt, J.H.: A trustworthy human-machine framework for collective decision making in food-energy-water management: the role of trust sensitivity. Knowl.-Based Syst. **213**, 106683 (2021)

43. Uslu, S., Kaur, D., Rivera, S.J., Durresi, A., Durresi, M., Babbar-Sebens, M.: Trustworthy acceptance: a new metric for trustworthy artificial intelligence used in decision making in food–energy–water sectors. In: Barolli, L., Woungang, I., Enokido, T. (eds.) AINA 2021. LNNS, vol. 225, pp. 208–219. Springer, Cham (2021). https://doi.org/10.1007/978-3-030-75100-5_19

44. Wang, L.: Support Vector Machines: Theory and Applications, vol. 177. Springer, Heidlberg (2005). https://doi.org/10.1007/b95439

45. Xiao, C., Freeman, D.M., Hwa, T.: Detecting clusters of fake accounts in online social networks. In: Proceedings of the 8th ACM Workshop on Artificial Intelligence and Security, pp. 91–101 (2015)

46. Xie, X., Duan, L., Qiu, T., Li, J.: Quantum algorithm for mmng-based dbscan. Sci. Rep. **11**(1), 15559 (2021)

Energy-Saving Multi-version Timestamp Ordering Algorithm for Virtual Machine Environments

Tomoya Enokido[1]([✉]), Dilawaer Duolikun[2], and Makoto Takizawa[3]

[1] Faculty of Business Administration, Rissho University, 4-2-16, Osaki, Shinagawa-ku, Tokyo 141-8602, Japan
eno@ris.ac.jp

[2] Department of Advanced Sciences, Faculty of Science and Engineering, Hosei University, 3-7-2, Kajino-cho, Koganei-shi, Tokyo 184-8584, Japan

[3] Research Center for Computing and Multimedia Studies, Hosei University, 3-7-2, Kajino-cho, Koganei-shi, Tokyo 184-8584, Japan
makoto.takizawa@computer.org

Abstract. In order to realize distributed applications, conflicting transactions are required to be serializable. The Multi-Version Concurrency Control (MVCC) is widespread to not only improve the concurrency of transactions but also serialize transactions. The Multi-Version Timestamp Ordering (MVTO) algorithm is one of algorithms to realize the MVCC. However, the MVTO algorithm does not consider the electric energy consumption of servers to perform transactions. In this paper, the Energy-Saving Multi-Version Timestamp Ordering with Virtual Machines (ESMVTO-VM) algorithm is newly proposed to not only improve the concurrency of transactions but also decrease the electric energy consumption of a cluster of servers which are equipped with virtual machines without performing meaningless write methods. In evaluation, we show the ESMVTO-VM algorithm can decrease the electric energy consumption of a cluster of servers than the MVTO algorithm.

Keywords: Transaction · Virtual machine · Electric energy consumption · Green computing · Multi-version concurrency control

1 Introduction

Various kinds and a large volume of data are required to realize current distributed applications. Data required by an application is collected from various types of devices like IoT (Internet of Things) devices [1,2]. Collected data and methods to manipulate the collected data is encapsulated as an object [3–5] like a database system. An object is a component to realize applications. A transaction [6,7] is an atomic sequence of methods to manipulate objects. Transactions are initiated on clients by application users to utilize provided application services. In current distributed application services, a huge number of transactions are

© The Author(s), under exclusive license to Springer Nature Switzerland AG 2023
L. Barolli (Ed.): CISIS 2023, LNDECT 176, pp. 13–20, 2023.
https://doi.org/10.1007/978-3-031-35734-3_2

initiated to utilize provided application services. Therefore, a scalable and high performance computing system like a cloud computing system [2] is required to provide application services. A cloud computing system is composed of a cluster of physical servers. Virtual machines are equipped with each physical server to efficiently utilize the computation resources of the server. Objects are deployed to virtual machines in a server cluster. Conflicting transactions issued by users have to be serialized [6–10] to maintain consistency among objects.

The *Multi-Version Concurrency Control (MVCC)* [8,9] is so far proposed to serialize conflicting transactions based on the "read data from" relation [8,9]. The latest committed version of each object is read by each read method in the MVCC. Therefore, each read method conflicts with neither any read nor write methods on different resources. Hence, the concurrency of transactions can be improved. The *Multi-Version Timestamp Ordering (MVTO)* algorithm [8,9] is one of algorithms to realize the MVCC. However, reduction of electric energy consumption of servers to perform methods on each object is not considered in the MVTO algorithm. In a server cluster, not only performance of a system but also reduction of electric energy consumed by servers have to be considered as discussed in Green computing [3–5,11].

In this paper, the *Energy-Saving Multi-Version Timestamp Ordering with Virtual Machines (ESMVTO-VM)* algorithm based on the MVTO algorithm is newly proposed to decrease the electric energy consumption of a server cluster equipped with virtual machines without performing *meaningless write methods* [11] on each object. We evaluate the ESMVTO-VM algorithm compared with the MVTO algorithm. In evaluation, we show the electric energy consumed by servers can be decreased in the ESMVTO-VM algorithm than the MVTO algorithm.

The system model of this paper is presented in Sect. 2. In Sect. 3, we newly propose the ESMVTO-VM algorithm. Evaluation results of the ESMVTO-VM algorithm are shown in Sect. 4.

2   System Model

2.1   Multi-version Concurrency Control

A server cluster SC is composed of multiple physical servers s_1, ..., s_n ($n \geq 1$). Each physical server s_t is equipped with one multi-core CPU. A CPU is equipped with a set $Core_t = \{c_{1t}, ..., c_{nc_t t}\}$ of cores where nc_t is the total number of cores. Let ct_t (≥ 1) be the total number of threads in a core c_{gt}. There is a set $Thread_t = \{th_{1t}, ..., th_{nt_t t}\}$ of threads in each server s_t where $nt_t = ct_t \cdot nc_t$. A set $VM_t = \{vm_{1t}, ..., vm_{nt_t t}\}$ of virtual machines are supported by each server s_t. Each virtual machine vm_{kt} occupies one thread th_{kt} in a server s_t. A set $O = \{o_1, ..., o_m\}$ ($m \geq 1$) of objects [3–5] are distributed to virtual machines in a cluster SC. Data d_h and methods to manipulate the data d_h is encapsulated as the object o_h. Methods supported by each object o_h are classified into *read* (r), *full write* (fw), and *partial write* (pw) methods. Data d_h in an object o_h is fully written by a full write method while a part of data d_h is written by a partial write method.

A *transaction* [6,7] is an atomic sequence of methods and issues write ($w \in \{fw, pw\}$) and read (r) methods to manipulate objects. Let $\mathbf{T}$ be a set $\{T^1, ..., T^v\}$ ($v \geq 1$) of transactions. The Multi-Version Concurrency Control (*MVCC*) [8,9] is so far proposed to *serialize* [6–9] conflicting transactions based on the "*read data from*" relation. A notation S shows a schedule of $\mathbf{T}$. Each object holds a totally ordered set $D_h = \{d_h^1, ..., d_h^l\}$ ($l \geq 1$) of versions of data d_h. Let $\ll_h$ ($\subseteq D_h^2$) be an order of versions in D_h. A version d_h^j is written after another version d_h^i in an object o_h if $d_h^i \ll_h d_h^j$. Notation $\ll$ shows an union of version orders $\ll_h$ in a schedule S, i.e. $\ll = \bigcup_{o_h \in O} \ll_h$. A transaction T^j *reads data from* another transaction T^i ($T^i \rightarrow_S T^j$) in a schedule S iff (if and only if) a version d_h^i of an object o_h written by the transaction T^i is read by the transaction T^j. $T^i \parallel_S T^j$ iff neither $T^i \rightarrow_S T^j$ nor $T^j \rightarrow_S T^i$.

[**One-copy serial schedule**] A schedule $S = \langle \mathbf{T}, \rightarrow_S \rangle$ ($\subseteq \mathbf{T}^2$) is *one-copy serial* [8] iff either $T^i \rightarrow_S T^j$, $T^j \rightarrow_S T^i$, or $T^i \parallel_S T^j$ for every pair of transactions T^i and T^j in $\mathbf{T}$.

Let ocS be an one-copy serial schedule $\langle \mathbf{T}, \rightarrow_{ocS} \rangle$ ($\subseteq \mathbf{T}^2$). In the ocS, $T^i \rightarrow_{ocS} T^j$ if $T^i \rightarrow_S T^j$ and the "read data from" relation $\rightarrow_{ocS}$ is acyclic.

Let mvS be a *multi-version schedule* $\langle \mathbf{T}, \rightarrow_{mvS} \rangle$ ($\subseteq \mathbf{T}^2$). In the mvS, the following conditions hold:

1. If $T^i \rightarrow_{mvS} T^j$, $T^i \rightarrow_{ocS} T^j$.
2. $d_h^i \ll_h d_h^k$ or $d_h^i = d_h^k$ if T^i writes a version d_h^i, T^j reads a version d_h^k, and $T^i \rightarrow_{mvS} T^j$,

[**One-copy serializability**] A multi-version schedule $mvS = \langle \mathbf{T}, \rightarrow_{mvS} \rangle$ ($\subseteq \mathbf{T}^2$) is *one-copy serializable* [8] iff either $T^i \rightarrow_{mvS} T^j$, $T^j \rightarrow_{mvS} T^i$, or $T^i \parallel_{mvS} T^j$ for every pair of transactions T^i and T^j in $\mathbf{T}$.

The *Multi-Version Timestamp Ordering* (*MVTO*) algorithm [8,9] is so far proposed to make transactions one-copy serializable. Let $r_{kt}^i(d_h^j)$ and $w_{kt}^i(d_h^i)$ be a read and write methods issued by a transaction T^i to read a version d_h^j and write a version d_h^i, respectively, in an object o_h supported by a virtual machine vm_{kt}. Each transaction T^i is given an unique timestamp $TS(T^i)$. In the MVTO algorithm, a method op issued by a transaction T^i is performed on an object o_h by the following procedure [8,9]:

1. If $op = r_{kt}^i(d_h^u)$, the read method $r_{kt}^i(d_h^u)$ reads a version d_h^u written by a transaction T^u whose timestamp $TS(T^u)$ is the maximum in $TS(T^u) < TS(T^i)$.
2. If $op = w_{kt}^i(d_h^i)$, the write method $w_{kt}^i(d_h^i)$ is performed iff there is no read method $r_{kt}^j(d_h^u)$ such that $TS(T^u) < TS(T^i) < TS(T^j)$. Otherwise, the write method $w_{kt}^i(d_h^i)$ is rejected.

2.2   Data Access Rate of Each Method

The *Data Access in Virtual Machine environments* (*DAVM*) model [12] is so far proposed to perform write and read methods on virtual machines based on our experiment. Let $r_{kt}^i(d_h)$ and $w_{kt}^i(d_h)$ be read and write methods issued by a transaction T^i to read and write version d_h of an object o_h supported by a virtual machine vm_{kt}, respectively. Let $Read_t(\tau)$ and $Write_t(\tau)$ be sets of read and write methods being performed on virtual machines in a server s_t at time τ, respectively. The maximum write rate $maxW_t$ and the maximum read rate $maxR_t$ of a server s_t depends on the performance of the server s_t. Let $rR^i_{kt}(\tau)$ and $wR^i_{kt}(\tau)$ [B/s] be the read rate of a read method $r^i_{kt}(d_h)$ and the write rate of a write method $w^i_{kt}(d_h)$, respectively, at time τ. The read rate $rR^i_{kt}(\tau)$ ($\leq maxR_t$) of a read method $r^i_{kt}(d_h)$ is $degr_t(\tau) \cdot maxR_t$ at time τ where $0 \leq degr_t(\tau) \leq 1$. The write rate $wr^i_{kt}(\tau)$ ($\leq maxW_t$) of a write method $w^i_{kt}(d_h)$ is $degw_t(\tau) \cdot maxW_t$ at time τ where $0 \leq degw_t(\tau) \leq 1$. Here, $degr_t(\tau)$ and $degw_t(\tau)$ are degradation ratios of read and write rates, respectively, at time τ. Here, $degr_t(\tau) = 1 \ / \ (|Read_t(\tau)| + rw_t \cdot |Write_t(\tau)|)$ where $0 \leq rw_t \leq 1$ and $degw_t(\tau) = 1 \ / \ (wr_t \cdot |Read_t(\tau)| + |Write_t(\tau)|)$ where $0 \leq wr_t \leq 1$.

2.3　Electric Power Consumption

The *Power Consumption for Data Access in Virtual Machine* environment (*PCDAVM*) model [12] is so far proposed to perform write and read methods on virtual machines based on our experiment. Let $BEP_t(\tau)$ [W] be the base electric power of a server s_t at time τ. The base electric power BEP_t [W] of a sever s_t depends on the number $aC_t(\tau)$ of active cores and the number $aT_t(\tau)$ of active threads in a server s_t at time τ. If $aC_t(\tau) \geq 1$, a server s_t consumes the electric power CE_t [W]. A server s_t consumes the electric power thE_t [W] and coE_t to make one thread and one core active, respectively. The maximum electric power $maxE_t$ [W] and the minimum electric power $minE_t$ [W] of a server s_t depends on the performance of the server s_t. The base electric power $BEP_t(\tau)$ [W] is given as Eq. (1):

$$BEP_t(\tau) = \begin{cases} minE_t + CE_t + aC_t(\tau) \cdot coE_t + aT_t(\tau) \cdot thE_t) & \text{if } aC_t(\tau) \geq 1. \\ minE_t, & \text{otherwise.} \end{cases}$$

(1)

The electric power $E_t(\tau)$ [W] to perform read and write methods is consumed by a server s_t at time τ as shown in Eq. (2):

$$E_t(\tau) = \begin{cases} BEP_t(\tau) & \text{if } |Write_t(\tau)| = |Read_t(\tau)| = 0. \\ BEP_t(\tau) + wE_t & \text{if } |Read_t(\tau)| = 0 \text{ and } |Write_t(\tau)| \geq 1. \\ BEP_t(\tau) + wrE_t(\alpha(\tau)) & \text{if } |Read_t(\tau)| \geq 1 \text{ and } |Write_t(\tau)| \geq 1. \\ BEP_t(\tau) + rE_t & \text{if } |Read_t(\tau)| = 1 \text{ and } |Write_t(\tau)| \geq 0. \end{cases}$$

(2)

If no method is performed in every virtual machine of a server s_t, $E_t(\tau) = BEP_t(\tau)$. Let wE_t [W] and rE_t [W] be the electric power consumed in the server s_t to perform only write and read methods, respectively. Let $wrE_t(\alpha(\tau))$ [W] be the electric power consumed in the server s_t to concurrently perform read and write methods. Here, $wrE_t(\alpha(\tau)) = \alpha(\tau) \cdot wE_t + (1 - \alpha(\tau)) \cdot rE_t$ where $\alpha(\tau) = |Write_t(\tau)| / |Read_t(\tau) \cup Write_t(\tau)|$. $rE_t \leq wrE_t(\alpha(\tau)) \leq wE_t$. The total processing electric energy (TPE) $TPE_t(\tau_1, \tau_2) = \Sigma_{\tau=\tau_1}^{\tau_2}(E_t(\tau) - minE_t)$ [J] is consumed in a server s_t to perform write and read methods from time τ_1 to τ_2.

3   ESMVTO-VM Algorithm

A method op^i *locally precedes* another method op^j ($op^i \rightarrow_{mvS_h} op^j$) in a local multi-version schedule mvS_h of an object o_h iff $op^i \rightarrow_{mvS} op^j$.

[**Absorption**] A write method op^j is *absorbed* by another full write method op^i in a local schedule mvS_h iff one of the following conditions holds:

1. $op^j \rightarrow_{mvS_h} op^i$ and there is no read method op' such that $op^j \rightarrow_{mvS_h} op'$ $\rightarrow_{mvS_h} op^i$.
2. op^i absorbs op' and op' absorbs op^j for some method op'.

A pair of multi-version schedules mvS^1 and mvS^2 are *equivalent* iff the same state is obtained by mvS^1 and mvS^2 for every state in the system. A write method w is *meaningless* iff a multi-version schedule mvS^1 is *equivalent* to another multi-version schedule mvS^2 even if the write method w is removed from mvS^1.

[**Meaningless Write Methods**] A write method w is *meaningless* on an object o_h iff the write method w is absorbed by another full write method fw in the local schedule mvS_h.

In this paper, the *Energy-Saving Multi-Version Timestamp Ordering in Virtual Machine environment (ESMVTO-VM)* algorithm is newly proposed based on the MVTO algorithm [8,9] to decrease the TPE consumption of a server cluster without performing meaningless write methods. A method op^i issued by a transaction T^i is performed on an object o_h by ESMVTO-VM algorithm shown in Algorithm 1. A variable $o_h.wcheck$ shows a write method $w_t^j(d_h^j)$ issued by a transaction T^j. A write method $w_t^j(d_h^j)$ shown by $o_h.wcheck$ waits for performing until a method op' to be performed just after the write method $w_t^j(d_h^j)$ is received by the object o_h. If the write method $w_t^j(d_h^j)$ shown by $o_h.wcheck$ is absorbed by the method op', the method op' is performed on the object o_h without performing the write method $w_t^j(d_h^j)$.

The TPE consumption of a server cluster can decrease in the ESMVTO-VM algorithm without performing meaningless write methods on each object. The ESMVTO-VM algorithm can also reduce the execution time of each transaction than the MVTO algorithm since each transaction can commit without performing meaningless write methods.

4   Evaluation

The ESMVTO-VM algorithm is evaluated compared with the MVTO algorithm [8,9]. There are ten homogeneous servers s_1, ..., s_{10} in a cluster CS of servers.

Algorithm 1. ESMVTO-VM Algorithm

---

procedure ESMVTO-VM(op^i)
  if $o_h.wcheck = \phi$ **then**
    if $op^i(d_h)$ = a *read* method **then**
      perform($op^i(d_h)$);
    else   ▷ $op^i(d_h)$ = a *write* method
      $o_h.wcheck = op^i(d_h)$;
    end if
  else   ▷ $o_h.wcheck \neq \phi$
    if $op^i(d_h)$ = a *read* method **then**
      perform($o_h.wcheck$); $o_h.wcheck = \phi$; perform($op^i(d_h)$);
    else   ▷ $op^i(d_h)$ = a *write* method
      if $op^i(d_h)$ absorbs $o_h.wcheck$ **then**
        perform($op^i(d_h)$);
      else
        perform($o_h.wcheck$); $o_h.wcheck = op^i(d_h)$;
      end if
    end if
  end if
end procedure

Every server s_t holds the same PCDAVM model and DAVM model [12]. Each server s_t is equipped with a dual-core CPU ($nc_t = 2$) and each core holds two threads ($ct_t = 2$). Hence, each server s_t holds four threads ($nt_t = 4$). There are forty threads and virtual machines in a cluster CS. The maximum read rate $maxR_t$ of every server s_t is 98.5 [MB/s]. The maximum write rate $maxW_t$ is 85.3 [MB/s]. $rw_t = 0.077$, $wr_t = 0.667$, $minE_t = 17$ [W], $CE_t = 1.1$ [W], $coE_t = 0.6$ [W], $thE_t = 0.5$ [W], $wE_t = 4$ [W], and $rE_t = 1$ [W]. The parameters of each server s_t are based on our previous studies [12]. One hundred objects o_1, ..., o_{100} are randomly allocated to forty virtual machines. The size of data for each object o_h is randomly selected between 10 to 100 [MB]. The number nt ($0 \leq nt \leq 3{,}000$) of transactions are initiated in a system. Each transaction issues five methods randomly selected on one hundred objects.

The TPE consumption [KJ] of the cluster CS to perform the number nt of transactions in the MVTO and ESMVTO-VM algorithms is shown in Fig. 1. For $0 \leq nt \leq 3{,}000$, the TPE consumption of the cluster CS in the ESMVTO-VM algorithm is lower than the MVTO algorithm since meaningless write methods are not performed on each object in the ESMVTO-VM algorithm.

The average execution time (average ET) [s] of each transaction in the ESMVTO-VM and MVTO algorithms are shown in Fig. 2. For $0 < nt \leq 3{,}000$, the average ET of each transaction in the ESMVTO-VM algorithm is shorter than the MVTO algorithm since each transaction can commit without performing meaningless write methods.

The TPE consumption of a homogeneous cluster CS and the average ET of each transaction can be more reduced in the ESMVTO-VM algorithm than the

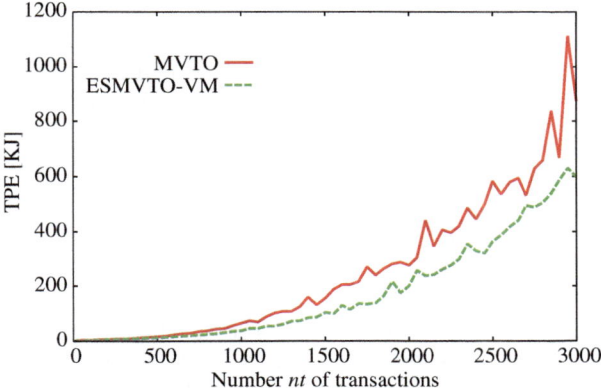

Fig. 1. TPE consumption [KJ].

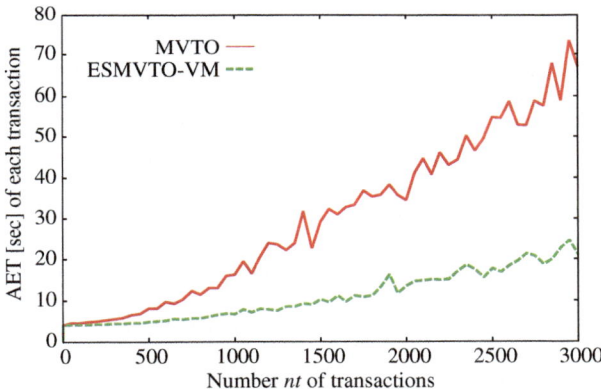

Fig. 2. Average ET [s].

MVTO algorithm. Therefore, the ESMVTO-VM algorithm is more useful than the MVTO algorithm.

5   Concluding Remarks

The ESMVTO-VM algorithm is newly proposed based on the MVTO algorithm to decrease the TPE consumption of a server cluster without performing meaningless write methods on each object. In the evaluation, the TPE consumption of a cluster of servers are shown to be more decreased in the ESMVTO-VM algorithm than the MVTO algorithm. The average ET in the ESMVTO-VM algorithm can be shorter than the MVTO algorithm. Following the evaluation, the ESMVTO-VM algorithm is more efficient than the MVTO algorithm for a homogeneous cluster of servers.

References

1. Nakamura, S., Enokido, T., Takizawa, M.: Time-based legality of information flow in the capability-based access control model for the Internet of Things. Concurr. Comput. Pract. Exp. **33**(23), e5944 (2021). https://doi.org/10.1002/cpe.5944
2. Enokido, T., Takizawa, M.: The redundant energy consumption laxity based algorithm to perform computation processes for IoT services. Internet Things **9** (2020). https://doi.org/10.1016/j.iot.2020.100165
3. Enokido, T., Duolikun, D., Takizawa, M.: An energy-efficient quorum-based locking protocol by omitting meaningless methods on object replicas. J. High Speed Netw. **28**(3), 181–203 (2022)
4. Enokido, T., Duolikun, D., Takizawa, M.: Energy-efficient concurrency control by omitting meaningless write methods in object-based systems. In: Proceedings of the 36th International Conference on Advanced Information Networking and Applications (AINA-2022), pp. 129–139, (2022)
5. Enokido, T., Duolikun, D., Takizawa, M.: Energy consumption laxity-based quorum selection for distributed object-based systems. Evol. Intell. **13**, 71–82 (2020)
6. Gray, J.N.: Notes on data base operating systems. In: Bayer, R., Graham, R.M., Seegmüller, G. (eds.) Operating Systems. LNCS, vol. 60, pp. 393–481. Springer, Heidelberg (1978). https://doi.org/10.1007/3-540-08755-9_9
7. Bernstein, P.A., Hadzilacos, V., Goodman, N.: Concurrency Control and Recovery in Database Systems. Addison-Wesley, Boston (1987)
8. Bernstein, P.A., Goodman, N.: Multiversion concurrency control - theory and algorithms. ACM Trans. Datab. Syst. **8**(4), 465–483 (1983)
9. Reed, D.: Naming and synchronization in a decentralized computer system. Technical Report. MIT/LCS/TR-205, Department Electrical Engineering and Computer Science, Massachusetts Institute of Technology (1978). http://hdl.handle.net/1721.1/16279
10. Garcia-Molina, H., Barbara, D.: How to assign votes in a distributed system. J. ACM **32**(4), 814–860 (1985)
11. Enokido, T., Duolikun, D., and Takizawa, M.: Energy-efficient multi-version concurrency control (EEMVCC) for object-based systems. accepted for publication. In: Proceedings of the 25th International Conference on Network-Based Information Systems (NBiS-2022) (2022)
12. Enokido, T., Takizawa, M.: The power consumption model of a server to perform data access application processes in virtual machine environments. In: Proceedings of the 34th International Conference on Advanced Information Networking and Applications (AINA-2020), pp. 184–192 (2020)

Towards a Blockchain-Based Crowdsourcing Method for Robotic Ontology Evolution

Wafa Alharbi[1,2(✉)] and Farookh Khadeer Hussain[1]

[1] School of Computer Science, University of Technology, Sydney, NSW, Australia
wafamatara.alharbi@student.uts.edu.au, walharbi@su.edu.sa,
farookh.hussain@uts.edu.au
[2] College of Computer Science and Information Technology, Shaqra University, Riyadh, Saudi Arabia

Abstract. All domain knowledge may change over time due to the addition or updating of new concepts, rules or requirements, especially when the domain is in its infancy such as robotics. This mandates the need to evolve the corresponding ontologies. In this paper, we propose a new blockchain-based robotic ontology evolution method using a crowdsourcing approach. The proposed method integrates the use of a crowdsourcing approach to allow a crowd of robotic experts to evolve the ontology and then use blockchain as a storage mechanism to store and manage the different versions of the ontology. We aim to solve some of the issues emanating from the methods proposed in the literature, such as preventing manipulation in ontologies, accessing different versions of ontologies, and involving a good number of domain experts in proposing changes and seeking their approval.

Keywords: Crowdsourcing · Blockchain · Ontology · Ontology Evolution

1 Introduction

Robotics technology is the science that incorporates designing, creation and operation of robots. It intersects with many disciplines such as electrical engineering, mechanical engineering and computer science [1]. As robotics field is rapidly growing, there is no agreement in the robotic knowledge terminologies. For example, the robot attributes are expressed differently depending on the manufacturers, the purpose of the robot, and the intended robot's users. Ontology comes to address similar issues by harmonizing the terminology of any domain and standardize them across all users to simplify gathering, managing, retrieving and presenting these data.

What Is Ontology?
An ontology is a knowledge representation that is used to share a common understanding of domain information among different users [2]. It is a data model that is used to reduce conceptual and terminological ambiguity, merge all relevant data from heterogeneous databases, enable knowledge sharing and much more [3]. It represents the knowledge of a domain as classes which describe the concepts and relationships between these classes.

L. Barolli (Ed.): CISIS 2023, LNDECT 176, pp. 21–29, 2023.
https://doi.org/10.1007/978-3-031-35734-3_3

A key issue with ontology is the evolution of ontologies [4]. As domain knowledge changes overtime, it necessitates a change in the corresponding domain ontology. Unfortunately, a number of ontologies are static in nature, in the sense that once engineered, they are unable to evolve or change over time to accommodate the new (or in some instances changed) knowledge. This has given rise to the notion of 'ontology evolution'.

Crowdsourcing has been proposed as one of the approaches to enable ontology evolution [5]. In the crowdsourcing method, a requester identifies a set of tasks that have to be undertaken and the crowd voluntarily undertakes these tasks. We apply the concept of crowdsourcing to evolve robotics ontology, where the crowd (robotic expert community) is involved in improving or updating the robotic ontology. Using approaches such as crowdsourcing for ontology evolution will result in ontology transitioning from a static semantic body of knowledge to a dynamic semantic body of knowledge.

The remainder of this paper is structured as follows: in Sect. 2, we review some of the methods proposed in the literature. In Sect. 3, we discuss our proposed approach, crowdsourcing blockchain-based robotic ontology evolution. In Sect. 4, we illustrate an example of the proposed method and we conclude the work in Sect. 5.

2   Ontology Evolution Related Work

Ontology evolution is necessary when an expert's perception of domain knowledge changes over time. Hence, there is a need for the corresponding domain ontology to evolve.

When reading through the existing literature on ontology evolution, it is observed that the studies in this field can be classified into three main groups: the first group proposes ontology evolution methods; the second group focuses on storing and versioning the ontology changes without paying any attention to the ontology evolution methodologies; and the third group measures the impact of ontology evolution on dependent applications which is not in the scope of this study.

Many studies have proposed new methodologies for ontology evolution. One of the proposed methods to enable ontology evolution is "Ontology Evolution Wiki" which is based on the crowdsourcing technique of Wikipedia, where different community users are involved in the evolution of ontologies by proposing changes/updates and discussing them with domain experts until reaching an agreement [6].

In a similar vein, the proposed method in [5] employed the use of a crowdsourcing technique to allow learners in an online learning environment to contribute to ontology generation and evolution. This approach has been tested on three learning activities: a semantic annotation activity, a knowledge graph activity, and a reverse quiz activity. In [7], the authors presented a semantic navigation support service and tool for ontology evolution called OntoAssist. It evolves a base ontology by aggregating knowledge from a large number of actual web users.

Other studies in the literature tried to solve the issue of ontology history access. [8] proposed a method to manage and simplify the versioning of an ontology by storing all the evolved versions of an ontology in a rational database. This involves identifying all types of primitive operations that are used to evolve an ontology and identifying each version of the ontology with a unique timestamp. Then, any version of the ontology can be manipulated, reconstructed or retrieved using SQL queries. In a similar premise, [9] proposed a hybrid approach which also used rational databases to store ontology versions. In their approach, they tried to build a space-efficient storage strategy that keeps a complete history of ontology evolution by capturing the semantics of a domain using a predefined reference version and the most recent version of the ontology. A formal framework was developed by [10] to track the changes in an ontology by assigning a timestamp to each version of the ontology. The framework then takes two states of an ontology as input and returns a description of the changes. The framework treats the ontology changes in four levels: concepts, concepts' relations, instances and their relations.

In robotics domain, there are number of ontologies that have been built in the literature. Some of these works employ ontologies to support robot capabilities in the working environment by defining a conceptualization of all instructions and situations of the required task [11], while other studies built ontologies for technical terminologies of robots and robot parts [12]. For instance, an IEEE standard ontology termed as "Core Ontology for Robotics and Automation (CORA)" is developed in [13], that ontology describes the physical design of robots and it is currently being used to build other robotic ontologies that cover other robotic sub-domains.

The authors of [14] built a domain ontology that is designed to facilitate human-robot collaboration by considering three contexts: the Environment, the Behavior and the Production contexts. However, none of these works discussed how to evolve the proposed robotic ontologies.

In this paper, we are aiming to address the limitations in the current ontology evolution methods such as:

1. All robotic ontologies in the literature are static. We aim to allow robotic ontology evolution to change the nature of robotic ontologies to be dynamic ontologies.
2. There is no automated method to recover the previous version of the ontology before applying the changes [4].

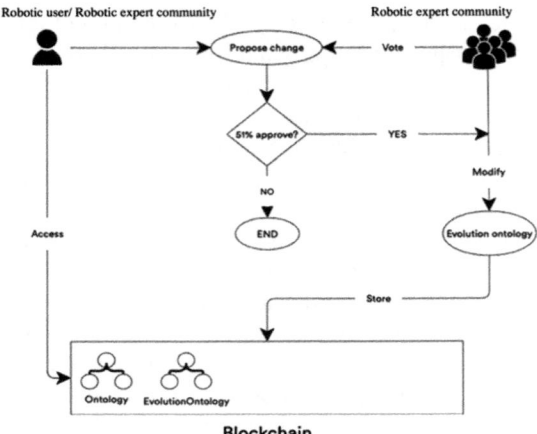

Fig. 1. Overview of the system architecture.

3   Crowdsourcing Blockchain-Based Robotic Ontology Evolution

Crowdsourcing Blockchain-based robotic ontology evolution is one solution to address some of the challenging issues facing the proposed ontology evolution methods in the literature. It is based on the crowdsourcing method proposed in [6] which allows the crowd to collaborate to evolve ontologies by sending change requests to domain experts who are required to approve the ontology change requests through a majority vote. We propose the use of blockchain as a platform for proposing ontology evolution requests and seeking the approval of the domain experts via a consensus vote on blockchain.

The system architecture is shown in Fig. 1. The crowd of people who are participating in ontology evolution in this method include any community member (robotic user in our case) or robotic expert who is using the robotic ontology, however, the robotic ontology can only be changed/updated after obtaining approval from the robotic expert community.

As shown in Fig. 1, the robotic expert and the robotic user community can access and use the ontology, and they can also submit a change proposal to the robotic expert community who are requested to vote on the proposal. Once the change proposal is validated and approved by the majority of robotic experts, the robotic ontology will be refined, updated and recorded in the blockchain ledger as a new block. Then, the robotic experts/ robotic user community members are able to access and use the evolved ontology again.

Using blockchain allows us to capture and control the crowdsourcing method and track changes in ontologies. The consensus protocols of blockchain assist in the process of reaching consensus between different people in the crowd. Moreover, blockchain's key property of transparency makes it an excellent option for this method since all previous versions of ontologies are stored and secured in the blockchain which enables a specific version to be recovered when needed.

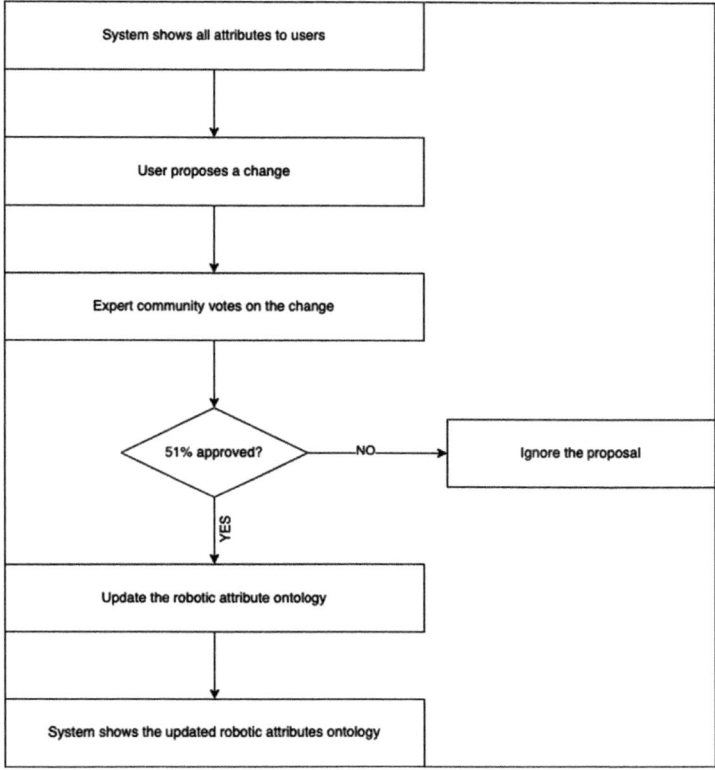

Fig. 2. System flow diagram.

4 Illustrative Example of Evolving Robotic Attribute Ontology Using Crowdsourcing Blockchain-Based Robotic Ontology Evolution Method

We built an ontological manifestation that encapsulates the attributes of robots and their relationship in attempt to standardize robotic attributes among all users and suppliers of robots. We termed that ontology as "Robot Attribute ontology (RAO)". RAO facilitates building any methodology that requires managing robotic attributes such as robot selection methods. Robot attributes ontology (RAO) is shown in Fig. 3. Then we allow the crowd to participate in the evolution of RAO. The system flow diagram of this example is illustrated in Fig. 2.

Figure 4 shows all initial attributes that are shown to the user community. In Fig. 5, a user proposes a change to the robotic attributes ontology. The expert community is allowed to vote on that change as shown in Fig. 6. 51% or more of the robotic experts have to approve the change before updating the robotic attributes as in Fig. 7.

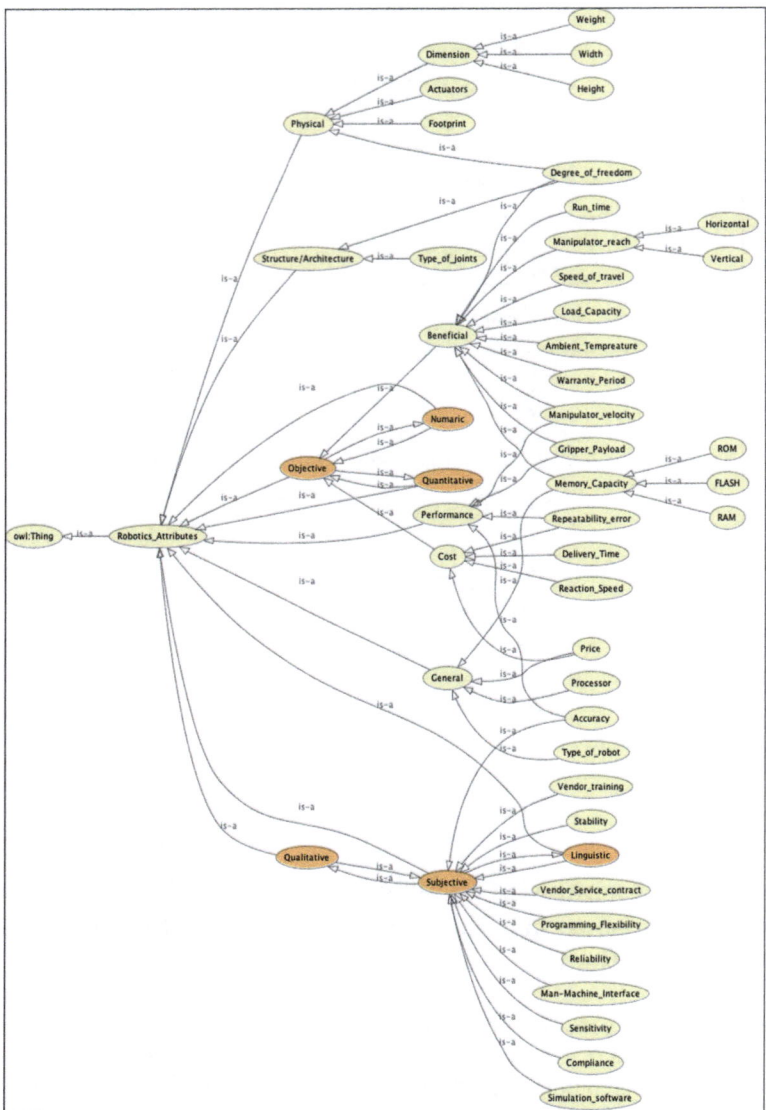

Fig. 3. Initial Robot Attribute Ontology.

Fig. 4. Initial attributes ontology.

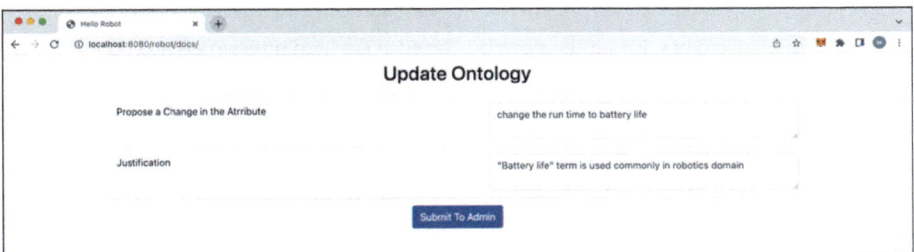

Fig. 5. Proposing a change by a user.

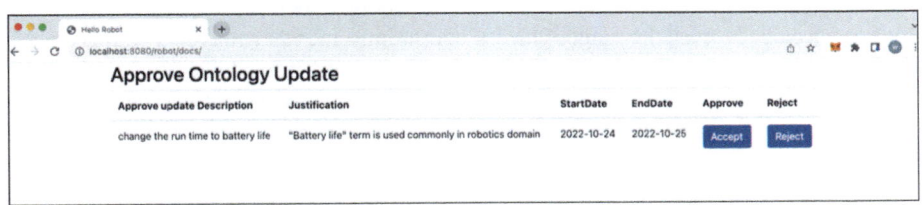

Fig. 6. Expert voting process.

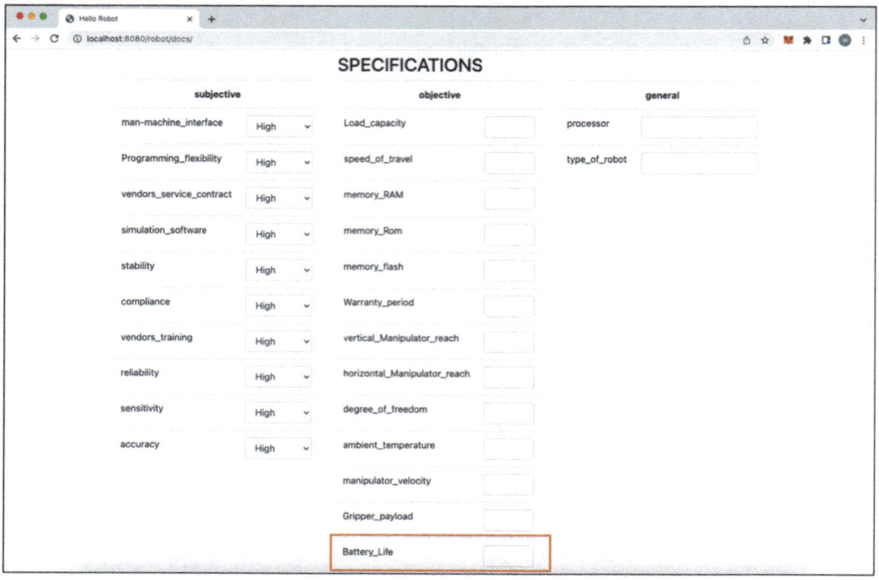

Fig. 7. Updated ontology after changing the attribute "run time" to "Battery life".

5 Conclusion

In this paper, we proposed a new method for robotic ontology evolution. The proposed method is based on using the crowdsourcing method that engages a community of robotic experts and robotic users to participate in proposing changes in ontologies. It also integrates the use of blockchain as the platform for proposing changes and seeking their approval (via a consensus vote). The approved changed will then be manually made to the ontology by the robotic experts. The robotic experts participate in the voting process for each proposed change. Once a community member makes a change to the ontology, this change will not be added to the chain until it has been approved by the majority of experts, then it will be added as a block at the end of the ontology chain. Using blockchain as a storage mechanism has numerous benefits: the immutability feature of blockchain ensures the stored ontologies are secure and the transparency of blockchain allows reliable and trusted old versions of ontologies to be accessed.

References

1. Vrontis, D., Christofi, M., Pereira, V., Tarba, S., Makrides, A., Trichina, E.: Artificial intelligence, robotics, advanced technologies and human resource management: a systematic review. Int. J. Hum. Resour. Manag. **33**(6), 1237–1266 (2022)
2. Kim, M., et al.: Efficient regression testing of ontology-driven systems. In: Proceedings of the 2012 International Symposium on Software Testing and Analysis, pp. 320–330 (2012)
3. Ahmad, M.N., Zakaria, N.H., Sedera, D.: Ontology-based knowledge management for enterprise systems. Int. J. Enterp. Inf. Syst. **7**(4), 64–90 (2011)

4. Khattak, A.M., Batool, R., Khan, Z.P., Mehmood, A., Sungyoung, L.E.E.: Ontology evolution and challenges. J. Inf. Sci. Eng. **29**(5), 851–671 (2013)

5. Wang, Q., Ding, G., Yu, S.: Crowdsourcing mode-based learning activity flow approach to promote subject ontology generation and evolution in learning. Interact. Learn. Environ. **27**(7), 965–983 (2019)

6. Aseeri, A., Wongthongtham, P., Wu, C., Hussain, F.K.: Towards social network based ontology Evolution Wiki for an ontology evolution. In: Proceedings of the 10th International Conference on Information Integration and Web-based Applications & Services, Linz, Austria (2008)

7. Lin, H., Davis, J., and Zhou, Y.: 'Ontological services using crowdsourcing', in Editor (Ed.)^(Eds.): 'Book Ontological services using crowdsourcing' (2010, edn.), pp.

8. Grandi, F.: Dynamic class hierarchy management for multi-version ontology-based personalization. J. Comput. Syst. Sci. **82**(1, Part A), 69–90 (2016)

9. Bayoudhi, L., Sassi, N., Jaziri, W.: A hybrid storage strategy to manage the evolution of an OWL 2 DL domain ontology. Proc. Comput. Sci. **112**, 574–583 (2017)

10. Kozierkiewicz, A., Pietranik, M.: A Formal Framework for the Ontology Evolution, pp. 16–27. Springer International Publishing (2019)

11. Olivares-Alarcos, A., et al.: A review and comparison of ontology-based approaches to robot autonomy. Knowl. Eng. Rev. **34** (2019)

12. Schlenoff, C., et al.: An IEEE Standard Ontology for Robotics and Automation, pp. 1337–1342. IEEE (2012)

13. Prestes, E., et al.: Towards a core ontology for robotics and automation. Rob. Autom. Syst. **61**(11), 1193–1204 (2013)

14. Umbrico, A., Orlandini, A., Cesta, A.: An ontology for human-robot collaboration. Proc. CIRP **93**, 1097–1102 (2020)

Performance Evaluation of DTAG-Based Recovery Method for DTN Considering a Real Urban Road Model

Shura Tachibana[1], Shota Uchimura[1], Makoto Ikeda[2](✉) [iD], and Leonard Barolli[2] [iD]

[1] Graduate School of Engineering, Fukuoka Institute of Technology, 3-30-1 Wajiro-Higashi, Higashi-Ku, Fukuoka 811-0295, Japan
{mgm23106,mgm21102}@bene.fit.ac.jp
[2] Department of Information and Communication Engineering, Fukuoka Institute of Technology, 3-30-1 Wajiro-Higashi, Higashi-Ku, Fukuoka 811-0295, Japan
makoto.ikd@acm.org, barolli@fit.ac.jp

Abstract. In this paper, we present the performance of a Dynamic Threshold-based Anti-packet Generation (DTAG) method for Delay-/Disruption-/Disconnection-Tolerant Networking (DTN) considering a real road map scenario. We imported road data around Kagoshima central station, which is located in South Kyushu region in Japan. We considered the combination of DTAG method and anti-packet with Epidemic protocol. Based on the simulation results, we observed that the combination of the proposed DTAG method with the Epidemic protocol reduces overhead regardless the number of nodes.

Keywords: DTAG · DTN · Anti-packet · Epidemic · Recovery Method

1 Introduction

Recently, cellular networks improved their high-speed handover technology, but it is hard to connect wireless LANs and ad-hoc nodes because they use high frequencies, move at fast speeds and only have a short amount of time to connect to each other. Thus, communication systems for storing messages are attracting attention as well. Delay-/Disruption-/Disconnection-Tolerant Networking (DTN) is one of stored-based message delivery methods for land and space when are many disconnections [5–7,9,14]. In DTN, nodes continuously duplicate the original message to neighbors until the destination receives the message. The delivery management is important because there are many duplicated messages scattered using DTN technique. The basic architecture and specification of bundle message can be checked in [4,11] and some implementations are under development [2,3]. Epidemic [8,13] is typical contact-based DTN protocol. It resembles COVID-19, which is widespread all over the world during recent years.

In [12], we proposed a Dynamic Threshold-based Anti-packet Generation (DTAG) method, which considers replication progress of the adjacent nodes. The DTAG performed well for utilizing storage states and reducing message overhead for a grid road model.

L. Barolli (Ed.): CISIS 2023, LNDECT 176, pp. 30–37, 2023.
https://doi.org/10.1007/978-3-031-35734-3_4

In this paper, we present the performance of the proposed DTAG method combined with Epidemic protocol considering a real road map scenario. For evaluation, we considered a road model around Kagashima central staion in Japan. Based on the simulation results, we observed that the combination of the proposed DTAG method with the Epidemic protocol reduces overhead regardless the number of nodes.

The paper is structured as follows. Section 2 introduces the contact-based DTN protocols. In Sect. 3, we explain the approach for combining the DTAG method with the Epidemic protocol. The simulation scenario and results are presented in Sect. 4. The paper concludes with Sect. 5.

2   Contact-Based DTN Protocols

The Epidemic protocol involves sending a group of messages along with two control messages throughout the network. To keep track of the messages stored by each node, a Summary Vector (SV) is periodically distributed by every node. The SV is checked by nodes against their stored data, and in case an unknown bundle message is detected, the node generates a REQUEST.

However, the constant copying and storing of bundle messages by nodes to their neighbors poses a significant challenge to network resources and storage capacity. The copied bundle messages may remain in the storage even after being received by destinations and being deleted too late by recovery methods such as anti-packet.

Typical anti-packets are transmitted by destinations and contain a list of bundles received. Nodes are able to delete the messages from their storage once they have been verified and distributed the anti-packet to other nodes. However, anti-packets may have adverse effects on network resources.

3   Combination of DTAG Method with Epidemic Protocol

We combine the proposed DTAG Method with Epidemic protocol. The flowchart of the proposed DTAG method is illustrated in Fig. 1. The method has a procedure for handling received messages, including SVs, REQUESTs, MESSAGEs and anti-packets. The DTAG approach generates anti-packets at intermediate nodes before the bundle messages reach their destination. To calculate the number of new and identical messages, the node counts the received SVs without any predetermined threshold in the DTAG procedure.

If the result of Eq. (1) is true, the DTAG method generates an anti-packet. Otherwise, if Eq. (1) is false, the message is stored in the node's storage.

$$\frac{\text{Count New Msgs}}{\text{Count All Msgs}} \geq \frac{\text{Count Same Msgs}}{\text{Count All Msgs}}, (\text{Count New Msgs and All Msgs} \neq 0). \quad (1)$$

4   Evaluation of Proposed Approach

To evaluate the effectiveness of the proposed approach, we use four metrics: mean delivery ratio, mean delay, mean overhead and mean storage usage. The mean values are

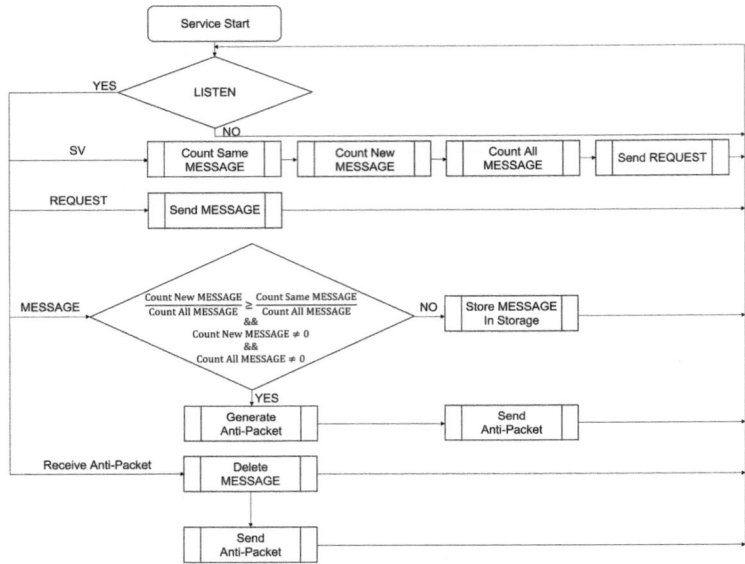

Fig. 1. Flowchart of the proposed method.

derived from ten different seeds and account for various vehicle movement patterns. To conduct the evaluation, we implement the DTAG method on the Scenargie platform [10], which is combined with the Epidemic protocol.

4.1 Simulation Environment

To evaluate the proposed approach, we employ a real-world road model in the vicinity of Kagoshima central station, Japan with vehicle densities ranging from 50 to 150 vehicles per square kilometer. The road data are obtained from OpenStreetMap. Table 1 presents the simulation parameters utilized in our experiments. The road model is depicted in Fig. 2. During simulations, the vehicles move on the roads based on the random waypoint mobility model. We evaluate the proposed approach under two different scenarios.

- I: DTAG method with Epidemic protocol.
- II: Conventional anti-packet with Epidemic protocol.

The bundle message are initially sent by the originator to the intermediate nodes, which then relay the message to other nodes until it reaches the destination. The simulations are conducted for a duration of 600 seconds. We use a radio frequency of 5.9 GHz and employ the ITU-R P.1411 [1] in our simulations.

4.2 Simulation Results

The simulation results of the mean delivery ratio are illustrated in Fig. 3. There is no significant difference in the results between the proposed and conventional methods at 50 vehicles/km^2. However, the performance difference is increased from 75 vehicles/km^2

Table 1. Simulation parameters for road model around Kagoshima central station.

Items	Setup Values
Number of Vehicles	50, 75, 100, 125, 150 [vehicles/km^2]
Vehicles Velocity	min: 8.333 [m/s], max: 16.666 [m/s]
Simulation Time	600.0 [s]
DTN: Recovery Method	Epidemic with DTAG Method and Anti-packet
Bundle Message Size	500 [bytes]
Bundle Message Generation Start and End Time	10.0 to 400.0 [s]
Bundle Message Generation Interval	10.0 [s]
Summay Vector Generation Interval	1.0 [s]

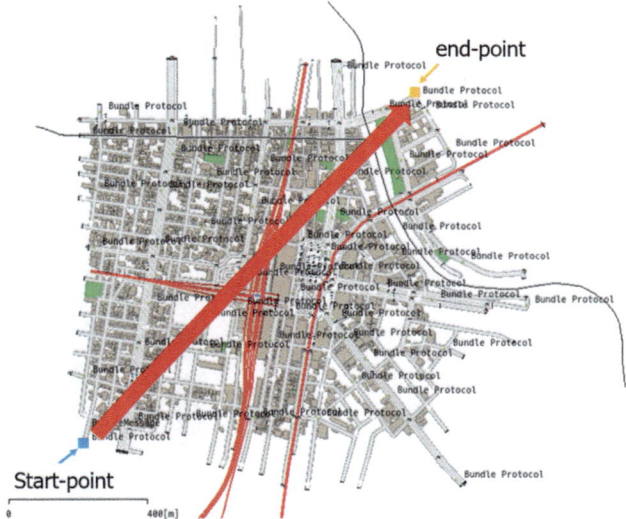

Fig. 2. Road model around Kagoshima central station in Japan.

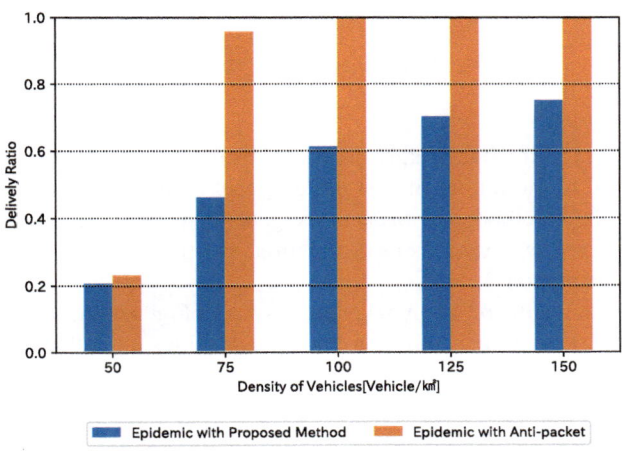

Fig. 3. Mean delivery ratio for different cases.

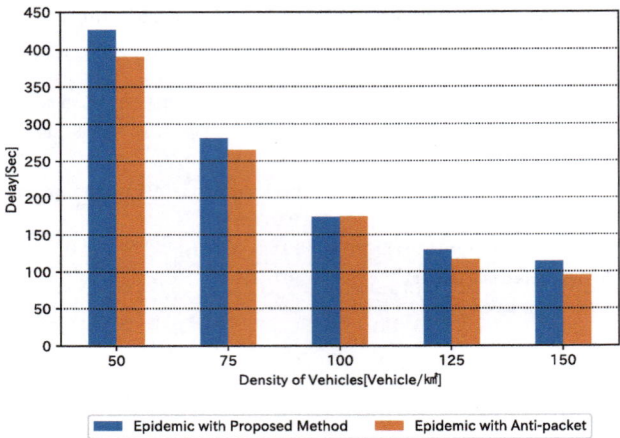

Fig. 4. Mean delay for different cases.

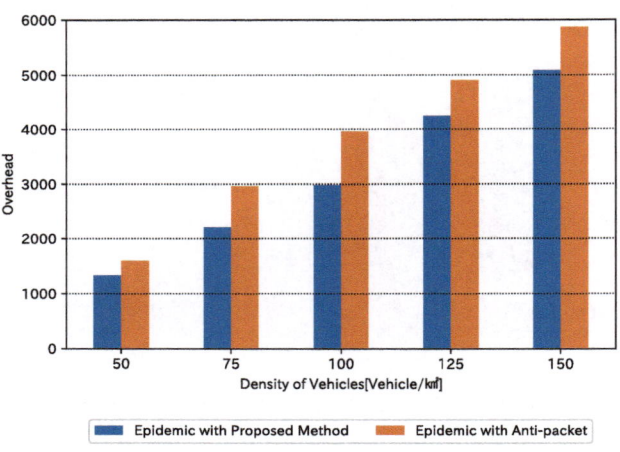

Fig. 5. Mean overhead for different cases.

to 150 vehicles/km^2, which gradually decreased as the vehicle density increased. This may be attributed to the timing of anti-packet generation in the proposed method. In the proposed DTAG method, anti-packets are generated when the condition of Eq. (1) is true before the message reaches its destination, and the messages are deleted before they reach the destination.

The simulation results of mean delay are shown in Fig. 4. When the proposed approach is used, the delay time is longer for vehicle densities other than 100 vehicles per km^2. This is because the condition for anti-packet generation is met before the messages reached the destination, causing the messages to try to reach th final destination and resulting in longer delays.

In Fig. 5, the average overhead results show that the DTAG method can reduce the number of duplicate messages. This reduction was consistently observed even when

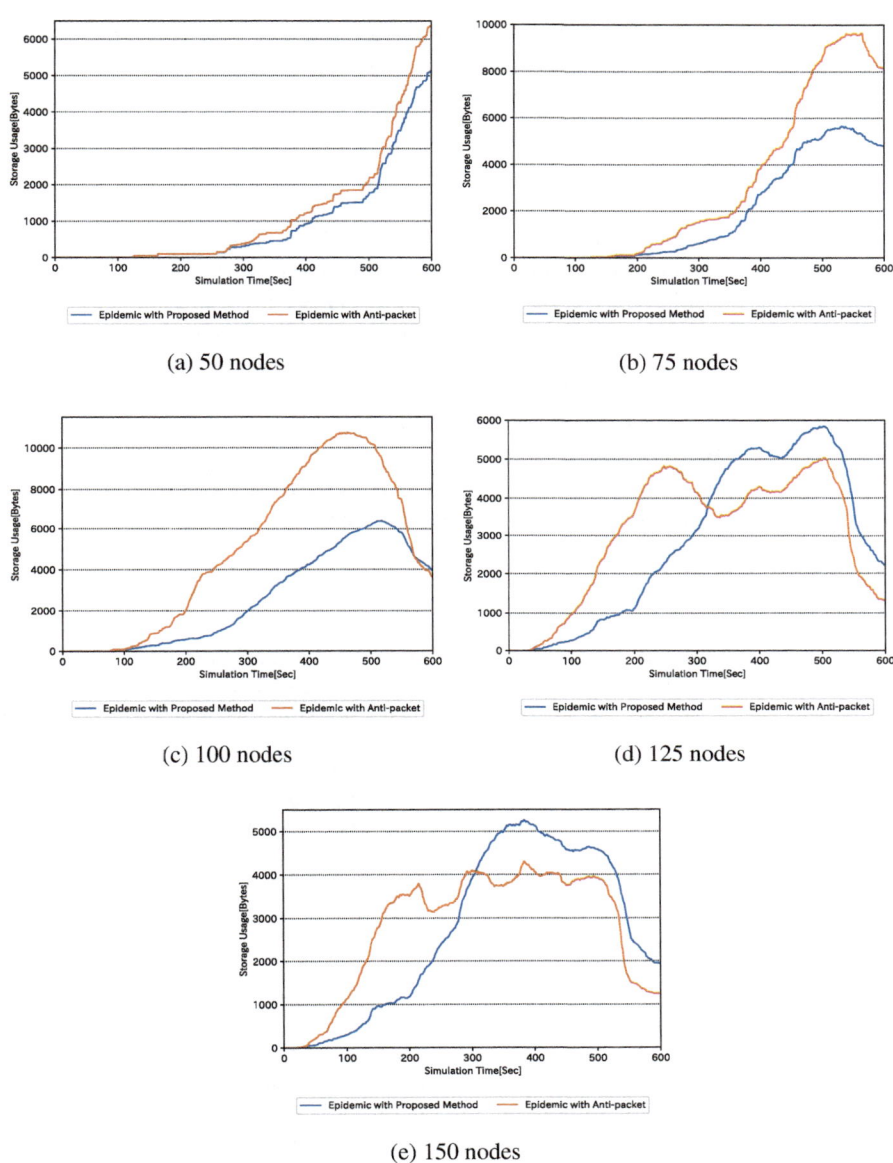

(a) 50 nodes

(b) 75 nodes

(c) 100 nodes

(d) 125 nodes

(e) 150 nodes

Fig. 6. Storage usage for different vehicles.

the number of vehicles increased, indicating the robustness of the proposed method for managing message redundancy.

The storage usage for different vehicles is illustrated in Fig. 6. The proposed approach showed lower storage usage from 50 vehicles/km² to 100 vehicles/km², but the situation reversed towards the end of the duration as the vehicle density increased. This is because the number of messages increased with more vehicles, and some vehicles were unable to receive anti-packets, leading to an increase in storage usage.

5   Conclusions

In this paper, we have presented the performance evaluation of the proposed DTAG method combined with Epidemic protocol considering a real-world road map scenario. Specifically, we considered the road model around Kagoshima central station in Japan and implemented two scenarios by combining the proposed DTAG method and conventional anti-packet method with Epidemic protocol. We conducted simulations to evaluate the implemented scenarios. Based on the simulation results, we observed that the combination of the proposed DTAG method with Epidemic protocol reduces overhead regardless the number of nodes.

In future work, we plan to explore other recovery methods and evaluate other parameters to improve the performance of proposed approach.

References

1. Recommendation ITU-R P.1411-11: Propagation data and prediction methods for the planning of short-range outdoor radiocommunication systems and radio local area networks in the frequency range 300 MHz to 100 GHz. ITU (2019). https://www.itu.int/rec/R-REC-P.1411-11-202109-I/en
2. Baumgärtner, L., Höchst, J., Meuser, T.: B-DTN7: browser-based disruption-tolerant networking via bundle protocol 7. In: Proceedings of the International Conference on Information and Communication Technologies for Disaster Management (ICT-DM-2019), pp. 1–8 (2019). https://doi.org/10.1109/ICT-DM47966.2019.9032944
3. Burleigh, S., Fall, K., E. Birrane, I.: Bundle protocol version 7. IETF RFC 9171 (Standards Track) (2022)
4. Cerf, V., et al.: Delay-tolerant networking architecture. IETF RFC 4838 (Informational) (2007)
5. Davarian, F., et al.: Improving small satellite communications and tracking in deep space - a review of the existing systems and technologies with recommendations for improvement. part ii: Small satellite navigation, proximity links, and communications link science. IEEE Aeros. Electron. Syst. Maga. **35**(7), 26–40 (2020). https://doi.org/10.1109/MAES.2020.2975260
6. Fall, K.: A delay-tolerant network architecture for challenged internets. In: Proceedings of the International Conference on Applications, Technologies, Architectures, and Protocols for Computer Communications, SIGCOMM 2003, pp. 27–34 (2003). https://doi.org/10.1145/863955.863960
7. Fraire, J.A., Feldmann, M., Burleigh, S.C.: Benefits and challenges of cross-linked ring road satellite networks: a case study. In: Proceedings of the IEEE International Conference on Communications (ICC-2017), pp. 1–7 (2017). https://doi.org/10.1109/ICC.2017.7996778
8. Ramanathan, R., Hansen, R., Basu, P., Hain, R.R., Krishnan, R.: Prioritized epidemic routing for opportunistic networks. In: Proceedings of the 1st International MobiSys Workshop on Mobile Opportunistic Networking (MobiOpp 2007), pp. 62–66 (2007). https://doi.org/10.1145/1247694.1247707

9. Rüsch, S., Schürmann, D., Kapitza, R., Wolf, L.: Forward secure delay-tolerant networking. In: Proceedings of the 12th Workshop on Challenged Networks (CHANTS-2017), pp. 7–12 (2017). https://doi.org/10.1145/3124087.3124094
10. Scenargie: Space-time engineering, LLC. http://www.spacetime-eng.com/
11. Scott, K., Burleigh, S.: Bundle protocol specification. IETF RFC 5050 (Experimental) (2007)
12. Uchimura, S., Azuma, M., Ikeda, M., Barolli, L.: DTAG: A dynamic threshold-based anti-packet generation method for vehicular dtn. In: Proceedings of the 37th International Conference on Advanced Information Networking and Applications (AINA-2023) (2023)
13. Vahdat, A., Becker, D.: Epidemic routing for partially-connected ad hoc networks. Duke University, Technical report (2000)
14. Wyatt, J., Burleigh, S., Jones, R., Torgerson, L., Wissler, S.: Disruption tolerant networking flight validation experiment on NASA's EPOXI mission. In: Proceedings of the 1st International Conference on Advances in Satellite and Space Communications (SPACOMM-2009), pp. 187–196 (2009). https://doi.org/10.1109/SPACOMM.2009.39

An Energy-Aware Dynamic Algorithm for the FTBFC Model of the IoT

Dilawaer Duolikun[1(✉)], Tomoya Enokido[2], and Makoto Takizawa[1]

[1] RCCMS, Hosei University, Tokyo, Japan
dilewerdolkun@gmail.com, makoto.takizawa@computer.org
[2] Faculty of Business Administration, Rissho University, Tokyo, Japan
eno@ris.ac.jp

Abstract. Various types of devices are scalably interconnected in addition to computers like servers in the IoT (Internet of Things) and the IoT consumes large electric energy. In the FC (Fog Computing) model of the IoT, application processes for calculating on sensor data from devices are executed on both servers and fog nodes. The TBFC (Tree-Based FC) model is proposed to reduce the total energy consumption of the IoT. The FTBFC (Flexible TBFC) model is also proposed, where operations for changing the tree structure. In this paper, a DC (Dynamical Change) algorithm is newly proposed to change the FTBFC model where a node is changed each time the node is congested Here, the tree structure and processes supported by nodes are changed to reduce the total energy consumption. In the evaluation, the total energy consumption of the IoT is shown to be reduced in the DC algorithm.

Keywords: Green computing · FTBFC model · IoT · Fog computing (FC) model · DC algorithm · Energy consumption

1 Introduction

The IoT (Internet of Things) [3] is a scalable information system where device nodes and clouds of servers are interconnected in networks. Accordingly, the IoT consumes large electric energy [1].

In the fog computing (FC) model [3] of the IoT, application processes to process sensor data sent by devices are executed on both servers and fog nodes differently from the CC (Cloud Computing) model [2]. First, sensor data is processed by fog nodes and the processed data is sent to other fog nodes. Thus, the processed data is finally delivered to servers. Because servers can process smaller data, smaller energy is consumed by servers than the CC model. On the other hand, fog nodes consume energy to process sensor data. The TBFC (Tree-Based FC) model [15,18–22] is proposed to reduce the total energy consumption of the FC model. In the TBFC mode, nodes are interconnected in a tree structure, whose leaf node shows a device node and root node is a cloud of servers. A device node is composed of sensors and actuators. Each node only communicates with the child nodes and parent node. Each node receives data from the child nodes. The node then processes the data and sends the processed data to the parent node. The FTTBFC (Fault-Tolerant TBFC) model [22–24] is proposed to be tolerant of node faults. In addition to node faults, traffic of sensor data is fluctuated. In order to adapt to the fluctuation of sensor data, the DTBFC (Dynamic TBFC) model [18, 24] is proposed. Processes supported by a node f are changed to reduce the energy consumption of congested nodes. In the TBFC model, nodes with which each node can communicate are fixed. In the NFC (network FC) model [25], each fog node f negotiates with another fog node g on whether or not g can process data sent by f. Operations to change the tree are proposed In the FTBFC (Flexible TBFC) model [16].

A DC (Dynamic Change) algorithm is newly proposed in this paper, by which parent-child relations and processes of nodes are changed each time more data units are queued in the nodes. The total energy consumption is shown to be reduced in the DC algorithm in the evaluation.

In Sect. 2, we present the FTBFC model. In Sect. 3, we propose the DC algorithm. In Sect. 4, we evaluate the DC algorithm.

2   The FTBFC Model

2.1   TBFC Model

Sensor data from devices are processed and actions to be executed by devices are decided by an application process in the IoT. In the CC model of the IoT, devices deliver sensor data to servers and the application process for the sensor data is executed on a server in a cloud. On the other hand, in the FC model [3], application processes are executed on both servers and fog nodes. A fog node receives sensor data from devices and sends the processed data to another fog node. Thus, processed data is delivered to servers by fog nodes.

In the TBFC (Tree-Based FC) model [15, 18–22], fog nodes are interconnected with clouds of servers and devices in a tree structure. A root node is a cloud of servers and a leaf node shows a device node. Each node only communicates with the child nodes and parent node. A notation f_I of a node means that I is a sequence $\langle i_1 i_2 \ldots i_{l-1} \rangle$ $(l \geq 1)$, i.e. a path $\langle f, f_{i_1}, f_{i_1 i_2}, \ldots, f_{i_1 i_2 \ldots i_{l-1}}(= f_I) \rangle$ from the root node f to f_I. A node f_I has b_I (≥ 0) child nodes $f_{I1}, \ldots, f_{Ib_I}$. Here, f_I is a *single-child* node and a *multi-child* node if $b_I = 1$ and $b_I > 1$, respectively. An edge node f_I has a device node d_I as a child node. A device node is an abstraction of a collection of devices. Let DV be a set of devices in the IoT.

We assume an application P is a sequence $\langle p_1, \ldots, p_m \rangle$ $(m \geq 1)$ of processes, where p_1 is a *top* and p_m are a *tail*. First, the tail process p_m receives a DU dt_{m+1} of sensor data from devices. Each process p_i calculates dt_i on dt_{i+1} from p_{i+1} and sends dt_i to p_{i-1} $(i = m, m-1, \ldots, 2)$. The top process p_1 finally obtains dt_1 which shows actions to be executed by actuators. dt_1 is sent to devices to activate actuators of the devices.

Each node f_I supports a subsequence P_I $(= \langle p_{t_I}, p_{t_I+1}, \ldots, p_{l_I} \rangle)$ of the application P, where $1 \leq t_I \leq l_I \leq m$, p_{t_I} is a top, and p_{l_I} is a tail. Here, let P_0

stand for a subsequence of processes supported by the root node f. In a path $\langle f, f_{i_1}, f_{i_1 i_2}, \ldots, f_{i_1 i_2 \ldots i_l} \rangle$ from f to each edge $f_{i_1 i_2 \ldots i_l}$, a concatenation of the subsequences P_0, P_{i_1}, $P_{i_1 i_2}$, $\ldots$, $P_{i_1 i_2 \ldots i_l}$ is P. A node f_I receives an input DU id_{Ii} from each child node f_{Ii} $(i = 1, \ldots, b_I)$. Let ID_I be $\{id_{I1}, \ldots, id_{Ib_I}\}$ of the input DUs of a node f_I.

2.2   Power Consumption and Computation Models

A fog node f_I follows the SPC (Simple Power Consumption) model [5–9] in the power consumption models [10–12]. Here, the power consumption NE_I [W] of f_I is maximum xE_I if at least one process is active, else minimum mE_I. A node f_I also consumes power RE_I and SE_I [W] to receive and send data, respectively.

The computation rate CR_I [bps] is the number of bits which f_I processes for one second. CR denotes the computation rate of the root node f. The computation ratio $cr_I = CR_I / CR$ of the Raspberry Pi3 node f_I [26] is 0.18 for a HP DL360 server node f [4,19]. It takes $PT_{Ih}(x)$ [sec] to process a DU of size x by a process p_h in the subsequence P_I. The computation residue R_h [bit] of a process p_h is assumed to be initially x or x^2. A process p_h whose R_h is x and x^2 is referred to as *linear* and *square*, respectively. The execution time T_h [sec] is initially zero. R_h is decremented by CR_I and T_h is incremented by one for each second. If $R_h \leq 0$, p_h terminates. Thus, T_h is the execution time $PT_h(x)$ to calculate on the input DU of size x. Here, the size $|od_h|$ of the output DU od_h of a process p_h is $\rho_h \cdot x$ where ρ_h is a *reduction ratio* of p_h. The total time $ET_I(x)$ of f_I to calculate a DU of size x is $PT_{l_I}(x) + PT_{l_I - 1}(\rho_{l_I} \cdot x) + \ldots + PT_{t_I}(\rho_{l_I} \cdot \ldots \cdot \rho_{t_I+1} \cdot x)$. The size $|od_I|$ is $\rho_I \cdot x$ where ρ_I is $\rho_{l_I} \cdot \ldots \cdot \rho_{t_I}$.

The total execution time of a node f_I for input DUs ID_I of size x is $TT_I(x) = RT_I(x) + ET_I(x) + ST_I(\rho_I \cdot x)$ [sec]. The energy $TE_I(x) = RT_I(x) \cdot RE_I + ET_I(x) \cdot NE_t + ST_I(\rho_I \cdot x) \cdot SE_I$ [J] is consumed by f_I.

2.3   Change Operations

If some node f_I is congested, f_I is a point of energy consumption bottleneck. Here, if some input DUs of f_I are sent to another node, the energy consumption of f_I gets smaller since the node f_I receives and calculates on smaller amount of input DUs. Each node f_I has a receipt queue RQ_{Ii} to receive input DUs from each child node f_{Ii} $(1 \leq i \leq b_I)$. ql_{Ii} shows the length, i.e. number of input DUs of a receipt queue RQ_{Ii}. The total queue length ql_I is the total number $\sum_{i=1}^{b_I} ql_{Ii}$ of input DUs in the receipt queues $RQ_{I1}, \ldots, RQ_{Ib_I}$. In this paper, a node f_I is referred to as *more congested* than another node f_J $(f_I > f_J)$ if $ql_I > ql_J$.

For a process sequence $P = \langle p_1, \ldots, p_m \rangle$, suppose a node f_I supports a process subsequence $P_I = \langle p_{t_I}, p_{t_I+1}, \ldots, p_{l_I} \rangle$ and its child node f_{Ii} supports a subsequence $P_{Ii} = \langle p_{t_{Ii}}, p_{t_{Ii}+1}, \ldots, p_{l_{Ii}} \rangle$. Here, $p_{l_I+1} = p_{t_{Ii}}$. For every pair of child nodes p_{Ii} and p_{Ij}, the top processes $p_{t_{Ii}}$ and $p_{t_{Ij}}$ are the same while the tail processes p_{l_I} and p_{l_I} may not be the same.

In order to change the tree T, we consider the following change operations on a node f_I for changing the tree structure and making the process migrate among parent and child nodes. Here, let f_H be a parent node of f_I, i.e. $I = Hi$ for some i $(\le b_H)$.

[Change operations]

1. Operations for process migration:
   (a) $MD(f_I)$.
2. Operations for structure change:
   (a) $f_N = SPT(f_I)$.
   (b) $f_N = XPD(f_I)$ and $\langle f_N, f_M \rangle = XPD2(f_I)$.

The more processes are supported by a node f_I, the more energy f_I consumes. In order to decrease the energy consumption of a node f_I, a tail process p_{l_I} migrates to every child node of f_I. A process on f_I can migrate by the live migration technologies of virtual machines [11–15]. Nodes which support only one process and multiple processes are referred to as *single-process* and *multi-process* nodes, respectively.

The tail process p_{l_I} of f_I migrates to every child node f_{Ii} in the *migration down* operation $MD(f_I)$. Here, f_I supports a subsequence $P_I = \langle p_{t_I}, \ldots, p_{l_I-1} \rangle$ and each child node f_{Ii} supports $P_{Ii} = \langle p_{l_I}, p_{t_{Ii}}, \ldots, p_{l_{Ii}} \rangle$, where $l_I = t_{Ii} - 1$. If f_I is a single-process node, no process can migrate from f_I to a child node f_{Ii}. Here, a multi-process ancestor node f_L of f_I is referred to as a *least upper (LU)* node, where f_L is a parent node of f_I or every node in a path from a child node of f_L to a parent node of f_I is a single-process one. If a parent node f_H of f_I is a single-process node, $MD(f_H)$ is performed if the node. On an LU node f_L, the tail process p_{L,l_L} first migrates to every child node. Then, the tail process from each child node migrates to every child node. Thus, the parent node f_H gets a multi-process one and the tail process of f_H finally migrates to f_I.

In the *split* operation $SPT(f_I)$, a new node f_N is created as a sibling node of f_I. Then, the half of the child nodes of f_I get child nodes of the new node f_N if f_I is not an edge node. If f_I is a single-child node, $SPT(f_{Ii})$ is performed on the child node f_{Ii}. If f_{Ii} is not a multi-child one, $SPT(f_{Ii})$ and $SPT(f_N)$ are performed after $SPT(f_I)$. Thus, descendant nodes of f_I are recursively split until descendant nodes are multi-child ones. If f_I is an edge node whose device node is d_I, a new device node d_N is created as a child node of the new sibling node f_N. Here, a half of devices of d_I are supported by the edge node f_N. This means, the half of the output DUs from d_I are sent to the edge node f_N by the new device node d_N. The energy consumption of f_I can be reduced because the number of input DUs decreases to the half.

In the *expand* operation $XPD(f_I)$, a new child node f_N is created for f_I and every child node f_{Ii} of f_I is a child node of f_N. Then, the tail process p_{I,l_I} of f_I migrates to f_N in $MD(f_I)$. Here, f_I has one child node f_N whose child nodes are $f_{I1}, \ldots, f_{Ib_I}$ and supports one process. In the operation $\langle f_N, f_M \rangle = XPD2(f_I)$, $f_N = XPD(f_I)$ and then $MD(f_N)$ are performed and $f_M = SPT(f_N)$ is performed. The energy consumption of f_I is reduced and the process p_{Il_I} is in parallel performed by a pair of the nodes f_N and f_M.

3   The DC Algorithm

A node f_I which has the longest total queue length ql_I is *most congested* in the tree T. For a most congested node f_I in T, migration down (MD), split (SPT), and expand $(XPD, XPD2)$ operations are applied to f_I in order to reduce the energy consumption of f_I (Fig. 1).

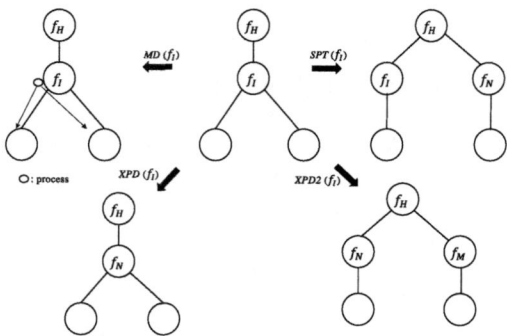

Fig. 1. Change operations.

In one way, the tail process p_{I,l_I} of f_I migrates to every child node f_{Ii} in $MD(f_I)$. Since f_I supports fewer processes, f_I consumes smaller energy. In addition, the process p_{Il_I} is in parallel performed on fewer input DUs by child nodes of f_I.

In another way, a new child node f_N is created by the split operation $f_N = SPT(f_I)$. Since the half of the input DUs to f_I are sent to f_N, the energy consumption of f_I can be reduced. Thus, sibling nodes of f_I increase and the input DUs in ID_I are in parallel processed by the nodes f_I and f_N.

In the other way, one new child node f_N of f_I is created by the expand operation $f_N = XPD(f_I)$. The tail process p_{Il_I} of f_I migrates to the new child node f_N. Every child node of f_I gets a child node of f_N. In the $XPD2(f_I)$ operations, a pair of new child nodes f_N and f_M are created for f_I and the tail process p_{Il_I} of f_I migrates to f_N and f_M. Then, the halves of the child nodes of f_I are connected to f_N and f_M as the child nodes. By the XPD and $XPD2$ operations, the height of the tree increases.

For each most congested node f_I, each operation $op(f_I)$ of the operations MD, SPT, XPD, and $XPD2$ is applied on the node f_I in the tree T by the following function $change\_node(T, f_I)$. A function $N = exec(T, op, f_I)$ means that an operation $op(f_I)$ is applied on a node f_I in the tree T and a tree N is obtained. A function $energy(T)$ gives the energy consumption of the tree T. The functions $exec$ and $energy$ are realized by the simulator which we implement

[17]. A tree N is obtained by changing the tree T in an operation op ($\in \{MD$, SPT, XPD, $XPD2\}$) on a node f_I where the energy consumption E_{op} is the smallest.

```
change_node(T, f_I) {
N_MD   = exec(T, MD, f_I);     E_MD   = energy(N_MD);
N_SPT  = exec(T, SPT, f_I);    E_SPT  = energy(N_SPT);
N_XPD  = exec(T, XPD, f_I);    E_XPD  = energy(N_XPD);
N_XPD2 = exec(T, XPD2, f_I);   E_XPD2 = energy(N_XPD2);
NE = min(E_MD, E_SPT, E_XPD, E_XPD2);
N   = N_op where E_op = NE;  /* changed tree */
return (N);
```

First, most congested nodes are found in the tree T. MC_T is a set of most congested nodes in T. If MC_T includes multiple nodes, nodes in MC_T are sorted in a descending order of levels of nodes. Nodes are sequentially taken in the order of MC_T. Nodes of lower levels are processed earlier in the algorithm. For each congested node f_I in the set MC_T, the function $change\_node(T, f_I)$ is applied on the node f_I and returns a tree N obtained by changing f_I. If the energy consumption of the changed tree N is smaller than T, another most congested node f_J following the node f_I in the set MC_T is taken and the function $change\_node(N, f_J)$ is applied on the node f_J in the tree N. Otherwise, another node f_J in the set MC_T is taken without changing the tree T and the function $change\_node(T, f_J)$ is applied on the node f_J.

```
change_tree(T) {
E    = maximum energy;   /* energy */
MC_T = ordered set of most congested nodes in T;
for each node f_I in MC_T, {
N   = change_node(T, f_I);    NE = energy(N);
if (NE < E) { E = NE;  T = N; } /* tree changed */
}; /* for end */
return(T);
};
```

4 Evaluation

We consider applications LP_k^m and SP_k^m ($k = 1, 2$) which are sequences of m processes $p_1, \ldots, p_m$, whose reduction ratios $\rho_1, \ldots, \rho_m$ are shown in Table 1. In LP_k^m and SP_k^m, every process p_i ($i = 1, \ldots, m$) is linear and square, respectively. In the evaluation, we consider $m = 5$ and $m = 10$. For $k = 1$, the reduction ratios are ordered as $\rho_i > \rho_j$ ($i > j$). That is, a process p_i generates a fewer output DUs than a process p_j for $i < j$. For $k = 2$, $\rho_i < \rho_j$ ($i > j$). That is, the lower level a process generates the fewer number of output DUs.

Table 1. Reduction ratios of processes.

Processes	p_1	p_2	p_3	p_4	p_5	p_6	p_7	p_8	p_9	p_{10}
LP_1^5, SP_1^5	0.2	0.4	0.6	0.8	1.0					
LP_2^5, SP_2^5	1.0	0.8	0.6	0.4	0.2					
LP_1^{10}, SP_1^{10}	0.1	0.2	0.3	0.4	0.5	0.6	0.7	0.8	0.9	1.0
LP_2^{10}, SP_2^{10}	1.0	0.9	0.8	0.7	0.6	0.5	0.4	0.3	0.2	0.1

The computation rate CR of the root node f is 800 [Kbps] and CR_I of a Raspberry pi3 fog node f_I is $0.185 \cdot CR = 148$ [Kbps]. The maximum and minimum power consumption $xE = 301.3$ and $mE = 126.1$ [W] for f while $xE_I = 3.7$ and $mE_I = 2.1$ for f_I. These are obtained through our experiment [19, 21].

A DU id_I of size sd_I is periodically sent by each device node d_I to the edge node f_I every $ist = 10$ [sec]. The total size tsd of DUs from all the device nodes is fixed while the size sd_I of a DU of each device node d_I gets fewer as the number of device nodes increases. Here, $tsd = 8,000$ [Kbit]. Each node f_I consumes energy $xE/5$ [W sec] to receives input DUs ID_I and no energy to send a DU.

If a DU du is sent by a device node at time τ, $du.time = \tau$. An input DU id_{Ii} from each child node f_{Ii} has the same time, i.e. $id_{Ii}.time = id_{Ij}.time$ for every pair of child nodes f_{Ii} and f_{Ij} of f_I. On receipt of input DUs ID_I, a node f_I obtains a output DU od_I where $od_I.time = ID_I.time$. Finally, the root node f obtains an output data od at time τ. The delivery time DT is defined as $\tau - od.time$.

In the simulation, the DC algorithm to change the TBFC tree T is evaluated in terms of the total energy consumption TE [W sec] of the nodes. A tree T is initially composed of a root node f which has a child device node d. Here, d denotes a set of all the devices. Let T_0 be an initial tree which stands for a CC model. A DU of size tsd is sent to the root node f by d every ist [sec]. Every xt [sec], each non-device node is checked and a set MC_T of most congested nodes is obtained from the tree T. Then, the function $change\_tree(T)$ is applied to the tree T. In the evaluation, xt is 100 [s].

TE_0 and DT_0 indicate the total energy consumption and delivery time of the initial tree T_0, respectively. Then, a tree T is obtained by applying the function $T = change\_tree(T_0)$ on the tree T_0. TE and DT show the total energy consumption and delivery time of a tree T, respectively. Table 2 shows the ratios $RTE = TE/TE_0$ and $RDT = DT/DT_0$, respectively. RTE_0 and RDT_0 are the ratios of TE_0 and DT_0 for $k = 2$ to $k = 1$. As shown in Table 2, TE and DT can be reduced in the DC algorithm compared with the CC model. Especially, if processes are square type, TE can be more reduced.

Table 2. Ratios of total energy consumption TE and delivery time DT.

Application process	RTE_0	RTE	RDT_0	RDT
LP_1^5	1.0	0.26	1.0	0.09
LP_2^5	0.54	0.41	0.43	0.56
SP_1^5	1.0	0.02	1.0	0.01
SP_2^5	0.02	0.26	0.35	0.001
LP_1^{10}	1.0	0.41	1.0	0.69
LP_2^{10}	0.46	0.31	0.36	0.20
SP_1^{10}	1.0	0.01	1.0	0.23
SP_2^{10}	0.01	0.01	0.02	0.001

5   Concluding Remarks

We discussed how to reduce the total energy consumption of the IoT by using the TBFC model in this paper. The tree structure of nodes has to be changed in order to adapt the IoT to the fluctuation of sensor data traffic from devices. Here, not only the parent-child relations on nodes but also processes of nodes have to be changed. In this paper, we newly proposed the DC algorithm by which a congested node is changed each time the node gets congested. In the evaluation, we showed the total energy consumption of the nodes and the delivery time can be reduced in the DC algorithm compared with the CC model.

Acknowledgment. This work is supported by Japan Society for the Promotion of Science (JSPS) KAKENHI Grant Number 22K12018.

References

1. Dayarathna, M., Wen, Y., Fan, R.: Data center energy consumption modeling: a survey. IEEE Commun. Surv. Tutor. **18**(1), 732–787 (2016)
2. Qian, L., Luo, Z., Du, Y., Guo, L.: Cloud computing: an overview. In: Jaatun, M.G., Zhao, G., Rong, C. (eds.) CloudCom 2009. LNCS, vol. 5931, pp. 626–631. Springer, Heidelberg (2009). https://doi.org/10.1007/978-3-642-10665-1_63
3. Rahmani, A.M., Liljeberg, P., Preden, J.-S., Jantsch, A.: Fog Computing in the Internet of Things, 1st edn., p. 172. Springer, Cham (2018). https://doi.org/10.1007/978-3-319-57639-8
4. HPE: HP server DL360 Gen 9. https://www.techbuyer.com/cto/servers/hpe-proliant-dl360-gen9-server
5. Enokido, T., Aikebaier, A., Takizawa, M.: Process allocation algorithms for saving power consumption in peer-to-peer systems. IEEE Trans. Ind. Electron. **58**(6), 2097–2105 (2011)
6. Enokido, T., Aikebaier, A., Takizawa, M.: A model for reducing power consumption in peer-to-peer systems. IEEE Syst. J. **4**(2), 221–229 (2010)
7. Enokido, T., Aikebaier, A., Takizawa, M.: An extended simple power consumption model for selecting a server to perform computation type processes in digital ecosystems. IEEE Trans. Ind. Inform. **10**(2), 1627–1636 (2014)

8. Enokido, T., Takizawa, M.: Integrated power consumption model for distributed systems. IEEE Trans. Ind. Electron. **60**(2), 824–836 (2013)

9. Kataoka, H., Duolikun, D., Sawada, A., Enokido, T., Takizawa, M.: Energy-aware server selection algorithms in a scalable cluster. In: Proceedings of the 30th International Conference on Advanced Information Networking and Applications, pp. 565–572 (2016)

10. Kataoka, H., Nakamura, S., Duolikun, D., Enokido, T., Takizawa, M.: Multi-level power consumption model and energy-aware server selection algorithm. Int. J. Grid Util. Comput. **8**(3), 201–210 (2017)

11. Duolikun, D., Enokido, T., Takizawa, M.: Energy-efficient dynamic clusters of servers. In: Proceedings of the 8th International Conference on Broadband and Wireless Computing, Communication and Applications, pp. 253–260 (2013)

12. Duolikun, D., Enokido, T., Takizawa, M.: Static and dynamic group migration algorithms of virtual machines to reduce energy consumption of a server cluster. In: Nguyen, N.T., Kowalczyk, R., Xhafa, F. (eds.) Transactions on Computational Collective Intelligence XXXIII. LNCS, vol. 11610, pp. 144–166. Springer, Heidelberg (2019). https://doi.org/10.1007/978-3-662-59540-4_8

13. Duolikun, D., Enokido, T., Takizawa, M.: Simple algorithms for selecting an energy-efficient server in a cluster of servers. Int. J. Commun. Netw. Distrib. Syst. **21**(1), 1–25 (2018). 145–155 (2019)

14. Duolikun, D., Enokido, T., Barolli, L., Takizawa, M.: A monotonically increasing (MI) algorithm to estimate energy consumption and execution time of processes on a server. In: Barolli, L., Chen, H.-C., Enokido, T. (eds.) NBiS 2021. LNNS, vol. 313, pp. 1–12. Springer, Cham (2022). https://doi.org/10.1007/978-3-030-84913-9_1

15. Duolikun, D., Nakamura, S., Enokido, T., Takizawa, M.: Energy-consumption evaluation of the tree-based fog computing (TBFC) model. In: Barolli, L. (ed.) BWCCA 2022. LNNS, vol. 570, pp. 66–77. Springer, Cham (2022). https://doi.org/10.1007/978-3-031-20029-8_7

16. Duolikun, D., Enokido, T., Barolli, L., Takizawa, M.: A flexible fog computing (FTBFC) model to reduce energy consumption of the IoT. In: Barolli, L. (ed.) EIDWT 2023. LNDECT, vol. 161, pp. 256–262. Springer, Cham (2022). https://doi.org/10.1007/978-3-031-26281-4_26

17. Duolikun, D., Enokido, T., Takizawa, M.: An energy-aware algorithm for changing tree structure and process migration in the flexible tree-based fog computing model. In: Barolli, L. (ed.) AINA 2023. LNNS, vol. 654, pp. 268–278. Springer, Cham (2023). https://doi.org/10.1007/978-3-031-28451-9_24

18. Mukae, K., Saito, T., Nakamura, S., Enokido, T., Takizawa, M.: Design and implementing of the dynamic tree-based fog computing (DTBFC) model to realize the energy-efficient IoT. In: Barolli, L., Natwichai, J., Enokido, T. (eds.) EIDWT 2021. LNDECT, vol. 65, pp. 71–81. Springer, Cham (2021). https://doi.org/10.1007/978-3-030-70639-5_7

19. Oma, R., Nakamura, S., Duolikun, D., Enokido, T., Takizawa, M.: An energy-efficient model for fog computing in the Internet of Things (IoT). Internet Tings **1–2**, 14–26 (2018)

20. Oma, R., Nakamura, S., Enokido, T., Takizawa, M.: A tree-based model of energy-efficient fog computing systems in IoT. In: Barolli, L., Javaid, N., Ikeda, M., Takizawa, M. (eds.) CISIS 2018. AISC, vol. 772, pp. 991–1001. Springer, Cham (2019). https://doi.org/10.1007/978-3-319-93659-8_92

21. Oma, R., Nakamura, S., Duolikun, D., Enokido, T., Takizawa, M.: Evaluation of an energy-efficient tree-based model of fog computing. In: Barolli, L., Kryvinska, N., Enokido, T., Takizawa, M. (eds.) NBiS 2018. LNDECT, vol. 22, pp. 99–109. Springer, Cham (2019). https://doi.org/10.1007/978-3-319-98530-5_9
22. Oma, R., Nakamura, S., Duolikun, D., Enokido, T., Takizawa, M.: A fault-tolerant tree-based fog computing model. Int. J. Web Grid Serv. **15**(3), 219–239 (2019)
23. Oma, R., Nakamura, S., Duolikun, D., Enokido, T., Takizawa, M.: Energy-efficient recovery algorithm in the fault-tolerant tree-based fog computing (FTBFC) model. In: Barolli, L., Takizawa, M., Xhafa, F., Enokido, T. (eds.) AINA 2019. AISC, vol. 926, pp. 132–143. Springer, Cham (2020). https://doi.org/10.1007/978-3-030-15032-7_11
24. Oma, R., Nakamura, S., Enokido, T., Takizawa, M.: A dynamic tree-based fog computing (DTBFC) model for the energy-efficient IoT. In: Barolli, L., Okada, Y., Amato, F. (eds.) EIDWT 2020. LNDECT, vol. 47, pp. 24–34. Springer, Cham (2020). https://doi.org/10.1007/978-3-030-39746-3_4
25. Guo, Y., Saito, T., Oma, R., Nakamura, S., Enokido, T., Takizawa, M.: Distributed approach to fog computing with auction method. In: Barolli, L., Amato, F., Moscato, F., Enokido, T., Takizawa, M. (eds.) AINA 2020. AISC, vol. 1151, pp. 268–275. Springer, Cham (2020). https://doi.org/10.1007/978-3-030-44041-1_25
26. Raspberry Pi 3 model B (2016). https://www.raspberrypi.org/products/raspberry-pi-3-model-b

A CCM, SA and FDTD Based Mesh Router Placement Optimization in WMN

Yuki Nagai[1], Tetsuya Oda[2(✉)], Kyohei Toyoshima[1], Chihiro Yukawa[1], Sora Asada[3], Tomoaki Matsui[3], and Leonard Barolli[4]

[1] Graduate School of Engineering, Okayama University of Science (OUS), 1-1 Ridaicho, Kita-ku, Okayama 700-0005, Japan
{t22jm23rv,t22jm24jd,t22jm19st}@ous.jp
[2] Department of Information and Computer Engineering, Okayama University of Science (OUS), 1-1 Ridaicho, Kita-ku, Okayama 700-0005, Japan
oda@ous.ac.jp
[3] Graduate School of Science and Engineering, Okayama University of Science (OUS), 1-1 Ridaicho, Kita-ku, Okayama 700-0005, Japan
{r23smk5vb,r23smf3vv}@ous.jp
[4] Department of Information and Communication Engineering, Fukuoka Insitute of Technology, 3-30-1 Wajiro-Higashi-ku, Fukuoka 811-0295, Japan
barolli@fit.ac.jp

Abstract. A Wireless Mesh Network (WMN) is expected to be used for temporary networks during disasters and networks for IoT devices in factories since it can provide a stable wireless sensor network over a wide area. Since the placement of mesh routers has a significant impact on the overall WMN communication, a mesh router placement problem is defined to decide efficient mesh router placement. In our previous work, we proposed various optimization methods for the mesh router placement optimization problem. However, these methods may result in a dense placement of mesh routers. Therefore, by spreading the interval between mesh routers, it is possible to spread the range that can be covered by the mesh routers in the entire subject area. In this paper, we propose an optimization method for mesh router placement considering electric field strength with the Finite Difference Time Domain Method (FDTD). We carried out some simulations and from the simulation results, the Area Ratio of Strong Electric field (ARSE) increases about 3 [%] by the proposed method from we proposed optimization methods in our previous work.

1 Introduction

A Wireless Mesh Network (WMN) [1–3] consists of multi-hop communication between multiple mesh routers. WMN has the advantage of enabling communication over other mesh routers even if some network links fail. Therefore, WMN is expected to be utilized for temporary networks during disasters and for IoT devices in factories, as they can provide networks with stable wireless connectivity over a wide area. In a wireless mesh network, the decision placement of mesh routers significantly impacts the connectivity and transmission loss of communication across the entire WMN. Therefore, the mesh router placement optimization problem is defined to decide efficient placement of mesh routers [4–8].

We previously proposed the Coverage Construction Method (CCM) [9], CCM-based Hill Climbing (HC) [10] and CCM-based Simulated Annealing (SA) [11] to solve the problem. In the mesh router placement optimization problem, the main objective functions are the Size of Giant Component (SGC), which indicates network connectivity and the Number of Covered Mesh Clients (NCMC), which indicates the number of mesh clients within the communication range of the mesh routers. However, these methods may results in a dense placement of mesh routers and even if the clients can be covered, the range in which the entire subject area can be communicated may be narrow range. By spreading the interval between mesh routers, it is possible to expand the range that can be covered by the mesh routers in the entire subject area. Therefore, we consider applying radio wave propagation by electromagnetic waves analysis to improve these proposed methods. There is the Finite Difference Time Domain (FDTD) method [12–17] as a method for electromagnetic waves analysis. The FDTD method divides the space into a mesh of small rectangular or cubic elements called cells and uses an updated formula formulated based on Faraday's law and Ampere's law. The electromagnetic field of each cell is successively updated at each time according to the updating formula formulated based on Faraday's law and Ampere's law.

In this paper, we propose a CCM and SA based FDTD approach as an optimization method for the placement of mesh routers considering the coverage area of the electric field strength. Also, we perform simulations to evaluate the proposed method.

The paper is organized as follows. In Sect. 2, we present an overview of the proposed system. In Sect. 3, we show the simulation results. Finally, in Sect. 4, we conclude the paper.

2   Proposed Method

Figure 1 shows the image of the proposed method. The proposed method that the CCM and SA expansive approach based on radio propagation analysis by FDTD method decides the optimal placement of mesh routers considering the electric field strength. The electric field strength is derived through FDTD based radio propagation analysis. Also, the method optimizes the placement of the mesh routers to maximize the stable communication range in the considered area.

2.1   CCM

The CCM [9] is a method that derives a solution to maximize SGC in the mesh routers optimization problem. One coordinate is randomly decided and a mesh router is placed. Then again the one coordinate is randomly decided and a mesh router is placed. Each mesh router considers the radio communication ranges as a circle and collision detection is performed between two mesh routers and repeated until the radio communication ranges overlap. The generated mesh router is detected for collision with the previously generated mesh routers. The new mesh router is generated when any of the mesh routers

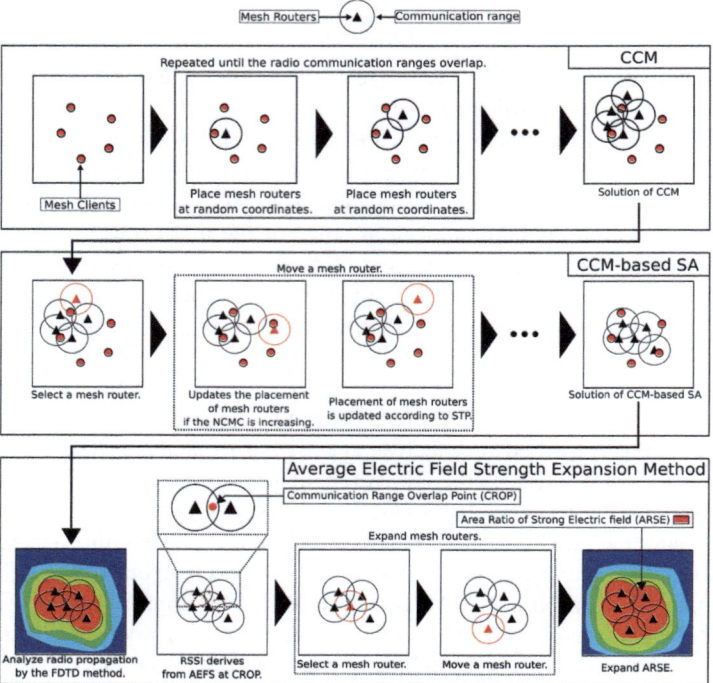

Fig. 1. Image of the proposed method.

have overlapping radio communication ranges. This process is continued until the number of mesh routers has been set. This process is repeated and the solution with the highest NCMC among generated solutions is the final solution for CCM.

2.2 CCM-Based SA

The CCM-based SA [11] derives the optimal solution by performing SA to the placement of mesh routers derived by CCM. The SA is the nearest neighbour search method that enables escape from local solutions by neighbourhood search and the state transition by random selection. Also, SA updates the solution according to the state transition probability (STP) which consists of the solution evaluation and temperature. In CCM-based SA, one mesh router is randomly selected and moved place. When the SGC is maximum, the NCMC derives and updates the placement of mesh routers if the NCMC is increasing. If the NCMC is decreased or un-updated, the placement of mesh routers is updated according to STP. This process is repeated in any number of iterations and derives the placement of mesh routers which is the results of the final iteration as the optimal solution.

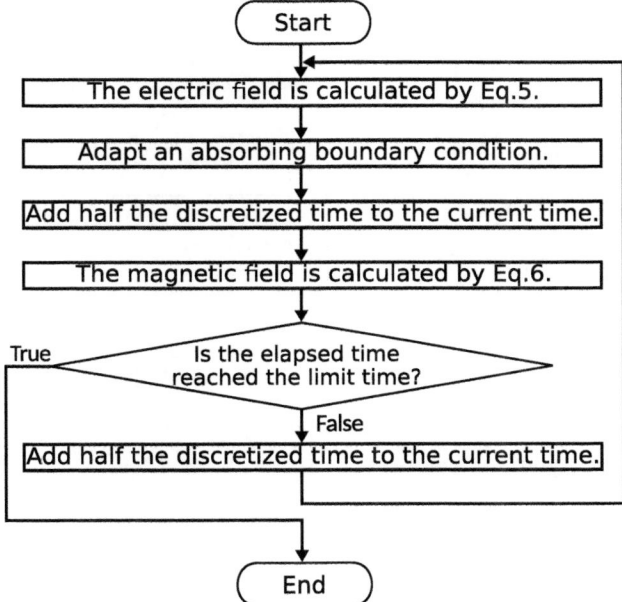

Fig. 2. Flowchart of FDTD method.

2.3　Average Electric Field Strength Expansion Method

Figure 2 shows the flowchart of the FDTD method. Figure 3 shows the flowchart of the proposed method. The proposed method aims to decide the optimal placement of mesh routers for covering a wide area and providing stability for wireless communication. The proposed method performs the radio propagation analysis based on the FDTD method considering the mesh routers as the source of radio waves to evaluate the stability of wireless communication.

The FDTD method is a numerical analysis technique for solving Maxwell's equations in the time domain. The method is discretizing both space and time and updates the values of the electromagnetic fields at each time step. E [V/m], H [A/m], D [C/m^2] and B [T] are the electric field, the magnetic field, the electric displacement and the magnetic flux density, respectively. J [A/m^2] and ρ [C/m^3] are the current density and charge density, respectively. Maxwell's equations shows in Eq. (1), Eq. (2), Eq. (3) and Eq. (4):

$$\nabla \times E = -\delta B/\delta t \tag{1}$$

$$\nabla \times H = J + \delta D/\delta t \tag{2}$$

$$\nabla \cdot D = \rho \tag{3}$$

$$\nabla \cdot B = 0 \tag{4}$$

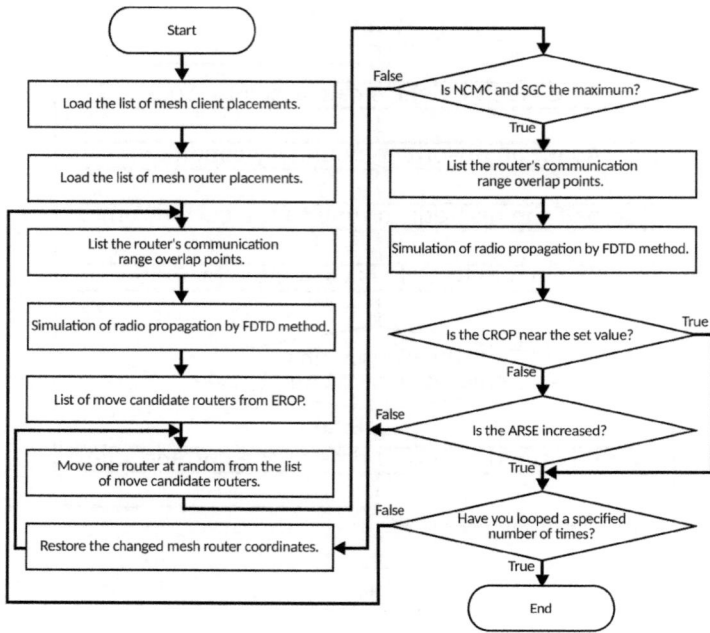

Fig. 3. Flowchart of proposed method.

The equations discretize in time and space with the central difference approximation and update the values of the electric and magnetic fields at each time step in two dimensions by the following Eq. (5) and Eq. (6). Where n indicates the current time step and Δt is the time step size. i and j indicates the indicate of the x and y coordinates and Δx and Δy are the grid spacings in each direction. $\varepsilon\ [F/m]$, $\mu\ [H/m]$ and $\sigma\ [H/m]$ are the permittivity, permeability and electrical conductivity of the medium, respectively.

$$
\begin{aligned}
E_z^n(i, j) =& \frac{1 - \frac{\sigma \Delta t}{2\varepsilon}}{1 + \frac{\sigma \Delta t}{2\varepsilon}} E_z^{n-1}(i, j) \\
&+ \frac{\delta t/\varepsilon}{1 + \frac{\sigma \Delta t}{2\varepsilon}} \frac{1}{\delta x} \{H_y^{n-\frac{1}{2}}(i+\frac{1}{2}, j) - H_y^{n-\frac{1}{2}}(i, j-\frac{1}{2})\} \\
&- \frac{\delta t/\varepsilon}{1 + \frac{\sigma \Delta t}{2\varepsilon}} \frac{1}{\delta y} \{H_x^{n-\frac{1}{2}}(i+\frac{1}{2}, j) - H_x^{n-\frac{1}{2}}(i, j-\frac{1}{2})\}
\end{aligned}
\tag{5}
$$

$$
\begin{aligned}
H_z^{n+\frac{1}{2}}(i+\frac{1}{2}, j+\frac{1}{2}) =& H_z^{n+\frac{1}{2}}(i+\frac{1}{2}, j+\frac{1}{2}) \\
&- \frac{\delta t}{\mu} \frac{1}{\delta x} \{E_y^n(i+1, j+\frac{1}{2}) - E_y^n(i, j+\frac{1}{2})\} \\
&+ \frac{\delta t}{\mu} \frac{1}{\delta y} \{E_x^n(i+\frac{1}{2}, j+1) - E_x^n(i+\frac{1}{2}, j)\}
\end{aligned}
\tag{6}
$$

Table 1. Parameters and values for simulation parameters.

Functions	Values
Width of Simulation Area	100 [m]
Hight of Simulation Area	100 [m]
Number of Mesh Routers	16
Radius of Radio Communication Range of Mesh Routers	10 [m]
Number of Mesh Clients	48
Number of Loops for CCM	3000 [unit]
Number of Loops for CCM-based SA	10000 [unit]
Number of Loops for ARSE expansion method	1000 [unit]

The proposed method set the simulation area as an open area and the second-order Mur Absorbing Boundary Condition (ABC) in the Mur ABC [18]. The Mur ABC is the one type of artificial boundary to limit the computational area in the numerical simulation of waves. Also, the Mur ABC include the first-order Mur ABC that considers only the perpendicular incident waves and the second-order Mur ABC also considers the oblique incident waves. The FDTD method applies ABC after performing electric field calculations and performs magnetic field calculations. By repeating this process for multiple time steps, we can simulate the behavior of electromagnetic waves.

We consider two factors to evaluate the stability of wireless communication: one is the Area Ratio of Strong Electric field (ARSE) and the other is the Received Signal Strength Indicator (RSSI) of the mesh routers Communication Range Overlap Point (CROP). The ARSE derives from the rate of area numbers with electric field strength above any value. The RSSI derives from the Average Electric Fields Strength (AEFS) at CROP of the mesh routers. The AEFS is calculated by taking the average of electric field strength for each time step in each area.

In the proposed method, the mesh routers placement is decided by CCM-based SA as the initial solution and derives the RSSI in each CROP. The electric field strength of the CROP and the AEFS in the simulation area calculates by the FDTD method. One CROP with an AEFS significantly different from the target value selects randomly within the list of CROPs. One of the mesh router comprising the selected CROP is moved to a neighbourhood. The target value set a number that stabilizes communication between the mesh routers. If the SGC and NCMC are not maximized when the mesh router is moved, the placement of the mesh routers is returned to the placement before the move. Also, the CROP and the AEFS with the mesh routers moved are derived by the FDTD method. When the mesh router is moved, if the AEFS of CROP is close to the target value or ARSE increases, the current solution is updated. This process is repeated in any number of iterations and derives the placement of mesh routers which is the results of the final iteration as the optimal solution.

Table 2. Parameters and values for FDTD method.

Functions	Values
Spatial Discretization Width	1 [m]
Speed of Light	2.998×10^8 [m/s]
Time Domain Interval	7.586×10^{-5} [s]
Frequency	$2.4\ [GHz]$
Permittivity	8.854×10^{-12} [F/m]
Magnetic Permeability	1.257×10^{-6} [H/m]
Electric Power	1 [mW]
Number of Loops	100 [unit]

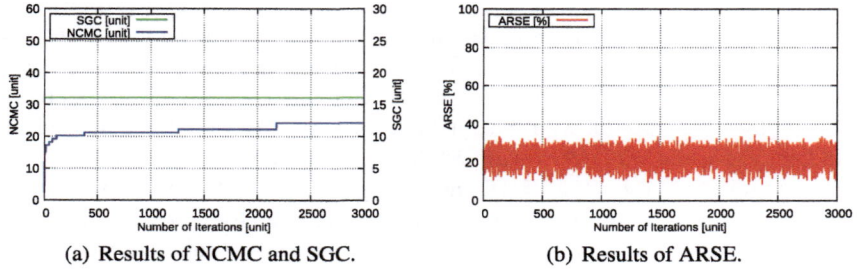

(a) Results of NCMC and SGC.   (b) Results of ARSE.

Fig. 4. Simulation results of CCM.

3   Simulation Results

Table 1 shows the simulation parameters. Table 2 shows the parameters of the FDTD method. The time domain interval is calculated based on Courant's stability condition. In this paper, ABC is set the second-order Mur ABC. Also, the electric power for RSSI calculation is set to 1 [mW]. The mesh clients are randomly placed by uniform distribution in the simulation area. Also, the distance to randomly move the mesh router is set to the range of ±5 [m] in the X and Y coordinate directions from the router's place. The target value of RSSI for the CROP to select the mesh router to be moved is set to ±3 [dBm] with −65 [dBm] as the reference value. The ARSE is derived from the area with RSSI above −65 [dBm] as the strong electric field.

Figure 4, Fig. 5 and Fig. 6 show the simulation results of the CCM, CCM-based SA and AEFS expansion method. Figure 4(a), Fig. 5(a) and Fig. 6(a) show the NCMC and SGC of the CCM, CCM-based SA and AEFS expansion method. Figure 4(b), Fig. 5(b) and Fig. 6(b) show the ARSE of the CCM, CCM-based SA and AEFS expansion method. From Fig. 4(a), Fig. 5(a) and Fig. 6(a), it can be seen the SGC is constantly maximized and is 16 [*units*], the NCMC is increased and covers all 48 [*units*] mesh clients. The results of ARES at the final iteration compare the CCM-Based SA shown in Fig. 5(b) and the AEFS expansion method shown in Fig. 6(b), the ARES is increased from about 78 [%] to about 81 [%] and improving about 3 [%].

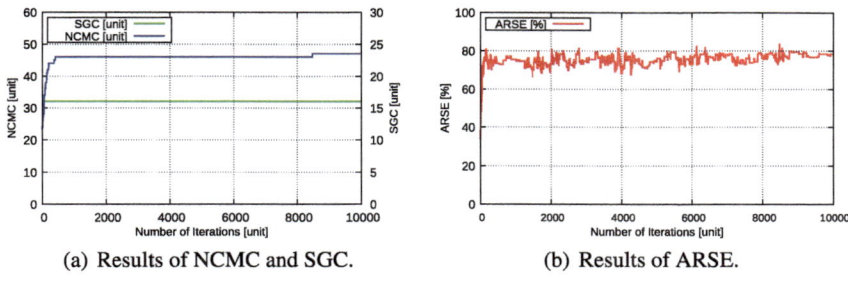

(a) Results of NCMC and SGC.   (b) Results of ARSE.

Fig. 5. Simulation results of CCM-based SA.

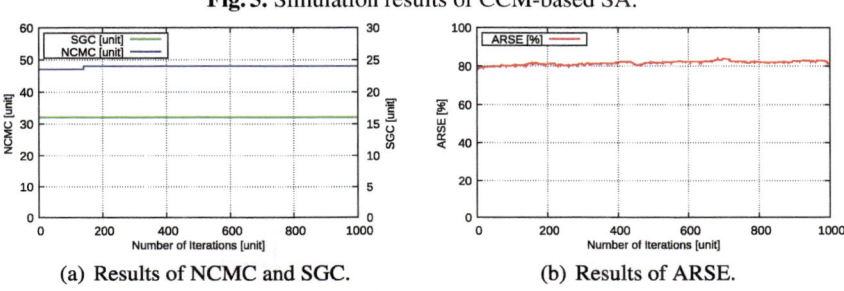

(a) Results of NCMC and SGC.   (b) Results of ARSE.

Fig. 6. Simulation results of AEFS expansion method.

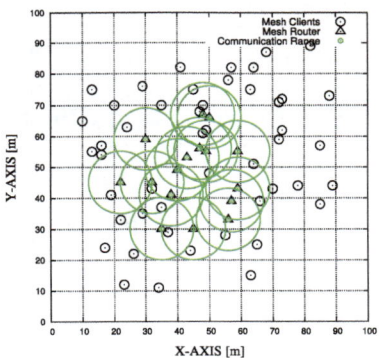

Fig. 7. Visualization results of CCM.

Figure 7 shows the visualization results of the CCM. Figure 8 shows the visualization results of the CCM-based SA. Figure 9 shows the visualization results of the AEFS expansion method. Figure 7 and Fig. 8, it can be seen that the placement of mesh routers density. On the other hand, From Fig. 9 are expanded to cover a large area and all clients are covered. Also, the ARSE of the AEFS expansion method in Fig. 9 is about 81 [%] and we can see the AEFS of the simulation area.

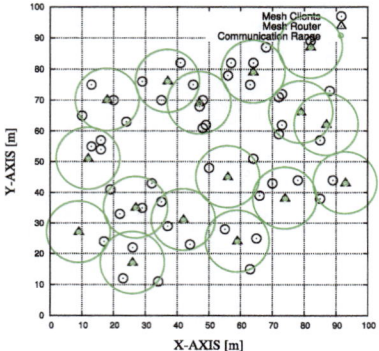

Fig. 8. Visualization results of CCM-based SA.

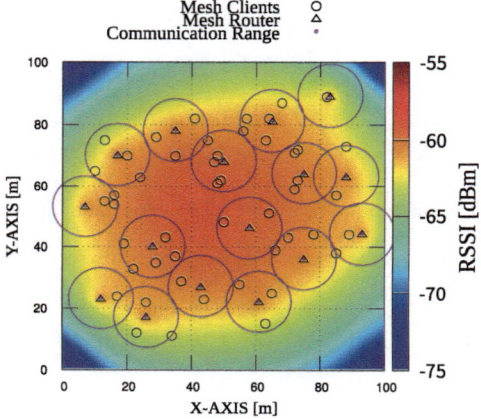

Fig. 9. Visualization results of AEFS expansion method.

4   Conclusions

In this paper, we propose a CCM and SA-based FDTD approach as an optimization method for the placement of mesh routers considering the coverage area of the electric field strength. From the simulation results, the ARSE is derived about 78 [%] by the CCM-based SA and about 81 [%] by the proposed method when the final solution. We can see that the ARSE improves about 3 [%] by the proposed method compared with the CCM-based SA. Therefore, the proposed system can derive the optimal placement of mesh routers and increase the ARSE.

In the future, we would like to simulate and evaluate extensive scenarios.

Acknowledgement. This work was supported by JSPS KAKENHI Grant Number JP20K19793.

References

1. Akyildiz, I.F., et al.: Wireless mesh networks: a survey. Comput. Netw. **47**(4), 445–487 (2005)
2. Jun, J., et al.: The nominal capacity of wireless mesh networks. IEEE Wirel. Commun. **10**(5), 8–15 (2003)
3. Oyman, O., et al.: Multihop relaying for broadband wireless mesh networks: from theory to practice. IEEE Commun. Mag. **45**(11), 116–122 (2007)
4. Oda, T., et al.: WMN-GA: a simulation system for WMNs and its evaluation considering selection operators. J. Ambient. Intell. Human. Comput. **4**(3), 323–330 (2013). https://doi.org/10.1007/s12652-011-0099-2
5. Ikeda, M., et al.: Analysis of WMN-GA simulation results: WMN performance considering stationary and mobile scenarios. In: Proceedings of the 28th IEEE International Conference on Advanced Information Networking and Applications (IEEE AINA-2014), pp. 337–342 (2014)
6. Oda, T., et al.: Analysis of mesh router placement in wireless mesh networks using Friedman test considering different meta-heuristics. Int. J. Commun. Netw. Distrib. Syst. **15**(1), 84–106 (2015)
7. Oda, T., et al.: A genetic algorithm-based system for wireless mesh networks: analysis of system data considering different routing protocols and architectures. Soft. Comput. **20**(7), 2627–2640 (2016). https://doi.org/10.1007/s00500-015-1663-z
8. Sakamoto, S., Ozera, K., Oda, T., Ikeda, M., Barolli, L.: Performance evaluation of intelligent hybrid systems for node placement in wireless mesh networks: a comparison study of WMN-PSOHC and WMN-PSOSA. In: Barolli, L., Enokido, T. (eds.) IMIS 2017. AISC, vol. 612, pp. 16–26. Springer, Cham (2018). https://doi.org/10.1007/978-3-319-61542-4_2
9. Hirata, A., et al.: Approach of a solution construction method for mesh router placement optimization problem. In: Proceedings of the IEEE 9th Global Conference on Consumer Electronics (IEEE GCCE-2020), pp. 1–2 (2020)
10. Hirata, A., Oda, T., Saito, N., Hirota, M., Katayama, K.: A coverage construction method based hill climbing approach for mesh router placement optimization. In: Barolli, L., Takizawa, M., Enokido, T., Chen, H.-C., Matsuo, K. (eds.) BWCCA 2020. LNNS, vol. 159, pp. 355–364. Springer, Cham (2021). https://doi.org/10.1007/978-3-030-61108-8_35
11. Hirata, A., et al.: A Voronoi edge and CCM-based SA approach for mesh router placement optimization in WMNs: a comparison study for different edges. In: Barolli, L., Hussain, F., Enokido, T. (eds.) AINA 2022. LNNS, vol. 451, pp. 220–231. Springer, Cham (2022). https://doi.org/10.1007/978-3-030-99619-2_22
12. Yee, K.S., Chen, J.S.: The finite-difference time-domain (FDTD) and the finite-volume time-domain (FVTD) methods in solving Maxwell's equations. IEEE Trans. Antennas Propag. **45**(3), 354–363 (1997)
13. Hwang, K.-P., Cangellaris, A.C.: Effective permittivities for second-order accurate FDTD equations at dielectric interfaces. IEEE Microwave Wirel. Compon. Lett. **11**(4), 158–160 (2001)
14. Mahmoud, K.R., Montaser, A.M.: Design of compact mm-wave tunable filtenna using capacitor loaded trapezoid slots in ground plane for 5G router applications. IEEE Access **8**, 27715–27723 (2020)
15. Chen, M., Wei, B., et al.: FDTD complex terrain modeling method based on papery contour map. Int. J. Antennas Propag. **1–10**, 2023 (2023)
16. Adao, R.M.R., Balvis, E., et al.: Cityscape LoRa signal propagation predicted and tested using real-world building-data based O-FDTD simulations and experimental characterization. Sensors **21**(8), 2717 (2021)

17. Choroszucho, A., Stankiewicz, J.M.: Using FDTD method to the analysis of electric field intensity inside complex building constructions. Poznan Univ. Technol. Acad. J. Electr. Eng. **97**, 39–48 (2019)
18. Gan, T.H., Tan, E.L.: Mur absorbing boundary condition for 2-D leapfrog ADI-FDTD method. In: Proceedings of the 1st IEEE Asia-Pacific Conference on Antennas and Propagation, pp. 3–4 (2012)

Design of Communication Protocol for Virtual Power Plant System in Distribute Environment

Yoshitaka Shibata[✉], Masahiro Ueda, and Akiko Ogawa

Iwate Prefectural University, Sugo, Takizawa 152-89, Iwate, Japan
{shibata,ueda,aki}@iwate-pu.ac.jp

Abstract. In order to reduce the use of fossil fuels, the renewable energy resources are important and should be developed on global scale. In this paper, virtual power plant system which is organized by various renewable energy resources is introduced to reduce the use of fossil fuels and CO_2 emissions. VPPs geographically distributed are aggregated and well controlled while balancing power demand and supply though communication protocol. In this paper, the communication protocol among those VPPs is designed to attain correct and efficient VPP operations. The protocol information and functions are discussed in local distributed environment.

1 Introduction

In recent, as increasing penetration of renewable energy sources (RES) such as solar power (PV), wind turbine (WT), geothermal power (GP), hydro power (HP), bio-mass power (BP) generations have been paid attentions as alternative power energy resources and replaced fossil fuels and atomic power generations around the world because they are clean without carbine emission and reproducible on the earth. Thus, the flow of electric power industry has been changing from large scale centralized power system to various small typed distributed power system. The automobile industries are also shifting from gasoline engine vehicles to product of electric vehicles based on battery power. Besides, as development of IoT technology, the power energy generated form those various renewable energy sources are precisely monitored in time and controlled to maintain the balance between the supply and demand.

Each size of renewable energy sources is relatively small and geographically distributed compared with conventional large scale centralized power system and each power generation characteristics is also different and varies depending on weather conditions. For example, VP generates energy efficiently at sunny day, but does not generate at night or bad weather such as cloudy or rainy days. WT generates more energy as increasing the window power regardless of sunshine time. GP, HP and BP always generate power at constant regardless weather conditions although their geographical construction areas are limited. Therefore, by aggregating those DERs and controlling to the power demand using IoT technology, a large power energy station can be realized as virtual power plant (VPP). VPP can provide more flexible usages than the conventional power plant. Using the Demand Response (DR) scheme which is one of the best method to control the generated power energy by DER, VPP can control the energy balance

L. Barolli (Ed.): CISIS 2023, LNDECT 176, pp. 59–67, 2023.
https://doi.org/10.1007/978-3-031-35734-3_7

between demand and supply by charging the temporal residual energy in battery storage and EV cars and discharging when weather conditions are bad for power generation. The residual power energy can be used for additional energy for regional area, such as electric power supply for EV car and automotive EV bus services. Furthermore, in case of blackout by disaster, VPP can provide the emergency power supply even the general power facilities are out of order. Thus, VPP can perform local production for local power consumption. In this paper, the communication network protocol to realize the distributed VP in regional area are discussed.

In the followings, the related works with VPP are introduced in section two. Then system configuration of power system with VPP is discussed in the section three. Next, communication protocol architecture among distributed VPP to monitor supply and demand and control the energy balance is explained in section four. After that, as example of VPP use case, geographically distributed DERS in local area is considered in section five. Finally, the concluding remarks of consideration and discussion of this research direction and future works are described in section six.

2   Related Works

So far, many researches have been made with VPP in 2010s. Awerbuch and Preston originally defined the concept of VPP [1] and established that individual participator can provide highly efficient power service to customers through virtually sharing of their private properties to improve individual utilization efficiency and avoid redundant construction using mathematical mode. Zhou and Yang, et al. defined VPP as an energy Internet hub, in which relies on remote control technology and central optimization [2]. Khodr and Halabi confined to a loose coalition between same types of energy resources such as microgrids (MGs) in [3]. Zamani and Zakariazadeh combined heat and power plants (CHPs) in [4], or active distribution networks (ADNs) in [5], and hybrid energy system in [6]. Through those researches above, the VPP is summarized as three key characteristics, Information gathering and processing, Geographical influence ignorance and Dynamic operation and optimization [7]. As internal control of VPP, Mashhour [8] established centralized control method to integrated distributed energy sources (DES). With communication protocol issue, Kolenc and Peter developed communication characteristics modelling for reliable system operation and evaluated communication system performance of actual VPP connected to two industrial distributed energy based on TCP/IP [9, 10]. Thus, most of the researches are concerned VPP issues with more industrial energy sources. In this paper, various different DERs which are widely distributed in local area and their communication protocols are considered.

3   System Configuration of VPP

So far, the conventional power supply and transmissions are monopolized by major regional electric companies which mainly operate power stations based on fossil fuels and partially atomic base stations in Japan. However, since the many local areas in Japan have higher potentials of RES, particularly northern parts with wide fields, volcanic mountains with trees, strong wind areas on top of mountains and coast lines, high potential of VPP which is equivalent energy size to the conventional power station can be realized as indicated in Fig. 1. Our objective VPP is organized by many different RESs including Phot-Voltaic (PV), Mega Phot-Voltaic (MPV), Offshore/Land Wind Turbines (WT), Hydraulic Power (HP), Geothermal Power (GT), Biomass (BM) as power supplies. Each RES has own temporal power storages and generates electric power on different locations and sends to consumers through power grid (PG).

By aggregating those RESs by a locally established VPP management company called by Aggregate Controller (AC), the power supply and demand to residents, offices/buildings can be managed and controlled to maintain power balance. Those generated power also supplies to the public transportation by electric vehicles and many other public services. Thus, precise and stable power demand-supply plan based on the big data of precise weather and predicted total production, precise supply and demand of power can be realized. Even though the case of emergency such as disaster, typhoon, heavy rain and snow occurred, blackout can be avoided because various different RESs can be mutually used and cooperated with back up line from the other areas.

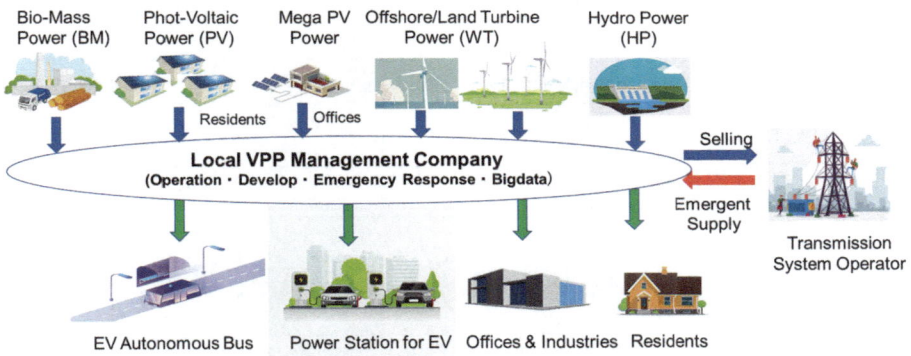

Fig. 1. System Configuration of VPP by Renewable Energy Resources

4   Communication Protocol and Architecture

As indicated in Fig. 2, VPP system mainly is classified as four different nodes including consumer, aggregator coordinator, resource aggregator and transmission system operator. The consumers include residents, offices and buildings and use electric power for various electric commodities and equipment. The consumers also generate power by PVs on their roofs. Both the generated and consumed energy states information is managed by Home Energy Management System (HEMS) for residents or Building Energy Management System (BEMS) for offices and buildings. The resource aggregator stands between the aggregate coordinator and the consumers, collects information with power states of Distributed Energy Resources (DERs) from HEMS and BEMS in regional consumers, controls the balance between the power supply and demand and maximizes the use of each customer's energy sources. The aggregate coordinator supervises all of the resource aggregators and controls energy balance in the whole area. The GW is a protocol interface to communicate between resource aggregator and Mega PV, Bio-mas and WT by OpenADR2.0b.

In order to perform VPP functions, communication protocol is important to realize efficient and correct deliver power among demand and supply of RERs. There are two standard protocols, OpenARD2.0b and ECHONET Lite in general VPP environment. OpenARD2.0b is defined as International Standard OpenADR Alliance and used for communication between the transmission system operator and aggregator controller or aggregator controller and resource aggregator as well as HEMS/BEMS as shown in Fig. 2. ECHONET Lite is used for communication among resident houses and offices.

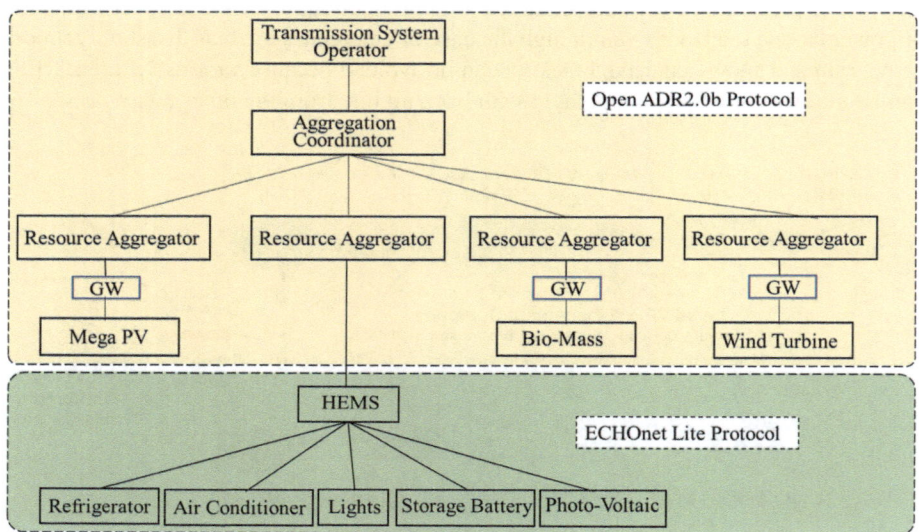

Fig. 2. VPP Components and Communication Protocol

OpenADR2.0 is modelled by communication nodes called Virtual Top Node (VTN) at message origin such as resource aggregator and Virtual End Node (VEN) at message destination such as HEMS/BEMS and performs event driven typed message exchange with demand request (DR) invocation and confirmation response, DR execution and periodical reporting measurement of current power generation quantity as shown in Fig. 3.

VTN and VEN perform message exchanging through Internet over wired network by PULL function by polling messages from VEN to VTN and PUSH function by passing messages from VTN to VEN. As the upper layer protocol on top of TCP/IP, HTTP and XMPP are used. The messages and their payloads for various services are defined by XML. As security functions, TLS (Transport Layer Security) and XML Signature are applied for message protection and security purposes.

Figure 3 shows a basic sequence and demand response event time chart of OpenADR2.0. In registration phase, the registrations of OpenADR2.0b are established among between transmission system operator, aggregator coordinator, resource aggregator and consumers. In DR invocation phase, the transmission system operator prepares invocation notice and sends its DR request to aggregate coordinator. When the aggregation coordinator received this request, it replies an acknowledgement for the DR request to the transmission system operator. Then this DR request message is delivered from the VTN of aggregator coordinator to the HEMS of consumer though resource aggregator. In measurement phase, when the HEMS received this request, it executes DR request and periodically measures the decreased power state and reports with to the aggregator coordinator and the transmission system operator as messages. This measurement process is repeated until the control period of DR is completed. Thus, the DR request to reduce consumption, such as "Reduce the current power 200 kW to 100 kW from 12:00 to 18:00" which is requested from the transmission system operator is transmitted and the equivalent measurement is periodically reported and eventually power balance is maintained.

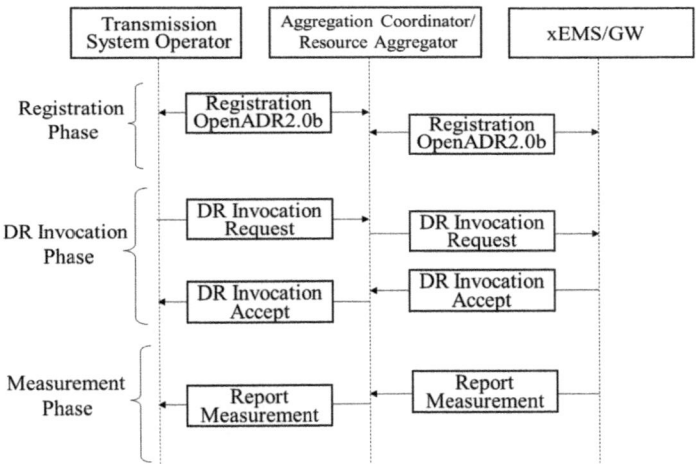

Fig. 3. Communication Protocol Sequence of Open ADR

The VTN of Transmission system operator is organized as DR programing a Demand Response Application Server (DRAS) to issue DR events to aggregator coordinator, resource aggregator and HEMS/BEMS. On the other hand, the VEN is implemented in HEMS or BEMS, receives the DR events from transmission system operator through aggregation coordinator and report a response.

On the other hand, ECHONET Lite is defined to communicate the HEMS and DERs including, air conditioners, refrigerators, lights, power storages and Photo-Voltaic, EV car in the house. HEMS has functions to monitor and control those EESs through home network. Figure 4 shows communication protocol stack of home network. As physical network, Ethernet and Wi-Fi, Bluetooth, Wi-SUN, G3-PLC are supported. As transmission and transport protocol, TCP/IP, UDP/IPv6 are used. As communication middleware on top of transmission and transport protocol, ECHONET Lite communication middle-ware and interface defined ISO/IEC 14543-4-3/301 are defined as International Standard. On top of those, application service applications and control function based on demand request from upper site, such as transmission system operator through aggregation coordinator or resource aggregator can be developed. The communication and protocol for offices and buildings are also performed as the same as for resident houses.

Thus, DR requests from transmission system operator can be transmitted from/to the HEMS and BEMS and eventually those ERSs can controlled, for example, reducing the power of air conditioner or turning off the lights to maintain the power balance in the region.

6~7	Application (Services)		
5	ECHOLite(ISO/IEC14543-4-3)		
4	TCP or UDP	TPC or UDP	UDP
3	IP or IPv6	IP or IPv6	IPv6
2	IEEE802.3	IEEE802.11n/ax	IEEE802.15.4g/e
1			
Standards	Ethernet	Wi-Fi, Bluetooth	Wi-SUN
Media	Wired	Wireless	

Fig. 4. Communication Protocol of ECHOnet Lite

5   Distributed VPP System for Local Area

In order to show the effects and superiority of VPP system we designed a prototype system of VPP in a local area, Yamagata area in Kuji city in northern part of Japan. Yamagata is located in wide mountain area, strong wind. Since the roads and power lines are longer, their alternative routes are not existed and isolated when the disaster such as earthquake, typhoon, heavy rain and snow occurred. For those reasons, there are higher potentials of renewable energy s including PVs, MPV, Offshore/Land WT, BM. As result, this area is selected as Preceding decarbonization areas by Japanese government in 2022. The local government tries to plan to construct decarbonization energy system by renewable energy. We design a proper VPP system based on this plan to provide electricity for the residents and offices/buildings and public transportation services such as on-demand and automotive electric vehicle bus services as shown in Fig. 5. By supplying electric power to the autonomous EV bus service as public transportation, the senor peoples, students, residents who do not have driver's licenses can easily go to hospitals, clinics and healthcare, schools, shops, office without driving. This autonomous EV bus service can also apply for delivering loads as logistics. Tourists can take the autonomous EV bus to go to tourist spots, historical places parking lots while using electric coupon and cashless payment. On emergence response, evaluation activity for senor people and handicapped can be realized. Thus, electric EV service system can perform public transportation facility for rural area in local production in local consumption manner.

The power demand energy in Yamagata area is analyzed by the statistical consumption energy from governmental portal site, REPOS. For those values, the supply energy by the distributed renewable sources is summarized in Table 1.

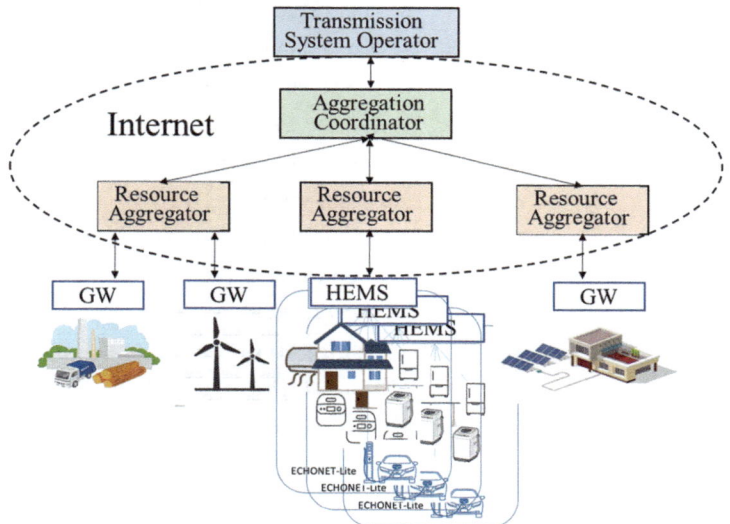

Fig. 5. VPP System by Various RESs for Local Area in Kuji

Yamagata area is consisted of 8 villages where 1059 people are living in 943 houses and 116 offices. We designed a VPP system by one aggregation coordinator and 8 resource aggregators on each village. As a communication protocol among the aggregation coordinator, 8 resource aggregations and 1059 HEMS/BEMS, OpenADR2.0b is used while ECHONET Lite is used within the houses and offices, The current power demand on all of the houses and offices is 11,782,248 kWh/year and managed by HEMS/BEMS, resource aggregators and aggregation coordinator.

On the other hand, total 26,228,168 kWh/year is generated by those DERs in 8 villages. Therefore, all of the power demand by residents and offices can be not only covered by those power energy supplies, but also 15,830,784 kWh/year as the residual energy can be used for various services in the area such as public transportation. Thus, even though the worst case such as black out by the serious disaster, all of the power energy required for the residents and offices in all villages can be provided in local production and local consumption manner.

Table 1. Power Energy Demand and Supply.

Power Energy Demand	Number	Annual Value	References
Resident Annual Demand	943	6,513,134 kWh	From Power Energy Co
Office Annual Demand	116	5727866 kWh	From Power Energy Co
Subtotal	1059	12,241,000 kWh	
Reduced Power Energy		458,752 kwh	By saving energy comodity
Total Demand		11,782,248 kWh	
Power Energy Supply		Annual Value	References
By On-Site PPA	312	2,055,881 kWh	By PVs
By Off-Site PPA	7	6,617,303 kWh	By Mega PVs
By Bio-Mass	2	624,000 kWh	
By Land Wind Turbine	1	16,830,784 kWh	
Total Supply		26,228,168 kWh	
Supply for Public Services		15,830,784 kWh	Pub. Trans., bus, charge. Sta
Real Supply for Resident		11,782,248 kWh	Equivalent to Total demand

6 Conclusions

This paper, VPP system which is organized by various RESs is introduced to reduce the use of fossil fuels and CO2 emissions and to effectively use the generated energy for customers. The general system configuration of VPP is defined to aggregate the geographically distributed RESs. The communication protocols including protocol information and interfaces for VPP system are precisely designed to well control while balancing power demand and supply using DR functions. As one of use case, a VPP system for potential RES in local area is designed as a prototype to support various public transportation service by the generated energy. Currently the precise functions and performance are evaluated for various load conditions such as changes of weather, consumer demand and system operation conditions by line failure, etc.

Acknowledgments. The research was supported by Japan Keiba Association, Grant Numbers 2021M-198 and Japan Science and Technology Agency (JST), Grant Numbers JPMUPF2003.

References

1. Awerbuch, S., Preston, A.: The Virtual Utility: Accounting, Technology and Competitive Aspects of the Emerging Industry. Springer, USA (1997)
2. Zhou, K., Yang, S., Shao, Z.: Energy internet: the business perspective. Appl. Energy **178**, 212–222 (2016)
3. Khodr, H.M., Halabi, N.E., García-Gracia, M.: Intelligent renewable microgrid scheduling controlled by a virtual power producer: a laboratory experience. Renew. Energy **48**, 269–275 (2012)

4. Zamani, A.G., Zakariazadeh, A., Jadid, S., et al.: Stochastic operational scheduling of distributed energy resources in a large scale virtual powerplant. Int. J. Electr. Power Energy Syst. **82**, 608–620 (2016)

5. Peik-Herfeh, M., Seifi, H., Sheikh-El-Eslami, M.K.: Two-stage approach for optimal dispatch of distributed energy resources in distribution networks considering virtual power plant concept. Int. Trans. Electr. Energy Syst. **24**(1), 43–63 (2014)

6. Abbassi, R., Chebbi, S.: Energy management strategy for a grid-connected wind-solar hybrid system with battery storage: policy for optimizing conventional energy generation. Int. Rev. Electr. Eng. **7**(2), 3979–3990 (2012)

7. Mashhour, E., Moghaddas-Tafreshi, S.M.: Bidding strategy of virtual powerplant for participating in energy and spinning reserve markets–part I: problem formulation. IEEE Trans. Power Syst. **26**(2), 949–956 (2011)

8. Mashhour, E., Moghaddas-Tafreshi, S.M.: Bidding strategy of virtual powerplant for participating in energy and spinning reserve markets – part II: numerical analysis. IEEE Trans. Power Syst. **26**(2), 957–964 (2011)

9. Kolenc, M., Nemček, P., Gutschi, C., Suljanović, N., Zajc, M.: Performance evaluation of a virtual power plant communication system providing ancillary services. Electr. Power Syst. Res. **149**, 46–54 (2017)

10. Kolenc, M., Suljanovic, N., Zajc, M.: Virtual power plant communication characteristics modelling for reliable system operation, pp. 1–8. In: The 41ast CIGRE Symposium 2021, Ljubljana Slovenia, 1–4 June 2021

Fine-Tuning VGG16 for Alzheimer's Disease Diagnosis

Huong Hoang Luong[2], Phong Thanh Vo[2], Hau Cong Phan[2],
Nam Linh Dai Tran[2], Hung Quoc Le[2], and Hai Thanh Nguyen[1(✉)]

[1] Can Tho University, Can Tho, Vietnam
nthai.cit@ctu.edu.vn
[2] FPT University, Can Tho, Vietnam

Abstract. Alzheimer's disease is a devastating neurological disorder that affects millions of people worldwide. It is characterized by the progressive death of brain cells, leading to brain shrinkage and a decline in cognitive abilities, behavior, and social skills. Therefore, early diagnosis is crucial for optimizing care, improving quality of life, and advancing our understanding of the disease. This study presents a model that utilizes transfer learning and fine-tuning techniques to distinguish between normal and Alzheimer-affected brain images. The dataset includes 6336 MRI images, categorized into three classes: non-demented, mild-demented, and very mild-demented. The results demonstrate that using the VGG16 model's transfer learning and fine-tuning techniques achieves promising outcomes, with accuracy and F1-score of 86%, 92%, 94%, and 98%, respectively. The proposed model performs better than some state-of-the-art methods, highlighting its potential to aid in early Alzheimer's diagnosis. Furthermore, this approach's ability to differentiate between various stages of Alzheimer's disease can improve patient outcomes, allowing for earlier interventions and better planning for patient care. Overall, the proposed model's performance demonstrates its potential as a valuable tool in diagnosing and treating Alzheimer's disease.

Keywords: brain · Alzheimer · fine-tuning · MRI images

1 Introduction

Alzheimer's disease (AD) is a neurological disorder that affects memory and cognitive abilities, with an estimated 200,000 people in the USA alone being impacted by the disease [1]. While most AD cases are sporadic, approximately 5% to 15% are familial, with early onset being linked to specific gene mutations [2]. In addition, the risk of developing AD can be increased by vascular risk factors such as smoking, hypertension, diabetes, and dyslipidemia [3]. Diagnosing AD typically involves various techniques and instruments, including memory, attention, problem-solving, language skill tests, and brain scans like MRI, CT, or PET [4]. However, accurately diagnosing AD can be challenging, which has

L. Barolli (Ed.): CISIS 2023, LNDECT 176, pp. 68–79, 2023.
https://doi.org/10.1007/978-3-031-35734-3_8

led to the development of machine learning models to assist with early detection and diagnosis. This study uses convolutional neural network (CNN) models to recognize AD patients and classify the disease stage using MRI scans. The researchers employed different CNN architectures [5], with VGG16 achieving the highest accuracy at 98.32%. The use of machine learning models for AD diagnosis shows promising results and may have implications for improving the early detection and treatment of the disease.

2   Related Work

Convolutional Neural Networks (CNNs) have been used to develop models for predicting and detecting Alzheimer's disease based on structural MRI scans. Suriya Murugan (2021) [6] developed a model that generates a high-resolution disease probability map and visually represents an individual's risk of Alzheimer's disease. Johannes Rieke (2020) [7] trained a 3D CNN and achieved a classification accuracy of 77% ± 6% and ROC AUC of 78% ± 4%. Samsuddin Ahmed (2019) [8] used ensembles of CNNs and achieved 90.05% accuracy by feeding three viewport patches to the CNN after manual localization of the left and right hippocampus using sMRI.

Several studies have used MRI image data to detect Alzheimer's disease. Yu Wang's research (2020) [9] improved the 3DPCANet model and achieved a classification accuracy of 95.00%, 92.00%, and 91.30% using CCA. Shangran Qiu's research (2021) [10] achieved high accuracy using multimodal inputs of MRI, age, gender, and Mini-Mental State Examination score on four datasets. Sheng Liu (2018) [11] developed a 3D CNN approach achieving an AUC of 85.12% for detecting mild Alzheimer's dementia. Finally, Abol Basher (2021) [12] proposed a method based on volumetric features from the left and right hippocampi achieving high classification accuracies and AUC values.

Various methods have been proposed to diagnose Alzheimer's disease using different modalities. Gupta (2019) [13] combined three features extracted from structural MR images achieving an accuracy of 96.42%. Cosimo Ieracitano (2021) [14] used the power spectral density of EEG traces, achieving an average accuracy of 89.8%. Donghuan Lu (2018) [15] used a multimodal and multiscale deep neural network achieving an accuracy of 82.4% for MCI. Jun Shi (2021) [16] used a multimodal stacked DPN algorithm achieving an accuracy of 55.34% for binary and multi-classification.

The training methodology employed in this study's proposed method differs from earlier methods. The model was trained on pre-existing models like VGG16 using transfer learning and fine-tuning, which led to a considerable decrease in training time and increased performance. Also, the suggested model attained high accuracy and F1 scores on par with or even superior to earlier state-of-the-art approaches. Thus, a possible method for identifying Alzheimer's disease using MRI images is the combination of transfer learning and fine-tuning with CNNs.

3   Proposed Model for Alzheimer

The implementation of our study was conducted through 9 main steps. They are illustrated in the diagram below (Fig. 1) and explained in detail below.

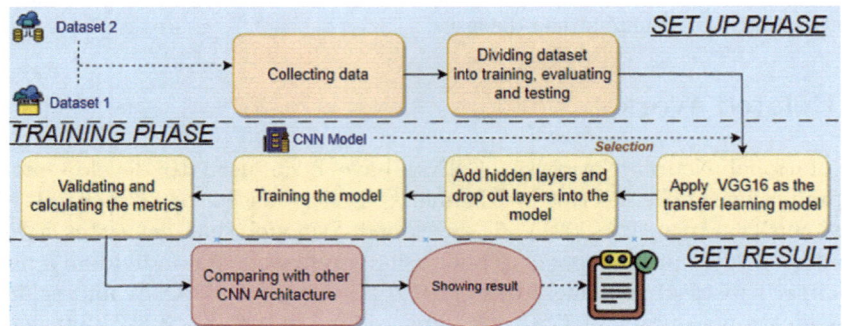

Fig. 1. The implementing procedure flowchart.

The data was hand collected by SARVESH DUBEY[1]. in 2019, with each label verified. It contains brain MRI images that have been converted into JPG files. The image is divided into Non-Demented, Mild Demented, and Very Mild Demented. After all the MRI images in the dataset are collected, they will be randomly chosen for the training, evaluation, and testing process with a ratio of 60-20-20.

We have changed some settings, like the epoch and batch size, then re-scaled the picture to fit the standard input of VGG16. We choose VGG16 [17] (Fig. 2) as a proposed model. Then, We added Flatten layer. The VGG16 model's convolutional layers produce feature maps as two-dimensional or three-dimensional arrays, which are unsuitable for input to fully connected layers. Flatten layers can convert these outputs into one-dimensional vectors, allowing the fully connected layers to take these outputs as input. This can be understood as Eq :
$output\_tensor_{i,j} = input\_tensor_{i,j_1,j_2,j_3}$.

Where i is the index for the i-th sample in the batch, j, j_1, j_2, and j_3 indicate the flattened and unflatten versions according to the input tensor. As a typical problem in deep learning, the sign of over-fitting is that the model performs well on training but fails to generalize effectively to new data. To avoid that, we have added a Dropout layer with a dropout rate of 0.25. That means 25% of the neurons will be dropped out during the training. So, all remaining neurons are forced to take on a bigger responsibility for representing the input data, making it more difficult for the model to over-fit the training data. This can be described as follows. Let I be the input to the dropout layer, X be the dropout rate, and Y be the output after applying the dropout layer. We can

---

[1] https://www.kaggle.com/datasets/tourist55/alzheimers-dataset-4-class-of-images.

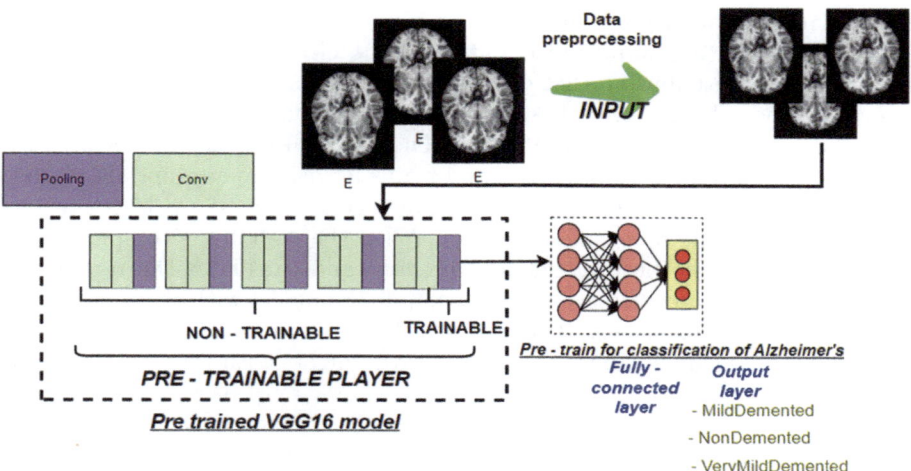

Fig. 2. The original architecture of VGG-16 architecture.

perform the following operation for each element in I to obtain the corresponding element in Y: first, generate a random number R between 0 and 1. If $R \leq X$ sets the element in Y to 0 (dropout the unit). Otherwise, if $R > X$, scale the element in Y by a factor $1/(1 - X)$ (keep the unit). The fully connected layer is also known as the Dense layer. They comprise a set of neurons connected to every neuron in the layer before it. Each neuron in the layer learns a weight for each input feature. To generate the neuron's output, the Rectified Linear Unit (ReLU) activation function can conduct a non-linear change to the weighted sum of the inputs. This allows the model to learn more complicated decision boundaries and overcome the limits of linear models. Which can be shown as Eq: output $= \mathrm{x} \cdot \mathrm{dot(w)} + \mathrm{b}$. Where x is the input tensor to the layer, w is the weight matrix, b is the bias vector, dot is the matrix multiplication operation, and relu stands for the Rectified Linear Unit activation function, which can be defined as $\mathrm{relu}(x) = \max(x, 0)$. Batch normalization can be used to normalize the activation of the preceding layers. By normalizing the activation of each layer, we can speed up the training process and reduce dependence on the initialization of the parameters. This can be explained as follow Eq: output $= \gamma \cdot \frac{\mathrm{x-mean}}{\sqrt{\mathrm{variance}+\epsilon}} + \beta$. Where γ and β are learn-able parameters, mean and variance are the mean and variance of the activation, and ϵ is a small constant added for numerical stability. After that, we added another dropout layer with a rate of 0.4; 40% of the units in the previous layer will be dropped out. The model can ensure that the predicted probabilities sum to 1 by applying the SoftMax activation function, and it can deliver a level of confidence for each prediction. This can be useful for determining how confident the model predictions are and can be used to guide downstream decision-making. This can be explained as Eq: $\mathrm{Softmax}(z_i) = \frac{e^{z_i}}{\sum_{j=1}^{k} e^{z_j}}$. Where z_i is the input to the i-th output unit of the SoftMax layer, k is the total number of output units. This function takes an input vector z of length k and produces a probability distribution over k classes. The resulting output is a probability distribution that equals 1 overall k classes. The Model training process will be split into two phases. We train fully connected layers in phase A while the base model weights are frozen. Then, it compiles the model with the specified loss, optimizer, and metrics and trains it with the appropriate method using the training and validation dataset. After that, we used a callback to save the weights with the best validation loss. When the training is completed, the method will load the best weight and predicts and calculates the metrics. In Phase B, all layers are unfrozen, and re-compile the model with the same hyper-parameters same in Phase A. The model will be fine-tuned during the training and validation. Then, a callback is used to save weights with the best validation loss to apply to the test set. If the validation accuracy in Phase A lower than in Phase B, the method will load the best weights and predicts the validation set again. After that, the method will predict the test set and calculates the final metrics. After training, we checked the number of correctly classified images and calculated the obtained data to compare the proposed model with other CNN architectures, especially the test. Again, check the accuracy and result in F1-score. Comparing with other CNN architectures: After the training, validating, and testing of the proposed model is finished, we compared the results to original CNN architectures like VGG19, InceptionV3, and MobileNet. After the data has been compared, tables and graphs will be displayed for comparison.

4   Experimental Results

4.1   Dataset

The data set for Alzheimer's disease was compiled from academic journals [8], including 6,336 MRI images corresponding to each stage of the disease (non-demented, mild demented, and very mild demented). The dataset has an image size of 176×208. Sample images of the three stages are shown in Fig. 3 (Classified Alzheimer's disease). In addition, Table 1 provides a dataset distribution with several images in the resulting data. Alzheimer's Dataset:[2].

Table 1. Three Dataset Class of Alzheimer's disease.

Class	Train	Test
Non-Demented	2560	640
Very Mild-Demented	717	448
Mild-Demented	1792	179
Total:		6336

---

[2] https://www.kaggle.com/datasets/tourist55/alzheimers-dataset-4-class-of-images.

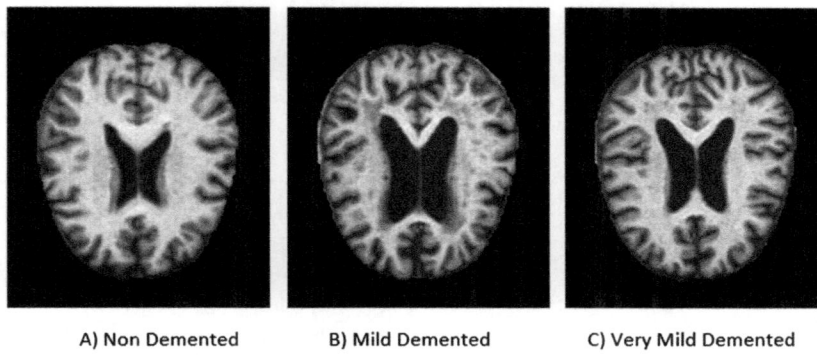

A) Non Demented   B) Mild Demented   C) Very Mild Demented

Fig. 3. Classified alzheimer disease.

4.2 Environmental Settings

With VGG16 serving as the foundational model, the suggested model was trained using transfer learning and fine-tuned on Google Colab Pro. The training was conducted over 40 epochs with a batch size of 32. The model's performance, including accuracy, AUC, and loss, was assessed during each epoch, and the results were recorded for analysis. In addition, a confusion matrix was created based on the test results after the trained model had been assessed on a split test set. The dataset was split into three groups for evaluation purposes, with a 60-20-20 ratio between training, validation, and testing. Using accuracy and F1-score in five different scenarios, the proposed model's performance was compared to that of VGG16, MobileNet, XCeption, and ResNet50.

4.3 Scenario 1 - Classify Non-demented and Mild-Demented

In this scenario, the proposed model was trained with data from two categories, non-Demented and Mild-Demented MRI images. The aim was to assess the model's performance in classifying these two categories and evaluate the results using test accuracy, F1-score, and the confusion matrix. The hyperparameters for all models were kept constant, with 40 epochs and 32 batch sizes. In addition, the hidden layers were configured to [1024,512,256]. This scenario's training and testing phase results are shown in Fig. 4. It was observed that fine-tuning provided better results than transfer learning. The accuracy and loss during the training phase of the model are illustrated in Fig. 4a and Fig. 4b. The confusion matrix of this scenario is presented in Fig. 4c, and it can be seen that the proposed model's training process has produced a promising result.

In conclusion, the results from Scenario 1 show that the proposed model has demonstrated its capability to classify between non-Demented and Mild-Demented MRI images with high accuracy and precision.

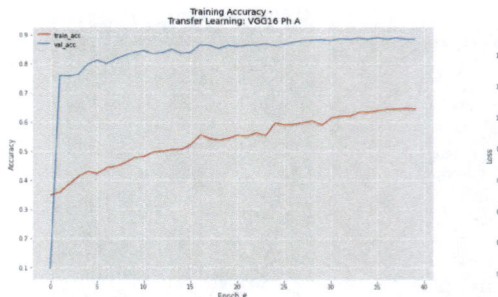
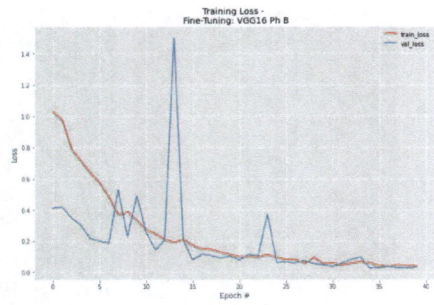

(a) The VGG-16 without fine-tuning: The accuracy on training and validation set during epochs

(b) The improved VGG-16: The accuracy on training and validation set during epochs

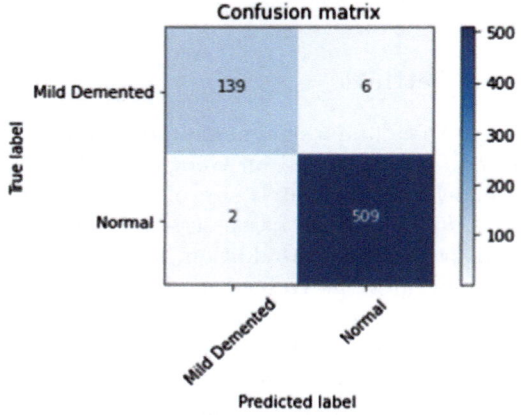

(c) Confusion matrix on the test set.

Fig. 4. The performance of VGG-16 - Scenario 1.

4.4   Scenario 2 - Classify Non-demented and Very Mild-Demented

In this scenario, we want to classify non-Demented and Very Mild Demented MRI images. To do this, we used to test accuracy, F1-score, and the confusion matrix as performance metrics. The hyperparameters used in this scenario were the same as those used in scenario 1. The experiment results showed that the VGG16 model with fine-tuning performed the best with an accuracy of 96.67%, as seen in Fig. 5a. This is significantly higher than the transfer learning model, which had an accuracy of 75%, as shown in Fig. 5b. It can be seen from the results of this scenario that the proposed model training process produced a promising outcome.

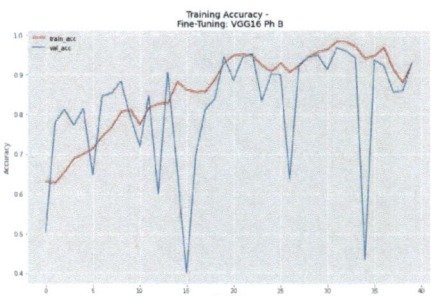

(a) The VGG-16 without fine-tuning: The accuracy on training and validation set during epochs

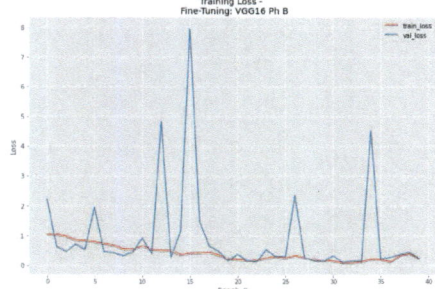

(b) The improved VGG-16: The accuracy on training and validation set during epochs

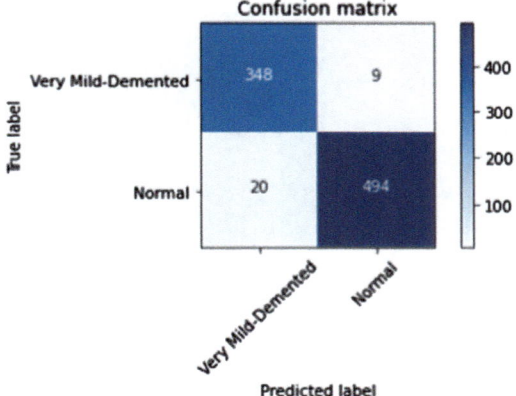

(c) Confusion matrix on the test set.

Fig. 5. The performance of VGG-16 - Scenario 2.

4.5 Scenario 3 - Classify Very Mild-Demented and Mild-Demented

In this scenario, we aimed to determine the model's ability to classify between Mild-Demented and Very Mild-Demented accurately. We used the same methods as in the previous scenarios to modify the architecture. The results support this model's use for predicting AD progression, especially at its mild and very mild stages (Fig. 6).

4.6 Scenario 4 - Classify Non-demented and Very Mild-Demented and Mild-Demented

In this scenario, we extended our study to a more complex classification task by gathering data from three categories: non-Demented, Mild-Demented, and Very Mild Demented. The goal was to test the model's capability in accurately

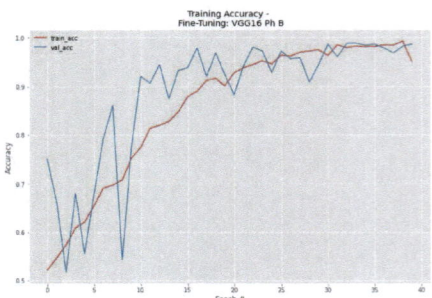 

(a) The VGG-16 without fine-tuning: The accuracy on training and validation set during epochs

(b) The improved VGG-16: The accuracy on training and validation set during epochs

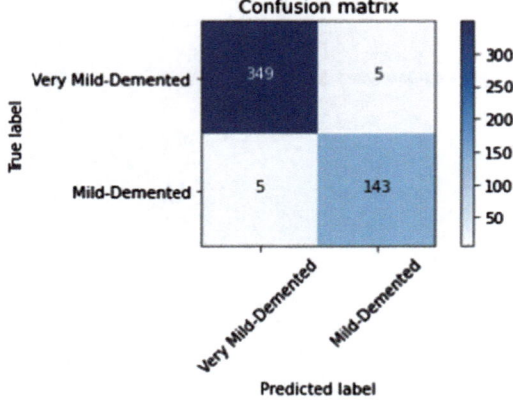

(c) Confusion matrix on the test set.

Fig. 6. The performance of VGG-16 - Scenario 3.

classifying and predicting Alzheimer's Disease (AD) across the three stages. In addition, the experiment's results were compared to other CNN architectures using various evaluation metrics such as accuracy, F1-score, and the confusion matrix. We applied the same hyperparameters as the previous scenarios. The study's results revealed that the VGG16 model with fine-tuning outperformed the transfer learning model with an accuracy of 98% compared to 70%. This promising result is reflected in the charts, graphs, and confusion matrix (Fig. 7).

Overall, this scenario further showcases the capability of the proposed model in tackling complex classification tasks. The results demonstrate that fine-tuning the VGG16 model yields better results than transfer learning in terms of accuracy, and it can be used for predicting the progression of AD (Table 2).

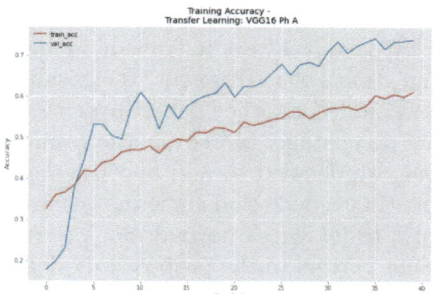 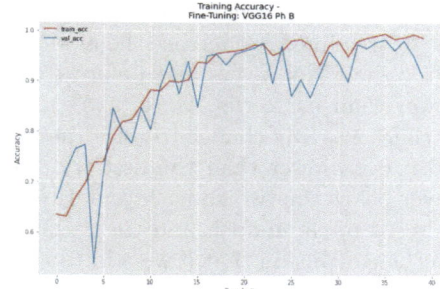

(a) The VGG-16 without fine-tuning: The accuracy on training and validation set during epochs

(b) The improved VGG-16: The accuracy on training and validation set during epochs

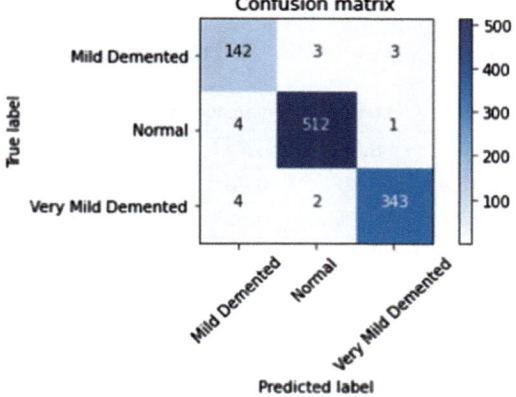

(c) Confusion matrix on the test set.

Fig. 7. The performance of VGG-16 - Scenario 4.

Table 2. Performance comparison with state-of-the-art method.

Ref.	DataSet	Architecture	ACC
Johannes Rieke [7]	sMRI	3D CNN	77%–83%
Samsuddin Ahmed [8]	sMRI	CNNs	90.05%
Shangran Qiu [10]	Framingham Heart Study	multimodal input	87.6%
Shangran Qiu [10]	NACC	multimodal input	95.4%
Jun Shi [16]	ADNI	multimodal stacked DPN	55.34%
Proposed model	sMRI	multimodal input	98.32%

4.7 Scenario 5 - Add More Dense Layers to the Model

In this scenario, we tried to add more dense layers to the proposed model. The purpose of this scenario is to experiment with various numbers of dense layers to optimize the architecture that balances between overfitting and underfitting. First, we aimed to increase the model accuracy percent by adding one dense layer after the CNN layers and before the classification layer. However, the result is lower than usual (98,1% to 98,3%). Then, we added another dense layer to see if there was any chance to raise the accuracy percentage. Moreover, the result is 96,1, even lower than the first try. Moreover, we decided to add another dense layer below the two above layers. However, it still needed to advance the accuracy percent of the model, and the result is 97,4% to 98,3% as normal. So, we went to the conclusion that if we add too many dense layers to the proposed model, it will break the operation of the model and lead to overfitting because the model becomes too complex and lazy to generalize new data but memorize the training data this can cause poor performance when encountering completely new data (Table 3).

Table 3. compare model VGG16 with other models.

Phase	Model	train_acc	val_acc	test_acc	train_loss	val_loss	F1
Transfer-learning	VGG16	0.62	0.74	0.71	0.97	0.56	0.69
	MobileNet	0.86	0.88	0.89	0.36	0.37	0.89
	Xception	0.74	0.82	0.82	0.68	0.56	0.82
	Resnet50	0.85	0.9	0.89	0.41	0.31	0.89
Fine-tuning	**VGG16**	**0.99**	**0.98**	**0.96**	**0.02**	**0.06**	**0.98**
	MobileNet	0.96	0.9	0.91	0.09	0.31	0.91
	Xception	0.97	0.85	0.84	0.08	0.56	0.84
	Resnet50	0.97	0.89	0.91	0.09	0.28	0.91

5 Conclusion

In this study, CNN architecture was used to classify AD disease stages. The model was trained and validated using open-source Kaggle data to classify stages of dementia. The vgg16 model is a famous model for classification; this model, after being trained and achieved an overall accuracy of up to 98% with fine-tuning machine learning techniques. Thus, it has the potential to identify AD-related brain regions and serve as an effective decision-support system for physicians in predicting AD severity based on the degree of dementia. In the future, we will train and test the proposed model on various datasets to improve model accuracy and add-on visualization to help with the forecast result and the explanation. In addition, we want to create better machine-learning models with better results and more disease-specific visualizations.

References

1. Qiu, C., Kivipelto, M., von Strauss, E.: Epidemiology of Alzheimer's disease: occurrence, determinants, and strategies toward intervention. Dialogues Clin. Neurosci. **11**, 111–128 (2022)
2. Giorgio, J., Landau, S.M., Jagust, W.J., Tino, P., Kourtzi, Z.: Modelling prognostic trajectories of cognitive decline due to Alzheimer's disease. NeuroImage **26**, 102199 (2020)
3. Silva, M., Loures, C., Alves, L., et al.: Alzheimer's disease: risk factors and potentially protective measures. J. Biomed. **33** (2019)
4. Kim, Y.: Are we being exposed to radiation in the hospital? Environ. Health Toxicol. **31** (2016)
5. Moser, E., Stadlbauer, A., Windischberger, C., Quick, H.H., Ladd, M.E.: Magnetic resonance imaging methodology. Eur. J. Nucl. Med. Mol. Imag. **36** (2009)
6. Murugan, S., et al.: Demnet: a deep learning model for early diagnosis of Alzheimer diseases and dementia from MR images. IEEE Access **9**, 90319–90329 (2021)
7. Rieke, J., Eitel, F., Weygandt, M., Haynes, J.-D., Ritter, K.: Visualizing convolutional networks for MRI-based diagnosis of Alzheimer's disease. In: Stoyanov, D., et al. (eds.) MLCN/DLF/IMIMIC -2018. LNCS, vol. 11038, pp. 24–31. Springer, Cham (2018). https://doi.org/10.1007/978-3-030-02628-8_3
8. Ahmed, S., et al.: Ensembles of patch-based classifiers for diagnosis of Alzheimer diseases. IEEE Access **7**, 73373–73383 (2019)
9. Jia, H., Wang, Y., Duan, Y., Xiao, H.: Alzheimer's disease classification based on image transformation and features fusion. Comput. Math. Methods Med. **2021**, 1–11 (2021)
10. Qiu, S., Joshi, P.S., et al.: Development and validation of an interpretable deep learning framework for Alzheimer's disease classification. Brain **143**, 1920–1933 (2020)
11. Liu, S., et al.: Generalizable deep learning model for early Alzheimer's disease detection from structural MRIs. Sci. Rep. **12**, 1–12 (2022)
12. Basher, A., Kim, B.C., Lee, K.H., Jung, H.Y.: Volumetric feature-based Alzheimer's disease diagnosis from SMRI data using a convolutional neural network and a deep neural network. IEEE Access **9**, 29870–29882 (2021)
13. Gupta, Y., Lee, K.H., Choi, K.Y., Lee, J.J., Kim, B.C., Kwon, G.R.: Early diagnosis of Alzheimer's disease using combined features from voxel-based morphometry and cortical, subcortical, and hippocampus regions of MRI T1 brain images. PLoS ONE **14**, e0222446 (2019)
14. Ieracitano, C., Mammone, N., Bramanti, A., Hussain, A., Morabito, F.C.: A convolutional neural network approach for classification of dementia stages based on 2D-spectral representation of EEG recordings. Neurocomputing **323**, 96–107 (2019)
15. Lu, D., Initiative, A.D.N., Popuri, K., Ding, G.W., Balachandar, R., Beg, M.F.: Multimodal and multiscale deep neural networks for the early diagnosis of Alzheimer's disease using structural MR and FDG-PET images. Sci. Rep. **8**, 1–13 (2018)
16. Shi, J., Zheng, X., Li, Y., Zhang, Q., Ying, S.: Multimodal neuroimaging feature learning with multimodal stacked deep polynomial networks for diagnosis of Alzheimer's disease. IEEE J. Biomed. Health Informat. **22**, 173 –183 (2018)
17. Simonyan, K., Zisserman, A.: Very deep convolutional networks for large-scale image recognition. In: International Conference on Learning Representations (2015)

Solving University Course Scheduling with Varied Constraints Using Integer Linear Programming

Seyed M. Buhari[1](✉) and Jyothi Manoj[2]

[1] UTB School of Business, Universiti Teknologi Brunei,
Bandar Seri Begawan, Brunei Darussalam
ismail.buhari@utb.edu.bn
[2] Accenture Solutions Pvt Ltd., Bengaluru, India

Abstract. Course scheduling is an optimization problem impacted by various constraints. In order to devise a course scheduling solution, the course structure should be modelled in terms of certain parameters like time, rooms, courses, instructors, etc. Several approaches and solutions exist for any given formulation of the course scheduling problem. Initially, the objective function along with hard and soft constraints were identified. A solution was obtained considering the accomplishment of hard constraints, which are mandatory to be accomplished. Further, the solution could be enhanced further by suitably handling various soft constraints, which are optional. Linear Programming approach used here is divided into two stages such as (1) Weekly lecture allocation and (2) Creating schedule so that further modifications could be easily addressed within a specific stage. Study reveals varying the number of days per week, number of periods per day and the number of medium rooms resulted in 37 to 44 solutions within a time frame of 0.57 to 0.86 s. Also, the utilization of the rooms largely varied between 60% to 80% at different time periods of the day. Room utilization reached more than 90% only when the number of medium rooms or the number of days is reduced.

1 Introduction

University timetabling problem is an intricate topic of research where a single universal solution is not likely. Formulation of a robust timetable as an optimization problem, while considering constraints is the objective. Research on timetabling and university class scheduling is recorded since 1960s [1]. The complexity of the problem is attributed to the difference in conditions and constraints of each institution. The methodology evolved from simple heuristic algorithms [2] to advanced algorithms [3, 4]. It includes linear models, non-linear mathematical models, heuristic and non-heuristic models. Assignment type problems, formulated as subproblems, are represented using Heuristic procedures [5]. Tahir et al. [6] worked on Dynamic Timetable Generation Conforming Constraints while determination of best students' sections is performed using applied fuzzy clustering, a fuzzy evaluator and a novel feature selection method [7]. Barkha Narang et al. [8] suggested the use of active rules and genetic algorithm for automatic timetabling and Oladipo [9] proposed an optimized technique to automate timetabling generation system using genetic algorithm.

L. Barolli (Ed.): CISIS 2023, LNDECT 176, pp. 80–91, 2023.
https://doi.org/10.1007/978-3-031-35734-3_9

A University timetabling problem entails that a set of courses is to be delivered to a group of students over a term or semester which is divided into a set of days, with each day divided into a set of time slots which can be referred to as a period. This group of students is called student group. A student group can be assigned the same course by the same lecturer in the given period and room. The other basic entities of the problem are the courses to be scheduled, the rooms available for the allocation of courses and teachers to deliver the lecture. The basic requirements of a timetable include:

- Lecture constraint (LC): a unique course must be assigned in a room at distinct periods.
- Room constraint (RC): not all rooms can be assigned to all courses; some may require lab or seminar halls and capacity of the room must be optimally utilised.
- Order constraint (OC): there may be some order of preference for some courses.
- Room occupancy (RO): at most one lecture can be assigned to each room in a period.
- Conflict Constraint (CC): the student group should not be assigned with more than one activity during the same period.

The above-mentioned constraints which should not be violated in the scheduling process are called hard constraints. The objective of the University Course Timetabling Problem (UCTP) is to develop a timetable which satisfies all LC, RC, OC, RO and CC by minimizing the penalty which are the weights assigned on violation of the soft constraints like:

- Room Capacity: when the room capacity is less than the size of the student group assigned leading to under utilization.
- Room stability: Total number of distinct rooms assigned.

The present research considers all the basic requirements of a timetabling problem and incorporates additional constraints such as

- Courses which require specific rooms like lab sessions
- Class sessions with different time duration.

The model solves timetabling problems and are amenable for extra constraints if any. The rest of the sections of the paper is presented as follows: Sect. 2 summarizes review of literature, Sect. 3 explains the Course Timetabling Problem (CTP), Sect. 4 elaborates the implementation and outcome of the proposed algorithm, and finally, Sect. 5 ends with conclusion.

2 Literature Review

We focus only on literature related to course and timetable scheduling in educational institutions. Linear programming, Genetic algorithms, heuristic approach etc. has gained wide popularity.

Integer Programming
Classroom scheduling is formulated as a LP model in Nakasuwan et al. [10]. The formulation objective is reducing the cost of scheduling the class where weights are penalty of

allocating course section of lesser strength in larger rooms. The implementation reduces the time consumed due to manual allocation and allows repeated modifications of time schedules. Incorporating Boolean nature of 0 – 1 variables in Integer Linear Programming Problems [11] includes pseudo Boolean constraints to reduce the complexity of expressing the constraints using Conjunctive Normal Form (CNF). A comparison with Electronic Design Automation is carried out and 0–1 ILP technique is found to outperform generic ILP as well as general EDA SAT problems. Scheduling of class and timetable fixation of Kuwait University [12] incorporates the required gender policies. Class-scheduling problem is solved using two-stages; Stage I determines classes based on gender and then in stage II availability of faculty members and their preferences are considered. The disintegrated class scheduling model derives the class schedules for males, females and joint sections by minimizing class conflicts. Naderi [13] presents three different algorithms - imperialist competitive algorithm, simulated annealing and variable modelling search for the scheduling of university course. The mathematical model based on integer linear programming is designed and then the three advanced algorithms were employed with novel procedures of encoding scheme, move operator and crossing operator. The comparative analysis shows that imperialist competitive algorithm outperforms the others. An integer programming model was adopted with objective of minimizing the number of academic terms, attempts to provide a roadmap to assist students in planning their courses completion [14]. Course-term-timeslot essential constraints type of decision variables were used for the integer linear programming model, simplified and solved using the Analytic Solver Platform. A different approach of optimal assignment is carried out [15] using mixed integer programming with three new objectives (i) assign course to faculty members based on their evaluation ratings, (ii) avoid overloading of teaching loads of regular faculty and (iii) hire the required number of part time faculty. The constraints also ensure that the faculty gets required time for preparation too. The model that pertains only to a single department of the institution achieves an overall average evaluation rating of 87.5% assigned for each faculty.

Genetic Algorithm
There is no dearth in literature related to application of genetic algorithm for resolving timetabling problem. Aldasht [16] presents a novel heuristic based on evolution algorithms providing optimal solution to one of the colleges. The evolutionary algorithm involving special mutation operator called Reordered Mutation proved to converge to optimal solution. Suitable search methodology is applied within a search space using a framework. The results indicate this method can give timetables with up to 97% modelling of soft constrains and up to 100% fulfilment of hard constraints. Wang [17] considers the extra requirements of a classroom scheduling like multimedia requirements or seat numbers, along with the regular soft and hard constraints. The authors use Service Oriented Architecture to develop SmartClass which has backstage designing space exploration algorithm and greedy algorithms. The presence of scheduling process in the SmartClass architecture at server side makes the academic scheduling process easier. The application of genetic algorithms [18] to optimize the process of university timetable scheduling with both hard and soft constraints concludes that genetic algorithm as a course scheduler could satisfy a good number of hard constraints. The architecture was designed with a population size of 100. Considering mutation rate as 0.01%, a crossover

rate of 0.9%, along with the number of elite individuals was 2 and the tournament size was 5, the resulting timetable had zero number of clashes with the fitness value of 1. Curriculum based Course timetabling problem is modelled as a bi-criteria optimization problem [19] where the two criteria are penalty function and robustness metric; a hybrid Multi-Objective Genetic Algorithm using Hill Climbing and Simulated Annealing algorithms in addition to the standard Genetic Algorithm approach is deployed. It aims to provide a suitable approximation to the Pareto-optimal front. Four hard and four soft constraints are considered along with the objective function. The weights are associated with the violation of the constraints. Robustness and solution with high quality are the advantages of the solution.

Meta-heuristic Algorithm
The meta heuristic procedure for solving optimization problems has gained wide acceptance. Tabu Search is used [4] to deal with a sub-optimal initial solution. The goal of obtaining a class as well as exam schedule is attained using a scoring method which will assign scores as positive and negative based on the feasibility of the allocation. Tabu will store the most recent optimal routine with least penalty by modelling the search space and averting inessential exploration. Parallel local search algorithm to assign schedule for staff meetings between classes [20] probes the efficiency of parallel competitive metaheuristics algorithm. Chen [21] proposed a novel Tabu search framework, which incorporates the controlled randomization strategy and threshold mechanism within the original Tabu search framework. Two dedicated move operators, swapping among and within rooms, are jointly employed. The approach outperformed 8 other algorithms and provided feasible solution to 60 test instances. The initial solution is improved by the Tabu search controlled randomized algorithm, which will stop only with the feasible solution or at reaching the time limit. In case of infeasibility in obtaining the best solution, the reconstruction procedure converts this situation into an incomplete timetable without any hard constraint violations, resulting in several unscheduled events. A mixed-integer programming formulation and a parallel metaheuristic based algorithm for solving high school timetabling problems (HSTP) with compactness and balancing requirements is proposed [22]. Column generation is exploited simultaneously using two pattern-based formulations and a solution algorithm; further, a team of metaheuristics is used to improve the obtained solutions. The parallel metaheuristic based algorithm is built on the diversification–intensification memory based (DIMB) framework which relies on a number of metaheuristic agents that work on parallel and share information via buffers of solutions. The proposed parallel algorithm DIMB-Color Generation outperformed previous HSTP variant methods and found near-optimal solutions. The main contribution of this algorithm is the introduction of new agents that exploit the developed column generation method for Extended Formulation to generate new feasible solutions.

3   Problem Description

This part of the paper describes the proposed model for optimization of university class scheduling. Class scheduling problem is based on the following key components: (1) number of days in a week, (2) number of working hours in a day, (3) number of courses, and (4) number of rooms and room types available. This work aims to assign classes to respective student groups and classrooms based on time, with due contribution towards the objective and constraints. Consideration is also given for objectives or constraints that could conflict each other.

The vast problem of scheduling the entire term of a university is broken down into two stages: (1) Weekly lecture allocation, and (2) Creating schedule. This will help simplify the task and the solution to be more comprehensible. Weekly schedule is to be obtained with the objective of optimizing the resource utilization. Furthermore, the results could be further optimized. Evaluation of the proposed approach is based on two main factors: (1) quality of the solution, and (2) algorithmic time complexity. These two-evaluation metrics also bring in trade-offs, thus obtaining a perfect solution could be impossible.

The University Class Scheduling problem considered in this study can be explained as follows. The scheduling is planned on for a term and weekly schedule is to be prepared. Resources that are to be allocated are teachers and classrooms, assuming that all classrooms are equally equipped with ICT and other requirements. Once the students decide the courses or programme for study, term wise schedule is to be provided. A group of students who have opted for a programme is referred to as student group, there is no restriction on the size of the student group. The duration of each class hour is constant, one hour for a lecture and two hours for a lab session. All other activities are also of one-hour duration.

General Elements

Any university class scheduling problem considers certain basic parameters like time, room, courses, students etc. (1) Time: Times indicate the number of weeks considered in a term, number of days within a week and the number of time slots per day. (2) Rooms: Rooms available for course allocation might vary on size, facilities, and time. Certain rooms like seminar hall or lecture theatre must be allocated for certain meetings, which occur at specific times during the week. (3) Courses: Courses might be structured in such a way that students are required to attend all offerings, whether a lab or lecture, of a specific course within a week. (4) Students: Students are grouped as student groups based on their term of study. The number of students per group will determine the number of sections required per course.

Constraints

Constraints can be classified as soft and hard constraints [23, 24]. The four basic parameters used in this research are total number of time slots, number of activities, number of student group and number of resources of different type available. Let the institution have n_p total number of time slots where p refers to the time period which will be engaged with n_a number of activities, where a being a specific activity which have n_s number of student groups where s represents different student group. The two resources considered r_i refers to rooms and teachers and let n_{ri} be the number of resource available of type i.

The hard constraints formulate the optimization parameters and are thus required to be satisfied. Not satisfying soft constraints will affect the quality of the solution.

The hard or required constraints include:

(1) Resource Related: In general, resources like rooms might vary in their capacities. Resource usage could be evaluated in terms of resource frequency (ratio of the number of used time periods over available time periods) and resource utilization (ratio of used to available resources considered over all the available time periods).

    i. The resources must be optimum utilized. Resource utilization cannot exceed a certain limit but must be utilized above a minimum. Instructors might have load limits on certain number of lectures/workload in a week [*InstructorWorkload*].

    ii. Every resource will have single booking at any point of time.

        a. A classroom can accommodate only one lecture at a time [*RoomOccupancy*].

        b. An instructor can teach only one class at a time [*InstructorTeach*].

(2) Activity Related:

    i. All activities must happen at least once in a week [*WeeklyOnce*].

    ii. Activity must be assigned according to a chosen order.

    iii. Some courses might need to be offered at the same time [*CourseOverlap*].

    iv. Every student and every resource may only have one activity booked at a time [*OneAtATime*].

    v. Activities have varied duration & required resource [*VariedAspects*].

Soft constraints include:

Based on student requirements: Student workload should be as even as possible; Daily workload must not violate the restrictions [*StudentWorkload*]. Certain number of breaks should be there in a day. There should not be gap between the schedules in the daily meetings of a course, i.e., two-hour lecture should be continuous.

Notation	Definition	Notation	Definition
n_a	total number of activities	n_{ri}	number of resource available of type i
n_p	total number of time slots where p refers to the time period	Periods	indicates the semester number; it varies from 1 to 8 for a 4 year program
n_s	number of student groups	P	set of all time periods, {1, 2, …n_p} where n_p is the number of time slots
p	time period which will be engaged	R	set of i type resource = {1, 2, 3….n_{ri}}, n_{ri} is the total number of resource i
S,SG,s	set of student groups. {1, 2, …n_s}	C,c	Courses
r_i	type of resource required for activity a_i	R	set of rooms that can be allotted
x[c][p]	indicates offering of a courses in a term. Here, term indicates semester number. x[c][p] is Boolean	P2	varies from 1 to nP2, where nP2 = (Number of terms)/(Number of years)
t-	total number of periods		

Notations used for Formulation of the Problem

In order to formalize the given problem, certain notations are required.

Terms for courses: Here, the decision is made with regards to which courses are offered in a specific term. Two options exist: (1) fixed courses per term, and (2) dynamically allocated due to resource [rooms, instructors, etc.] constraints. If the number of courses per term is fixed, there is no need for this process.

Objective Function

The objective function, in (1), minimizes the deviations from the allowed allocations.

$$\text{Min} \sum_{p \in Periods} \sum_{s \in StudentGroup} y1[p][s] + 2y2[p][s] + 0.5y3[p][s] \tag{1}$$

$y_1[p][s]$, $y_2[p][s]$ and $y_3[p][s]$ are integers; where y_2 punishes for deviation of 1 credit, y_1 is not enough (deviation of five credits is considered far worse than 5 times a deviation of one credit) and y_3 takes care of negative deviation.

Classroom Constraint: All courses must be held

$$\sum_{c} \sum_{p} x[c][p] = 1 \tag{2}$$

The variables used in the objective functions:

$$\forall s \in S, \forall p \in P,$$

$$y_1[p][s] \geq \sum_{c} StudyModule[c][s] * x[c][p] * CourseCredit[c] - 30 \tag{3}$$

$$y_1[p][s] \geq 0$$

where, 30 indicates the credits in a semester/term

Room Constraint: Room utilization is given an upper bound of 80%. Courses from different terms are considered for allocation. Courses for the 1st, 3rd, 5th and 7th semester should be offered at the old semester and the rest should be offered in the even semester.

$$\forall i, \forall r \in Roomtype \sum_{c \in courses} (x[c][i] * Room[c][r] + x[c][i+2] * Room[c][r]$$
$$+ x[c][i+4] * Room[c][r] + x[c][i+6] * Room[c][r]) \tag{4}$$
$$\leq 0.8 * Available[r]$$

where, i stands for term number in a year (Generally, two terms per year. This the value of P2, which is either one or two).

Output: Output of this stage includes X[c][p], y1[p][s], y2[p][s] and y3[p][s]. X[c][p] is Boolean and others are integers.

Weekly lecture allocation: Here, the decision is made with regards to how many offerings are needed for a specific course within a week. The input to this stage requires the courses to be offered in a specific term.

Terms Used:

ClassLength [c] → Number of hours at a stretch for a course

ClassLessons[c] $\rightarrow$ Number of hours per semester for the course; 45 h for a three credit course

CourseCredit[c] $\rightarrow$ Credit of the course

MinHrs[c][n] $\rightarrow$ Minimum hours per week; n represents weeks

MaxHrs[c][n] $\rightarrow$ Max hours per week; n represents weeks

z[c][n] $\rightarrow$ number of offerings or sessions on a week for a course

Objective Function

$$min \sum_{s \,\epsilon\, student} \sum_{n \,\epsilon\, weeks} y_1[s][n] + 2y_2[s][n] \tag{5}$$

Constraints

All lessons must be held

$$\forall_c \sum_{n \,\epsilon\, weeks} z[c][n] = \frac{CourseLessons[c]}{CourseLength[c]} \tag{6}$$

Minimum lessons held for all courses over the weeks of the semester.

$$\forall_c \sum_{n \,\epsilon\, weeks} z[c][n] \geq \frac{MinHour[c][n]}{ClassLength[c]} \tag{7}$$

Teacher constraint (<25 lecture hours per week for each teacher): Teacher constraint considers the maximum load of any instructor to be 25 per week.

$$\forall_{n \,\epsilon\, week} \forall_{t \,\epsilon\, teacher} \sum_c z[n][n] * ClassLength[c] * ClassTeacher[c][t] \leq 25 \tag{8}$$

Outputs: This stage outputs z[c][weeks], y1[s][weeks], y2[s][weeks] as integers.

Creating schedule: Creating schedule is to plan the final schedule for the semester.

Terms Used:

Ndaytime = number of sessions per day = 9
ClassLectures[c] – number of lecture hours per term

Objective Function: Minimize the objective function to reduce the cost involved in conducting any event at a time.

$$Min \sum_c \sum_{d \,\epsilon\, day} \sum_{h \,\epsilon\, daytime} \mu_u[h]$$
$$* (w[c][h + (d-1) * 9] + y[c][h + (d-1) * 9]) \tag{9}$$

$\mu_u[h]$- cost involved in conducting any event at a time
w[c][t] $\rightarrow$ class at time t; y[c][t] $\rightarrow$ 1 for more than 1h lecture slot

So, w[1][1] = 1 and y[1][2] = 1 for two hours lecture at a stretch

All Classes must be Held: Class hours equal to total class lectures

$$\forall_c \sum_{t \in time} w[c][t] + y[c][t] = ClassLectures[c] \qquad (10)$$

Courses that have two hours lecture/lab at a stretch cannot start at the last hour of a day. For courses that have two consecutive hours of lecture/lab, the value of w[c][t] and that of y[c][t + 1] are the same.

$\forall_c ClassLength[c] = 2 : w[c][45] = w[c][36] = w[c][27] = w[c][18] = w[c][9] = 0$

$\forall_c \forall_{t \in time}(Number 9, 18, 27, 36, 45)$

$ClassLength[c] = 2 : w[c][t] = y[c][t + 1]$

$$(11)$$

Similarly, courses that have three hours lecture/lab at a stretch cannot start at the last two hours of a day.

Only One Class Lecture at a Time: With two variables w and y used to indicate allocations, w and y cannot be 1 at the same time. This will make sure that a class is only offered once at a time.

$$\forall_c \forall_c w[c][t] + y[c][t] \leq 1 \qquad (12)$$

Student Group Constraint: Any student group should not have more than one lecture at a time.

$$\forall_t \forall_{s \in StudentGroup} \sum_c StudyModule[c][s] * (w[c][t] + y[c][t]) \leq 1 \qquad (13)$$

Outputs: This stage outputs y[classes][time] and w[classes][time] as Boolean variables.

4   Implementation Details

There is always a possibility to assign every course to the respective rooms within the given time frame, without violating any hard or required constraints. Meanwhile, the satisfactory nature of the soft constraints needs to be evaluated in terms of quality of the solution, along with computation time. Generally, computation time might not be considered as a salient feature as long as the quality of the solution meets the needs of the university. In order to test the feasibility of the proposed approach, real-life data for a semester from our faculty was obtained and tested. In this section, data description is followed by the results and the appropriate findings.

Dataset Characteristics
Different elements of the dataset constitutes towards the complexity of the problem. These elements include:

Problem size: Problem size is one of the aspects of problem complexity. Our dataset has 16 student groups, 168 courses and 5 room types. Number of rooms of each type are 4, 15, 1, 15 and 1 respectively.

Courses: The number of slots per course per week depends on whether the course is a lecture or a lab and the number of credits assigned to it. With regards to time slots for courses, generally it is considered to provide lab slots after the lecture slots.

Time: Time duration for a lecture could be determined by the working days of the week. With five working days (Sunday to Thursday, as in our case) in a week, three credits (three hours per week) means one hour for three days (Sun, Tue and Thursday) but with one-and-half hours in the remaining two days (Mon and Wednesday).

This study considers 16 student groups, 107 teachers, and 168 courses as its dataset.

Results and Findings

Different types of rooms are present in our setup. Number of rooms available are: Small (4), medium (15), large room (1), and network lab (1). In order to evaluate the proposed algorithm, number of days in a week, number of periods per day and number of medium rooms were varied as shown in Table 1. The number of courses, number of instructors and the number of student groups were maintained the same in all cases.

Table 1. Simulation Options and Outcome

	D5P9SR4 MR15	D5P8SR4 MR15	D5P7SR4MR15	D5P7SR4 MR10	D5P7SP4MR15 [25]
Number of days in a week	5	5	5	5	5
Number of periods per day	9	8	7	7	7
Number of periods	45	40	35	35	5
Number of Medium Rooms	15	15	15	10	15
Number of solutions found	37	38	44	41	43
Time taken to solve (Sec)	0.75	0.813	0.859	0.719	0.297

Considering the number of days in a week as 5, number of periods per day as 8 or 7 and the number of medium rooms as 8, does not return a feasible solution. Comparing with that referred work [25], facility utilization is better in the proposed case. When using [25], certain constraints are not considered and thus the number of rooms required to accomplish the scheduling is more than that of the available rooms. This shows the infeasibility in obtaining a solution that satisfies both the soft and hard constraints using [25]. At the same time, number of solutions (referred work has lesser soft constraints) found and the average number of consecutive lectures both for teachers and students are comparatively similar for both the proposed and the referred approaches.

5   Conclusion

University classroom scheduling problem considers a wide range of features for optimization. Consideration is given both for hard and soft constraints. Solution quality and time consumption are used as criteria to validate any obtained solution. The presence of diversified constraints could conflict each other and also the order of providing constraints to the specified formulation affects the resultant solution obtained. Considering the number of instructors as 107 and total number of rooms as 21, the utilization of the rooms varies between 60% to 80% at most of the time periods of the day. Similarly, utilization of medium rooms ranges between 50% to 60% with 15 rooms and increases to 80% to 90% with 10 rooms. Small rooms utilization varies between 70% to 80% considering the number of days and periods of work per day. Further enhancement can be made to the system in case of real-time changes to the availability of the facilities or instructors or student groups.

References

1. Gotlieb, C.: The construction of class-teacher time-tables | semantic scholar. In: Proceedings of the IFIP Congress 62. North Holland Pub. Co., Munich, Amsterdam. https://www.semanticscholar.org/paper/The-Construction-of-Class-Teacher-Time-Tables-Gotlieb/c6a16c404ccce6c8e8b7ce294c5c00bc3a80428b (1962). Accessed 26 Jun 2021
2. Almond, M.: An algorithm for constructing university timetables. Comput. J. **8**(4), 331–340 (1966). https://doi.org/10.1093/comjnl/8.4.331
3. Wasfy, A., Aloul, F.: Solving the University Class Scheduling Problem Using Advanced ILP Techniques. https://www.semanticscholar.org/paper/Solving-the-University-Class-Scheduling-Problem-ILP-Wasfy-Aloul/13e365cda5a392e6c882c182e496e9ef1b6b230c (2007). Accessed 26 Jun 2021
4. Islam, T., Shahriar, Z., Perves, M.A., Hasan, M.: University timetable generator using tabu search. J. Comput. Commun. **4**(16), 28–37 (2016). https://doi.org/10.4236/jcc.2016.416003
5. Aubin, J., Ferland, J.A.: A large scale timetabling problem. Comput. Oper. Res. **16**(1), 67–77 (1989). https://doi.org/10.1016/0305-0548(89)90053-1
6. Tahir, A.M., Hikmat, U.K., Sajjad, S.: Dynamic time table generation conforming constraints a novel approach. In: ICCIT (2012)
7. Amintoosi, M., Haddadnia, J.: Feature selection in a fuzzy student sectioning algorithm. In: Burke, E., Trick, M. (eds.) PATAT 2004. LNCS, vol. 3616, pp. 147–160. Springer, Heidelberg (2005). https://doi.org/10.1007/11593577_9
8. Barkha, N., Ambika, G., Rashmi, B.: Use of active rule and genetic algorithm to generate automatic time-table. Int. J. Adv. Eng. Sci. **3**(3) (2013)
9. Oladipo, W.K., Bamidele, A.O., Olalekan, A.M.: Automatic timetable generation using genetic algorithm. Int. J. Appl. Inform. Syst. **12**(19), 1–3 (2019). https://doi.org/10.5120/ijais2019451779
10. Nakasuwan, J., Srithip, P., Komolavanij, S.: Class scheduling optimization. Sci. Technol. Asia **4**, 88–98 (1999)

11. Aloul, F.A., Ramani, A., Markov, I.L., Sakallah, K.A.: Generic ILP versus specialized 0-1 ILP: an update. In: IEEE/ACM International Conference on Computer Aided Design, 2002. ICCAD 2002, pp. 450–457 (2002). https://doi.org/10.1109/ICCAD.2002.1167571

12. Al-Yakoob, S.M., Sherali, H.D.: A mixed-integer programming approach to a class timetabling problem: a case study with gender policies and traffic considerations. Eur. J. Oper. Res. **180**(3), 1028–1044 (2007). https://doi.org/10.1016/j.ejor.2006.04.035

13. Naderi, B.: Modeling and scheduling university course timetabling problems. Int. J. Res. Ind. **5**(1–4), 1–15 (2016)

14. Kumar, R.: A spreadsheet-based scheduling model to create individual graduation roadmaps. J. Supply Chain Oper. Manage. **15**(2), 165 (2017)

15. Ongy, E.E.: Optimizing student learning: a faculty-course assignment problem using linear programming. J. Educ. Hum. Resource Dev. **5**, 1–14 (2017)

16. Aldasht, M., Alsaheb, M., Adi, S., Qopita, M.A.: University course scheduling using evolutionary algorithms. In: 2009 Fourth International Multi-Conference on Computing in the Global Information Technology, pp. 47–51 (2009). https://doi.org/10.1109/ICCGI.2009.15

17. Wang, C., Li, X., Wang, A., Zhou, X.: A classroom scheduling service for smart classes. IEEE Trans. Serv. Comput. **10**(2), 155–164 (2017). https://doi.org/10.1109/TSC.2015.2444849

18. Herath, A.K.: Genetic Algorithm For University Course Timetabling Problem. University of Mississippi. https://egrove.olemiss.edu/etd/443 (2017)

19. Akkan, C., Gülcü, A.: A bi-criteria hybrid Genetic Algorithm with robustness objective for the course timetabling problem. Comput. Oper. Res. **90**, 22–32 (2018). https://doi.org/10.1016/j.cor.2017.09.007

20. Saviniec, L., Santos, M.O., Costa, A.M.: Parallel local search algorithms for high school timetabling problems. Eur. J. Oper. Res. **265**(1), 81–98 (2018). https://doi.org/10.1016/j.ejor.2017.07.029

21. Chen, M., Tang, X., Song, T., Wu, C., Liu, S., Peng, X.: A Tabu search algorithm with controlled randomization for constructing feasible university course timetables. Comput. Oper. Res. **123**, 105007 (2020). https://doi.org/10.1016/j.cor.2020.105007

22. Saviniec, L., Santos, M.O., Costa, A.M., dos Santos, L.M.R.: Pattern-based models and a cooperative parallel metaheuristic for high school timetabling problems. Eur. J. Oper. Res. **280**(3), 1064–1081 (2020). https://doi.org/10.1016/j.ejor.2019.08.001

23. Smith-Miles, K., Lopes, L.: Measuring instance difficulty for combinatorial optimization problems. Comput. Oper. Res. **39**(5), 875–889 (2012). https://doi.org/10.1016/j.cor.2011.07.006

24. Müller, T.: ITC2007 solver description: a hybrid approach. Ann. Oper. Res. **172**(1), 429 (2009). https://doi.org/10.1007/s10479-009-0644-y

25. Sandstrom, V.: Scheduling of teaching resources and classes using mixed integer linear programming. Aalto University, Helsinki, Independent Research Project in Applied Mathematics Mat-2.4108 (2012)

A Novel Hybrid Model Based on CNN and Bi-LSTM for Arabic Multi-domain Sentiment Analysis

Mariem Abbes[1,2(✉)], Zied Kechaou[2,3], and Adel M. Alimi[2,4]

[1] University of Sousse, 4011 ISITComSousse, Tunisia
`mariem.abbes@ieee.org`
[2] REGIM-Lab.: Research Groups on Intelligent Machines, University of Sfax, National Engineering School of Sfax (ENIS), BP1173, 3038 Sfax, Tunisia
`{zied.kechaou,adel.alimi}@ieee.org`
[3] Higher Business School of Sfax (ESCS), University of Sfax, Sfax, Tunisia
[4] Department of Electrical and Electronic Engineering Science Faculty of Engineering and the Built Environment, University of Johannesburg, Johannesburg, South Africa

Abstract. Since Web 2.0 and the freedom to share information, perspectives, and opinions on global events, services, goods, etc., most of the content on social media platforms comes from users. Social media data includes user sentiment-related subjects. Due to its complexity, ambiguity, dialects, shortage of resources, and morphological diversity, little work has been done on Arabic sentiment analysis. This study introduces a novel word embedding model, AraWord2Vec, based on Wor2Vec trained on Wikipedia and the Sentiment Benchmark dataset. We also examined how Wor2Vec embedding models improved classification tasks. Second, a novel CM_BiLSTM deep learning model called Convolutional Max Pooling and Bidirectional LSTM is presented to enhance Arabic sentiment analysis. Our experimental findings revealed the performance of all models on diverse datasets with an accuracy of 98.47% in binary classification; in multi-classes classification, we obtained a high performance of 98.92%.

Keywords: Deep Learning · Bi-LSTM · Convolutional Neural Network · word2vec · AraWord2Vec · Social Media · Arabic Sentiment Analysis

1 Introduction

Sentiment analysis has become one of the most important areas of study in natural language processing (NLP) because there is so much user-generated material on social media [1]. Users are increasingly likely to communicate with each other when making purchase choices, and policymakers and business leaders are now investing in finding out what people think of their goods and services. They spend money on mood research to make their customers happy, improve their goods and services, and even get new customers. Other studies have evaluated the impact of polarity detection on decision-making in different fields, including politics, business, marketing, tourism, and e-learning

[2–7]. Recently, with the increasing availability of Arabic social media platforms, the amount of Arabic text on the Internet has jumped significantly. Despite this, studies on Arabic sentiment analysis (ASA) have not yet seen significant developments [8], mainly because there aren't many Arabic emotion resources [9].

In an attempt to address the limitations of existing approaches, this article introduces novel deep learning (DL) based on "Bi-LSTM and CNN." The following are the research's main contributions to this article: we have presented a hybrid DL architecture based on CNN and Bi-LSTM. Secondly, we have created an AraWord2Vec model, based on Word2Vec embedding architecture, with different parameters trained with two large datasets, such as LABR [10] and Arabic Wikipedia offline.

The remainder of the paper is structured as follows: Sect. 2 provides a literature review and a survey of related models. Section 3 describes the novel DL model, including the background, proposed model parameters, configurations, and embedding scheme. The experimental procedures, results, and analyses are presented in Sect. 4. Conclusions are offered in Sect. 5, along with future work directions.

2 Related Works

Only a few recent works have explored deep-learning models for Arabic sentiment analysis. Abbes et al. proposed a novel approach for integrating the training of two modes of deep neural architectures: deep neural networks and recurrent neural networks. The experiment result shows that RNN (LSTM) outperforms DNN in terms of precision using the LABR dataset [11]. Authors in [12] studied several DL models for Arabic opinion mining, including DNN, Deep Belief Networks (DBN), Deep Automatic Encoder (DAE), and Recursive AutoEncoder (RAE). Results demonstrated that RAE outperformed all competing models, followed by DAE, presenting the advantage of recursive models over single models for learning accurate semantic representations. Another study by Al Sallab [13] addressed some limitations of RAE for ASA, such as its poor handling of Arabic morphological complexity. They introduced a recursive deep-learning model for Arabic opinion mining. The results showed that AROMA outperformed basic RAE and other classical machine learning approaches applied to ASA.

The authors of [14] presented their Twitter dataset of opinions about health services. For classification, they conducted several experiments using various DNNs and additional machine-learning techniques (NB, LR, and SVM). The DL approaches showed promising results (90% for CNN and 85% for DNN) using Word2Vec for unsupervised learning. However, they were outperformed by other classifiers like SVM, which achieved 90.88% accuracy, and in [15], they obtained a score of 95% for classifying sentences into two classes (positive and negative) using CNN + LSTM models.

In [16], the authors focused on sentiment analysis of dialectical tweets from highly unbalanced datasets. They extracted the main features by the Word2Vec model for word embedding. To solve the problem of unbalanced datasets, they oversampled the minority class by adding synthetic samples using the SMOTE technique [17]. In this work [18], authors applied the character-level deep CNNs for ASA, which significantly improved the performance over other machine learning classifiers, namely KNN, SVM, DT, Random Forest (RF), NB, and LR. CNN attained the optimal accuracy of 94.33% because deep CNNs are powerful in extracting features from the pre-processed and raw user-generated text. The authors of [19] proposed an ensemble model that combines CNN and LSTM, leading to significant accuracy.

In this work [20], the authors proposed a differential evolution (DE) algorithm to increase the performance of CNN. The DE algorithm was used to find the optimal configuration of the CNN architecture and network parameters. On five Arabic sentiment datasets, the DE-CNN framework obtained greater accuracy and required less time than existing algorithms. More recent work by [21] proposed an ABSA system using a bidirectional LSTM (Bi-LSTM) and a conditional random field (CRF). Integrating a Bi-LSTM with character-level integration features significantly improved the performance of the aspect opinion target expression. The overall accuracy obtained for identifying aspect sentiment polarity was 82.7%.

Elnagar et al. [22] constructed and tested nine deep-learning models for single and multi-label Arabic text classification without pre-processing. They examined how Word2Vec embedding models improved classification tasks. They trained CNN, Bi-LSTM, BiGRU, and HANGRU, the top four single-label categorization models, for the multi-label challenge. All models performed well on the SANAD corpus, with CGRU achieving 93.43% accuracy and HANGRU 95.81%. CNN had the greatest accuracy with confidence > 50% of 70.34% for a maximum subset of 8 categories in the SkyNewsArabia dataset, in contrast, HANGRU had 88.68% for a whole subset of 10 classes in the Masrawy dataset. In work proposed by [23], classification performance is compared using various WE vectors pre-trained in Arabic. The LSTM-RNN model was used for classification. The pre-trained embedding vector FastText won the three WE methods we compared for sentiment classification. It achieved an accuracy score of more than 90% on the training and testing set. The ensemble model scored the highest accuracy of 96.7% on the test set.

This article [24] provides Bidirectional Encoder Representations from Transformers (BERT) with a CNN classification head for Arabic sentiment analysis with few data resources. With 50% less batch size, training, and epochs, the model beats state-of-the-art models. Based on Google Play Store mobile app comments, this study introduces a novel sentiment analysis method for Arabic. The tests were done by comparing the accuracy of the Naive Bayes (NB) and the Levenshtein Distance (LD) algorithms. The NB algorithm's accuracy was 95.80%, while the LD algorithm's accuracy was 96.40% [25].

3 Proposed Approach

In this section, we will examine the preprocessing procedures, features, language resources, and ML approaches inspired by the literature we use for our comparative experimentation.

3.1 Words Embedding Model Based on Word2vec

The input data for DL models must be transformed into numerical representations from sentences. With word embedding, low-dimensional real-valued vectors can represent phrases or entire documents like reviews or tweets. Therefore, the input to CM_BiLSTM will be text representations generated from word embedding models. Mikolov [26] develops Word2Vec embeddings, which is an unsupervised model that converts a vast set of words into a numeric vector. Word2vec has two main variants: the Continuous Bag-of-Words (CBOW) model and the Skip-Gram model (SG). In the CBOW model, the goal is to predict the target word from the surrounding context words. In the SG model, the goal is to predict the surrounding context words from the target word.

There are Arabic word vectors like AraVec [27] are trained on three separate data sources (Twitter, the World Wide Web, and 2019 Wikipedia dumps). AraVec has noisy data, and unread symbols damage learning also needs a sentiment dataset.

For this work, we created our word embedding based on Word2Vec. AraWord2Vec models are used to develop efficient distributed word representation models using four datasets, such as Wikipedia offline 2022 [28], HARD [29], BRAD [30], and LABR [10].

Wikimedia Dump is Wikipedia offline in many languages and formats. The September 2022 Arabic Wikidata dump is 20 GB (a JSON file) with 5,448,137 articles. We used WikiExtractor [31] to line-by-line extract the dump and parse the JSON text to represent the current tools. The LABR has the most Arabic SA data—a book review website provided over 63 k imbalanced MSA and dialect reviews. The BRAD has 510,600 Arabic book reviews in MSA and dialect (a LABR extension). The HARD has 409,563 Arabic hotel reviews. Booking.com provided hotel reviews.

AraWord2Vec provides six different word embedding models, where each text domain has two different models, one using the CBOW technique and the other using the Skip-Gram technique. To build these models, we ran many experiments to tune the hyperparameters (Fig. 1).

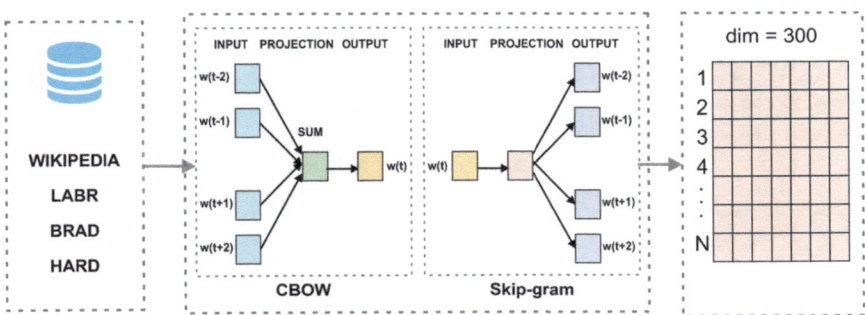

N=vocabulary size, dim = dimensionality

Fig. 1. General architecture of the Word2Vec proposed models

3.2 Arabic Sentiment Benchmark Datasets

This paper conducts experiments using seven Arabic benchmark datasets written in Egyptian dialects and Modern Standard Arabic (MSA). Moreover, it demonstrates the adaptability of our model across diverse domains. We have used two sentiment classes, i.e., Positive and Negative, and multi-sentiment classes, i.e., $--$, $-$, 0, $+$, $++$. The characteristics of these datasets are presented in Table 1.

Table 1. Benchmark tested datasets

Dataset		Classes	Data size
LABR	[10]	5($--$,$-$,0,+,++)	63257
ASTD	[32]	4($-$,0,obj,+)	10006
HTL	[33]	3 ($-$,0,+)	15572
RES	[33]	3 ($-$,0,+)	10970
MOV	[33]	3 ($-$,0,+)	1524
PROD	[33]	3 ($-$,0,+)	4272

3.3 The Proposed Hybrid Architecture Based on CNN and Bi-LSTM

In this section, we present our proposed approach to improving sentiment analysis. Figure 2 shows a graphical representation of the proposed architecture, named bidirectional LSTM, with a convolution layer and max pooling layer (CM_BiLSTM). The architecture consists of an embedding layer, two convolutional layers, two pooling layers, and Bi-LSTM, which combines a forward hidden layer and a backward hidden layer to access both preceding and subsequent contextual features.

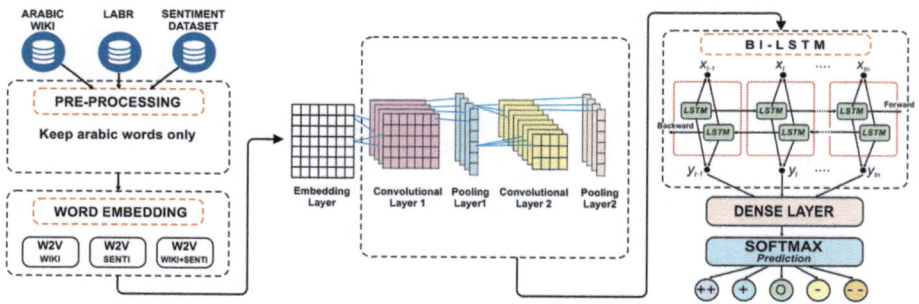

Fig. 2. CM_BiLSTM proposed architecture

3.3.1 Data Preprocessing

Arabic is one of the world's most frequently spoken languages. Semitic Arabic has 28 right-to-left letters. Data preprocessing is an initial step applied to the input data, aiming to modify the data for relevant feature extraction. We apply the following preprocessing steps to all datasets:

– Filtering Non-Arabic letters.

- Removing Arabic Stop Words (empty words).
- The normalization stage is to Standardising letters forms of the words, for example, replacing the different forms of the letter Alif written as by bare "alif," (إ,أ,آ,ا) were converted into (ا), the letter (ة) into (ه), and the letter (ى) into (ي).
- Cleaning stage: a step that contains the general cleaning of empty strings, elongation, numerical data, URLs, punctuation marks, extra spaces, and diacritics.

3.3.2 Embedding Layer Based on AraWord2Vec

The embedding layer encodes sentences as numbers. We chose this representation model because the Word2Vec model initializes word weights well. This phase prepares the datasets for learning the embedding and applies our pre-trained word embedding to a large text corpus.

The system imports the training set data into a lengthy list of tokens using the NLTK tokenizer [34] before fitting the Word2Vec model. Each phrase is an n × d (2-dimensional) vector, where n is the sentence's word count and d is the vector representation's dimension.

Six 300-dimensional models are used. We pad tweets with zeros to make them all the same size. We picked a maximum length of n = 200. Therefore, each sentence would be 200 × 300. The network learns meaning from these random vectors during training.

3.3.3 1D – Convolutional Neural Network

CNN has been widely used for image processing and has recently proven a powerful tool for text classification. It could pick up on sentiment reviews' syntactic and semantic features and make the input data smaller.

For text classification, we use a one-dimensional convolutional layer (1D-Conv), a well-known CNN architecture for dealing with sequences. The 1D-Conv applies a nonlinear function called a filter, which transforms a window into scalar values. After applying a filter, the 1D-Conv makes m vectors, each representing a filter. Local information about texts is stored in the text area while the filters move. Hence, the convolutional layer is an effective way to reduce dimensionality.

After the convolutional layer, a pooling layer is often used to gradually reduce the representation size, reducing the number of features and making the network easier to understand. So, a pooling task is needed to put all the m-vectors together into a single m-dimensional vector. Each word in a sentence of length nn is associated with an MM-dimensional vector. In this study, we apply two convolution layers and two max pooling layers.

3.3.4 Bidirectional LSTM and Dense Layer

Because LSTM architecture has proven effective in processing sequential data with fewer problems in vanishing gradients and memory saving, Bidirectional LSTM is placed after the last layer of Max-pooling for deep processing.

The hidden vectors are computed for each input sentence based on past and future data. After getting the output from the max-pooling layer, Eq. (1) is used to figure out the hidden state (ht). W is the weight matrix, W × h is the weight connecting input x to hidden layer h, b is the bias vector, and tanh is the activation function.

The outputs of Bi-LSTM are concatenated into a single vector using the sigmoid activation function in a range between 0 and 1. The dense layer has 256 vector sizes, the same as the layer size of the Bi-LSTM. The following equations can represent the output vector of the dense layer:

$$h_t = \tanh(W_{xh}x_t + W_{hh}h_{t-1} + b_h) \tag{1}$$

$$y_{t=}W_{xh}x_t + b_h \tag{2}$$

Bi-LSTM is specialized for sequential modeling and can further extract contextual information from the feature sequences obtained by the convolutional layer. However, the feature sequences obtained from the convolutional layer do not contain sequence information. The effect of Bi-LSTM is to build the text-level word vector representation. Assigning various weights to words is a common solution to this problem because each word contributes differently to the context's meaning. Finally, a dense sigmoid layer is used to predict the labels in the binary classification.

4   Experimental Results and Evaluation

This section shows the experimental results used to evaluate the performance of the proposed approach, CM_BiLSTM, of a sentence-level sentiment classification system based on CNN and Bi-LSTM structure on various benchmarking datasets.

4.1   Environment Setup and Parameters

Our implementation language is Python because it has many supporting APIs and libraries that make the workflow easier. For implementing our architecture and the deep learning experiments, we used Keras [35] on top of a TensorFlow [36] back-end in Google Collaboratory. For machine learning, we used the Scikit-Learn library [37], which provides a variety of machine-learning algorithms.

A neural network generally has more parameters than other machine learning algorithms, making optimizing hyperparameters critical to achieving the expected performance in DL. The hyperparameters can relate to network structure or training, such as the number of epochs (50 epochs) and the batch size (128). For all experiments, we used the Rectified Linear Unit (ReLU) in the convolutional layer, the dropout set to 0.5, and the Adam optimizer. We used a grid search estimator and specified ranges of values for possible parameters to optimize neural network performance and obtain the best selection of hyperparameters in this work.

4.2   Results Analysis and Discussion

The CM_BILSTM model was first applied separately to each dataset as a binary classification. Our goal was to check our precision against that of the most recent state-of-the-art baselines. For the embedding layer, we are developing Ara_word2vec, which provides six different word embedding models based on CBOW and SG Word2Vec architectures. These models are trained on Wikipedia and sentiment datasets. We tested all of them, and the best results were obtained using the SG architecture and sentiment (W2v_Senti_SG) and all datasets (W2v_all_SG). Table 2 displays the CM_BiLSTM model accuracy results for the binary classification datasets and Table 3 for the multi-classes classification datasets.

The CM_BiLSTM model clearly outperforms all published baselines, with accuracy ranging from 82.73% to 98.47% for binary classification and 72.88% to 98.92% for tertiary classification. This performance is anticipated given that the CM_BiLSTM model generates state-of-the-art results on international public datasets such as those listed in Table 1.

Table 2. CM_BiLSTM results with binary classification.

Word2vec	Datasets accuracy					
	HTL	PROD	RES	MOV	ASTD	LABR
W2v_all_SG	**98.47%**	**90.98%**	**93.76%**	**96.91%**	**83.93%**	**87.56%**
W2v_Senti_SG	98.17%	90.67%	93.50%	96.42%	82.74%	87.22%
[32]	88.7%	75.9%	84.6%	74.3%	-	81.2%

Table 3. CM_BiLSTM results with multi-classification

Word2vec	Datasets accuracy					
	HTL(3)	PROD(3)	RES(3)	MOV(3)	ASTD(4)	LABR(5)
W2v_all_SG	**98.92%**	**94.00%**	93.03%	97.46%	**87.53%**	**73.45%**
W2v_Senti_SG	98.68%	93.30%	**93.44%**	**97.95%**	87.40%	72.88%
[33]	67.1%	53.7%	60.6%	52.6%	-	60.6%
[32]	-	-	-	-	68.9%	-

5   Conclusion

This paper presents an enhanced architecture, CM_BiLSTM, to optimize Arabic text sentiment classification. There are three steps to developing our approach: data preprocessing, the creation of the word vector model, and the creation of the classifier. CM_BiLSTM was used for sentence-level classification, with features extracted using different proposed Word2Vec models. AraWord2Vec is trained on four large datasets (Wikipedia, LABR, HARD, BRAD), and the overall number of sentences is 594 k.

We test our system on several benchmarks consisting of MSA and dialectal Arabic reviews. Those benchmarks were used for training and testing across multiple polarities: very positive, positive, very negative, negative, and neutral.

We performed our experiments over multiple data splits and used cross-validation to ensure confidence in our classification performance. The accuracy for all tested models ranged between 72.88% and 98.92%.

Although the CM-BiLSTM has outperformed previous methods, we recommend the following changes to the approach in a future study to improve the CM_BiLSTM performance. (a) We want to go more into managing critical sentences that contain both positive and negative sentiment (ambiguous case). (b) Other word embedding models and datasets will be evaluated.

Acknowledgments. The research leading to the recorded results achievements has received funding from the Ministry of Higher Education and Scientific Research of Tunisia, under grant agreement number: LR11ES48.

References

1. Dashtipour, K., et al.: Multilingual sentiment analysis: state of the art and independent comparison of techniques. Cogn. Comput. **8**(4), 757–771 (2016)
2. Kechaou, Z., Ben Ammar, M., Alimi, A.M.: Improving e-learning with sentiment analysis of users' opinions. In: 2011 IEEE Global Engineering Education Conference (EDUCON), pp. 1032–1038 (2011)
3. Kechaou, Z., Ben Ammar, M., Alimi, A.M.: A new linguistic approach to sentiment automatic processing. In: IEEE ICCI, pp 265–272 (2010)
4. Kechaou, Z., Wali, A., Ben Ammar, M., Alimi, A.M.: Novel hybrid method for sentiment classification of movie reviews. In: DMIN, pp 415–421 (2010)
5. Kechaou, Z., Kanoun, S.: A new-arabic-text classification system using a Hidden Markov Model. KES J. **18**(4), 201–210 (2014)
6. Kechaou, Z., Wali, A., Ben Ammar, M., Karray, H., Alimi, A.M.: A novel system for video news' sentiment analysis. J. Syst. Inform. Technol. **15**(1), 24–44 (2013)
7. Kechaou, Z., Ben Ammar, M., Alimi, A.M.: A multi-agent based system for sentiment analysis of user-generated content. Int. J. Artif. Intell. Tools **22**(02), 1350004 (2013)
8. Al-Moslmi, T., Albared, M., Al-Shabi, A., Omar, N., Abdullah, S.: Arabic senti-lexicon: constructing publicly available language resources for arabic sentiment analysis. J. Inform. Sci. **44**(3), 345–362 (2018)
9. Zaghouan, W.: Critical survey of the freely available Arabic corpora. Comput. Sci. arXiv: 1702.07835 (2017)
10. Aly, M., Atiya, A.: LABR: large-scale arabic book reviews dataset. In: Association of Computational Linguistics (2013)
11. Abbes, M., Kechaou, Z., Alimi, A.M.: Enhanced deep learning models for sentiment analysis in Arab social media. In: Liu, D., Xie, S., Li, Y., Zhao, D., El-Alfy, E.-S.M. (eds.) ICONIP 2017. LNCS, vol. 10638, pp. 667–676. Springer, Cham (2017). https://doi.org/10.1007/978-3-319-70139-4_68
12. Al Sallab, A., Hajj, H., Badaro, G., Baly, R., El Hajj, W., Shaban, K.B.: Deep learning models for sentiment analysis in Arabic. In: Proceedings of the second workshop on Arabic natural language processing, pp. 9–17 (2015)

13. Al Sallab, A., Baly, R., Hajj, H., Shaban, K.B., El-Hajj, W., Badaro, G.: AROMA: a recursive deep learning model for opinion mining in Arabic as a low resource language. ACM Trans. Asian Low-Resour. Lang. Inf. Process. **16**(4), 25 (2017)
14. Alayba, A.M., Palade, V., England, M., Iqbal, R.: Arabic language sentiment analysis on health services. In: 1st international workshop on Arabic script analysis and recognition (Asar), pp. 114–118 (2017)
15. Alayba, A.M., Palade, V., England, M., Iqbal, R.: A combined cnn and lstm model for arabic sentiment analysis. In: Holzinger, A., Kieseberg, P., Tjoa, A.M., Weippl, E. (eds.) Machine Learning and Knowledge Extraction, pp. 179–191 (2018)
16. Al Azani, S., El-Alfy, E.S.: Emojis-based sentiment classification of Arabic microblogs using deep recurrent neural networks. In: International Conference on computing sciences and engineering, ICCSE, pp. 1–6 (2018)
17. Chawla, N.V., Bowyer, K.W., Hall, L.O., Kegelmeyer, W.P.: Smote: synthetic minority over-sampling technique. J. Artif. Intell. Res. 321–357 (2015)
18. Omara, E., Mosa, M., Ismail, N.: Deep convolutional network for Arabic sentiment analysis. In: International Japan–Africa Conference on Electronics, Communications and Computations, JAC-ECC, pp. 155–159 (2018)
19. Heikal, M., Torki, M., El-Makky, N.: Sentiment analysis of Arabic tweets using deep learning. Procedia Comput. Sci. **142**, 114–122 (2018)
20. Dahou, A., Abd Elaziz, M., Zhou, J., Xiong, S.: Arabic sentiment classification using convolutional neural network and differential evolution algorithm. Comput. Intell. Neurosci. **2019**, 2537689 (2019)
21. Al Smadi, M., Talafha, B., Al Ayyoub, M., Jararweh, Y.: Using long short-term memory deep neural networks for aspect-based sentiment analysis of Arabic reviews. Int. J. Mach. Learn. Cybern. **10**(8), 2163–2175 (2019)
22. Elnagar, A., Al Debsi, R., Einea, O.: Arabic text classification using deep learning models. Inform. Process. Manage. **57**(1), 102121 (2020)
23. Alwehaibi, A., Roy, K.: Comparison of pre-trained word vectors for Arabic text classification using deep learning approach. In: IEEE International Conference on Machine Learning and Applications, ICMLA (2019)
24. Fawzy, M., Fakhr, M.W., Rizka, M.A.: Sentiment Analysis for Arabic Low Resource Data Using BERT-CNN. In: International Conference on Language Engineering, pp. 24–26 (2022)
25. Al-Hagree, S., Al-Gaphari, G.: Arabic sentiment analysis on mobile applications using Levenshtein distance algorithm and naive Bayes. In: 2nd International Conference on Emerging Smart Technologies and Applications, pp. 1–6 (2022)
26. Mikolov, T., Chen, K., Corrado, G., Dean, J.: Efficient estimation of word representations in vector space. In: 1st International Conference on Learning Representations, pp. 1–12 (2013)
27. Soliman, A.B., Eisa, K., El-Beltagy, S.R.: AraVec: a set of arabic word embedding models for use in Arabic NLP. In: International Conference on Arabic Computational Linguistics (2017)
28. https://dumps.wikimedia.org/
29. Elnagar, A., Khalifa, Y.S., Einea, A.: Hotel arabic-reviews dataset construction for sentiment analysis applications. In: Shaalan, K., Hassanien, A.E., Tolba, F. (eds.) Intelligent Natural Language Processing: Trends and Applications. SCI, vol. 740, pp. 35–52. Springer, Cham (2018). https://doi.org/10.1007/978-3-319-67056-0_3
30. Elnagar, A., Einea, O.: BRAD 1.0: Book reviews in Arabic dataset. In: IEEE/ACS International Conference of Computer Systems and Applications, pp. 1–8 (2016)
31. Attardi, G.: WikiExtractor. GitHub repository. https://github.com/attardi/wikiextractor (2015)
32. Nabil, M., Aly, M., Atiya, A.: ASTD: arabic sentiment tweets dataset. empirical methods in natural language processing, pp. 2515–2519 (2015)

33. ElSahar, H., El-Beltagy, S.R.: Building large arabic multi-domain resources for sentiment analysis. In: Gelbukh, A. (ed.) CICLing 2015. LNCS, vol. 9042, pp. 23–34. Springer, Cham (2015). https://doi.org/10.1007/978-3-319-18117-2_2
34. Steven, B., Loper, E., Klein, E.: Natural Language Processing with Python. O'Reilly Media Inc. (2009)
35. Chollet, F., et al.: Keras. GitHub. https://github.com/fchollet/keras (2015)
36. Abadi, M., et al.: TensorFlow: Large-scale machine learning on heterogeneous systems. https://www.tensorflow.org/ (2015)
37. Pedregosa, F., et al.: Scikit-learn: machine learning in Python. J. Mach. Learn. Res. **12**, 2825–2830 (2011)

A Cost-Sensitive Ensemble Model for e-Commerce Customer Behavior Prediction with Weighted SVM

Jing Ning[1], Kin Fun Li[1(✉)], and Tom Avant[2]

[1] University of Victoria, Victoria, Canada
{jingning,kinli}@uvic.ca
[2] VINN Automotive, Victoria, Canada
tom@vinnauto.com

Abstract. In today's highly competitive environment, effective customer relationship management (CRM) is critical for every company, especially e-commerce businesses. Analyzing customer behavior is an initial step in marketing strategies and revenue generation, and then companies can predict purchasing behavior to enhance efficiency and boost profits. Many practitioners have noted that in customer behavior prediction, some special requirements, such as minimizing false predictions, need to be implemented to prevent the churn of potential customers. On the other hand, many studies on customer behavior prediction do not leverage unstructured data, such as conversation records, to improve a prediction model's performance. We propose a prediction model based on weighted support vector machines (SVM) that can reduce the churn of customers while maintaining reasonable accuracy. We validate our approach using customer data from an e-commerce company and conduct multi-dimensional comparisons to demonstrate the practicality of our method.

1 Introduction

VINN is an online automotive e-commerce platform committed to enhancing its core business by providing better customer support and improving its web functionalities and services. VINN has been collecting information to understand consumer behavior in automobile purchasing, to plan and launch new value-added services and promote its brand to potential customers. The data collected by VINN includes personal information filled out by a customer on their website, including name, phone number, email, etc., as shown in Table 1, in the form of structured data. Also, there is the description of the customer's family, occupation, interests, and so on, recorded by the marketing department personnel through phone and text message communication with customers and stored as unstructured text data.

L. Barolli (Ed.): CISIS 2023, LNDECT 176, pp. 103–115, 2023.
https://doi.org/10.1007/978-3-031-35734-3_11

In a competitive e-commerce scenario, companies would rather incur more administrative and communication costs than lose potential users. So, they must minimize the chance of detecting potential users as non-purchasers. However, this leads to decreased overall accuracy and more non-purchasers identified as purchasers [3]. For our customer behavior prediction model, we regard the classification of purchasers as true and the classification of non-purchasers as false. We aim to reduce the false negatives of the prediction results. Much research has addressed these issues, typically by building a model that adds penalties to a particular class, also known as cost-sensitive algorithms [17]. Our approach uses weighted SVM to implement a cost-sensitive prediction model.

Many practical cost-sensitive algorithms have been used to solve the problem of data imbalance [10,12,14], an often encountered issue in data mining and prediction models. In these solutions, cost-sensitive consumer behavior prediction has not been sufficiently validated in real e-commerce scenarios. Furthermore, there are many research works related to CRM, but most of them are based on structured data [1,2,4]. For example, most existing user churn models are based on structured data such as user name, phone number, and retention time [6,8,9]. Unstructured data, such as phone conversations, email content, and text messages, contain richer information than structured data. Processing and analyzing unstructured data are more complex but can often improve prediction performance. Some studies have shown that combining unstructured and structured data performs better than using structured data alone for executing a prediction model [5,7]. Our work aims to improve customer behavior prediction performance by combining structured and unstructured data and using ensemble learning.

Our contribution is twofold: First, we analyze some real e-commerce customers' data and design a cost-sensitive customer behavior prediction model for business demands. Secondly, using unstructured data with more information and structured data, we achieve a better customer behavior prediction model, with its effectiveness verified by Vinn's customer data. The results show that the single SVM model has a false negative rate of 28%, while the weighted SVM-based ensemble model has a false negative rate of 13% and maintains the same accuracy rate of 75%.

The remaining parts of this paper are organized as follows: The second section introduces the processing and analysis of the raw e-commerce data and then the extraction and selection of the features from the structural and unstructured data, respectively. Section 3 introduces the core algorithms of this work, including the ensemble model and the individual classifiers that make up the model. Finally, the fourth section describes the experiment and verification process and validates our model through various comparisons.

Table 1. Structured Data and Feature Extraction

Structured Data	Feature Extraction
Name	Offensive Check
Name	Name Entity Recognition
City	Name Entity Recognition
Phone number	Authenticity Check
Age	Adult Check
Email	Offensive Check
Email	Format Check
Priority Vehicle Type	Reasonable Budget
Payment Options and Budget	Reasonable Budget

2   Data Mining and Featurization

In the cases of our focus, structured data comes from the basic personal information filled in by the customer online, as shown in the left column of Table 1. In addition, we extract relevant features from this raw data, as shown in the right column of Table 1. Unstructured data comes from the voice or text communication between the marketing personnel and customers, which could include background description of the customers, including family, job, recreational activities, etc. This information is essential for us to predict customer purchase behavior.

2.1   Structured Data

Structured data includes basic customer information, which can be used to extract additional derived features for the prediction models. This data is obtained from customers' entered personal information on the company's web page. Often, some customers may enter offensive words which do not indicate buying intent. We also identify the relationships between different data categories to increase the number and quality of features. Various analytical methods are used to analyze the structured data as follows.

1. Offensive Word Detection We use databases containing words that are commonly regarded as offensive to identify the potentially inappropriate language in customer profiles. Users who include such terms in their profiles may be deemed non-genuine or not serious, potentially lowering the chances of successful transactions.
2. Name Entity Recognition Providing a real name, brand, or product represents the desired shopping needs of a customer, thus significantly improving the probability that the user is a purchaser. We utilize the most widely used NER (Name Entity Recognition) System [18] to determine whether the description contains a real name, brand, etc., to determine the likelihood that the user will be a purchaser. This approach is also used for unstructured data.

3. Phone Number and Email Address Check Often, many data items have the same phone number, which could result from a massive cyber attack. In other scenarios, hackers may fill in the same email address to increase the platform's workload. These users are meaningless to the business, so they have the lowest priority.
4. Correlation between Phone Number and City Suppose the area code in the phone number matches the city entered by the user. In that case, it indicates that the user's information is relatively reliable, which can be used to judge whether the user will become a purchaser.
5. Reasonability of Budget Comparing the budget and priority vehicle type filled in by a user can serve as a reference to determine whether the customer has the financial ability to make a potential purchase.

2.2   Unstructured Data

Vinn provides unstructured data in text paragraphs describing their customers' families, occupations, recreation, and other preferences, which are more diverse and richer than the structured data entered by the customers. To build an efficient prediction model, we need to identify the unique features that distinguish potential purchasers from non-purchasers, which we do by following these processes.

2.2.1   Pre-processing
Since transcribed telephone conversations are unstructured and contain redundant information, we need to do the following pre-processing to provide a baseline for subsequent feature extraction.

1. Filtering Stop Words - Stop words are the most frequently used words but contain the least information, which should be removed before further processing.
2. Lemmatization - There are multiple word forms, and we shall reduce these words to their most basic form to allow a prediction model to understand the text better.

2.2.2   Bag-of-Words
Bag-of-Words is widely used in text classification because of its simplicity and effectiveness. Vectorizing words involves transforming unstructured data into structured data that can be used in subsequent procedures. This transformation is achieved by creating individual lists of words and counting the frequency of their occurrence in different categories. By doing so, we can represent textual data in a numerical format that machine learning algorithms and other data analysis techniques can process.

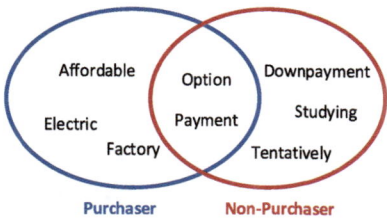

Fig. 1. Bag-of-Words

1. Split Word List - Some words only appear in a specific class, and the appearance of these words can significantly affect the results, which is a good source of features (Fig. 1).
2. Frequency Word List - For words that exist in both classes, their frequency of occurrence can also indicate the difference between the two classes. We choose 100 words with the most significant frequency difference of common words as the feature word list.
3. Priority Word List - Through consultation with Vinn's marketing department, there is a correlation between the frequency of some specific business-related words and purchasing behavior. We use this correlation to construct a priority word list, and the frequencies of these words are also used as features.

2.2.3   Sentiment Analysis

Sentiment analysis has been a popular research topic in recent years and a significant advancement to enable machines to understand human language. For CRM, sentiment analysis is critical. We analyzed four sentiment scores for the conversation text: positive, negative, neutral, and compound. These four scores help us to understand the users' emotions fully and to use them as features for prediction.

2.3   Feature Selection

We cannot guarantee that each one is effective for the multiple features that have been collected, so we need to carry out feature selection. For a binary prediction model, the relation between labels and features is usually nonlinear, so Spearman's and Kendall's correlations should perform better than Pearson's correlation. We use Kendall's method to calculate the correlation because it is more robust than Spearman's [19] for the many ties in our data set.

$$\tau = \frac{n_c - n_d}{n_c + n_d} = \frac{n_c - n_d}{n(n-1)/2} \tag{1}$$

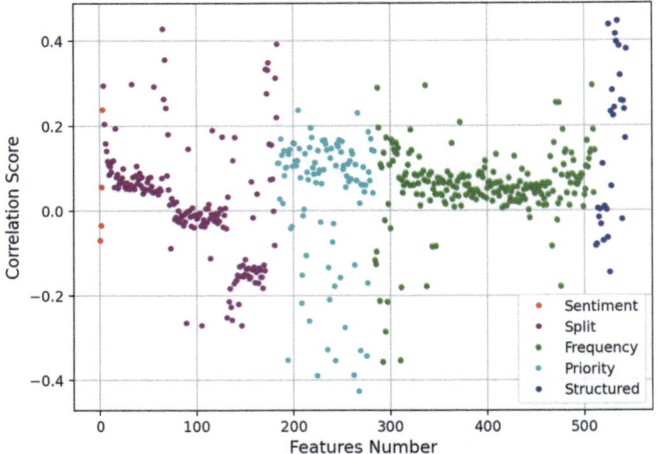

Fig. 2. Correlation between Features and Labels

where, n_c = number of concordant pairs, n_d = number of discordant pairs, and n = number of pairs.

As shown in Fig. 2, the abscissa is the feature number, and the ordinate is the correlation score between features and labels. A total of 544 features are shown in the figure, of which 4 features come from sentiment analysis, 280 from Bag-of-Words, 180 from the split word list, 100 from the frequency word list, and 228 from the priority word list. Additionally, there are 32 features extracted from the structured information. The appropriate number of features for prediction is selected based on the threshold of correlation scores determined through experimentation.

It is important to note that compound emotion scores in sentiment analysis show a stronger correlation with purchasing behavior. The priority word list outperforms the other features in the split word list because it is a manually selected list of business-related words. For the split word list, the classification is primarily associated with the most frequently used words, while the less frequently used words have minimal effect on the classification. For the frequency word list, most of the words have some correlation with prediction, but the correlation is not strong. Most structural data produce highly predictive features.

3 Prediction Methodology

A well-designed prediction model distinguishes between classes and can adapt to prioritize significant categories based on real-world situations. Our proposed model is illustrated in Fig. 3. We perform information extraction and natural language processing on structured and unstructured data, respectively, and complete feature selection for all the obtained features. These features are then fed

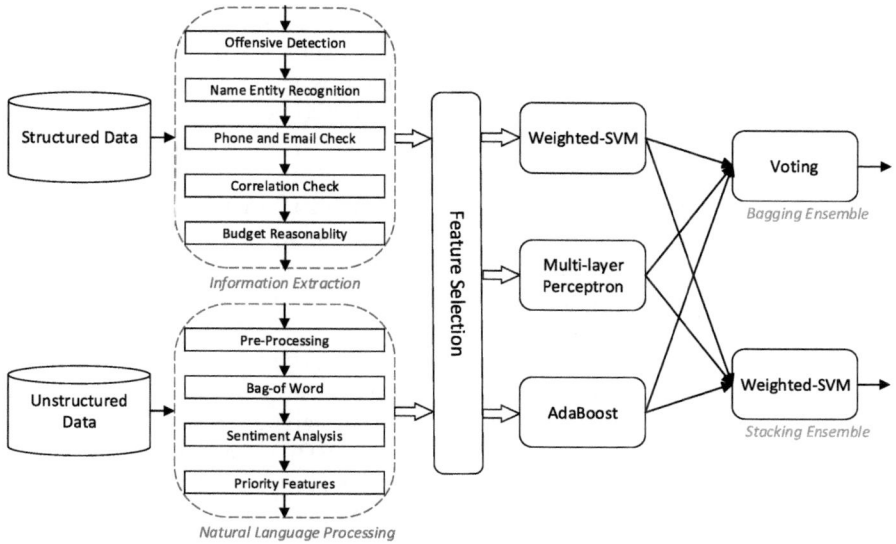

Fig. 3. Cost-Sensitive Bagging and Stacking Ensemble Model

into the three single individual models, and their outputs are combined using two different strategies (Bagging and Stacking) to achieve better performance.

3.1 Ensemble Model

Ensemble learning is a type of machine learning that seeks better predictive performance by combining multiple models. The three kinds of ensemble learning are bagging, stacking, and boosting. The bagging model is a simple and effective method to average or votes the predicted results. However, simple averaging or voting for a cost-sensitive model will lose the superiority of the single classifiers [10,12]. Therefore, we proposed a cost-sensitive stacking ensemble model, which combines individual models to achieve good performance on both accuracy and cost-sensitive.

3.2 Single Classifier

1. Weighted SVM False negatives can be detrimental to the success of any e-commerce platform. These occur when a customer is mistakenly categorized as not fitting the criteria for a particular product or service and, therefore, not shown that product or service. This misclassification results in a missed opportunity to make a sale and can lead to losing potential customers. To prevent misclassification, e-commerce platforms need a reliable and effective method to reduce false negatives. The mathematical foundation of SVM is robust and has been effectively employed in several classification applications.

SVM works by optimizing the separation between classes using hyperplanes [15]. However, its traditional form prioritizes accuracy and may not be suitable for cost-sensitive requirements [14]. This is where a weighted SVM comes into play. Weighted SVM is a cost-sensitive version of SVM that assigns different weights to each class based on their relative importance. The weighted SVM algorithm can prioritize specific categories and increase the penalty for false negatives within those categories. This helps to ensure that every potential customer is given appropriate consideration and is not overlooked due to a misclassification. By implementing a weighted SVM approach, e-commerce platforms can reduce the occurrence of false negatives and improve the prediction model's performance.

The kernel is the mathematical function that determines how support vector machines operate. The kernel's role is to process the input data to the desired format. Different SVM algorithm uses different type of kernel, such as linear, nonlinear, polynomial, radial basis function (RBF), and sigmoid [16]. Our model uses the polynomial kernel as it has demonstrated better prediction results in our experiments. The formula of the polynomial kernel is:

$$K(X_1, X_2) = (a + X_T^1 X_2)^b \tag{2}$$

where b = degree of kernel, a = constant term.

2. Multi-layer Perceptron Classifier

   As one of the most popular machine learning models, a neural network has excellent performance in classification and prediction. In our work, we use a multi-layer perception neural network because it is powerful and flexible [13].

3. AdaBoost

   Boosting learning improves performance by correcting the errors of a previous iteration. Because of its adaptability, boost has been proven to be an efficient classifier [11]. We use AdaBoost as one of the classifiers to improve the overall accuracy of our ensemble model.

4   Experiment

We used commercial data to validate the model's effectiveness through a series of experiments. After introducing the data sets used, we compared the performance of different SVM models and two other individual models and the ensemble models. In addition, we employed different validation methods to compare the algorithms used.

4.1   Data Sets

Real-world data best demonstrate the model's effectiveness. The data set we used has 14,599 users' structured data, 3333 of which contain unstructured data. Using the featurization described in Sect. 2, we processed the data in various ways and obtained good features and valid binary labels for training. Two-thirds of the data were used for training, and the rest were for testing.

4.2   Validation

A confusion matrix is used to summarize the performance of the prediction model. Divided into four parts as shown in Fig. 4 are True Positive, False Positive, False Negative, and True Negative, respectively.

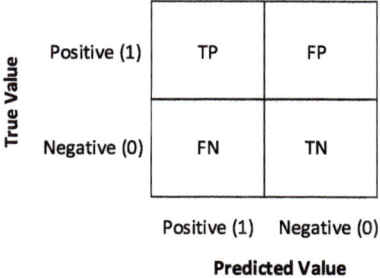

Fig. 4. Confusion Matrix

We examined different models' accuracy and confusion matrices and conducted various analyses. We first compared the performance of SVM and weighted SVM to verify the effectiveness of controlling false negatives. Secondly, we compared the predictive performance of two single models with that of ensemble models. Finally, we contrasted the strengths and weaknesses of two different ensemble models.

1. Class Weighted SVM
   As shown in Fig. 5, the accuracy of the SVM model is 73.45%, while the weighted SVM model achieves an accuracy of 72.23%. However, the difference in false negative rates is quite noticeable, with rates of 28.27% and 12.06%, respectively. It suggests that the weighted SVM model is more effective at controlling false negatives.
2. MLP and AdaBoost
   We used the MLP and AdaBoost models to improve the overall accuracy of our ensemble model. The MLP model achieves an accuracy of 75.7% with a

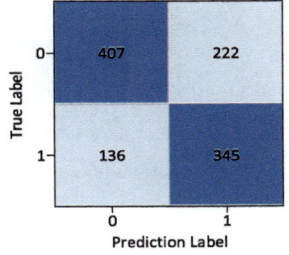
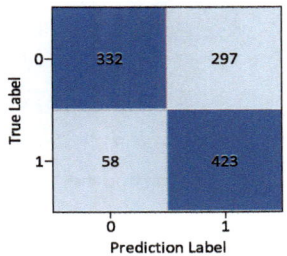

Fig. 5. SVM and Weighted SVM

false negative rate of 22.4%, while the AdaBoost model has an accuracy of 73.71% with a false negative rate of 27.65%, as shown in Fig. 6.

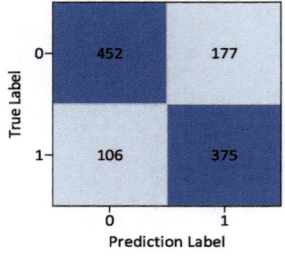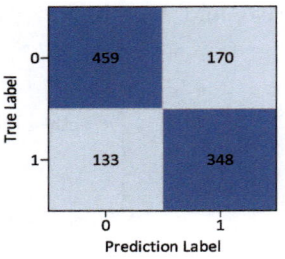

Fig. 6. MLP and AdaBoost

3. Ensemble Model
   We compared two different ensemble models, bagging and stacking. As shown in Fig. 7, we used the voting strategy for the bagging ensemble to make decisions. Due to the averaging characteristics of the decision-making method, the effectiveness of cost sensitivity is reduced, and the false negative is 18.31%. On the other hand, the bagging model combines the advantages of each model and achieves the highest accuracy of 76.33%. The SVM-based stacked ensemble model maintains the cost sensitivity feature well and obtains the best accuracy (75.61%) with low false negatives (13.31%), reflecting the model's effectiveness.

4.3   Technology Comparison

We used various validation methods to compare the performance of different algorithms in Table 2.

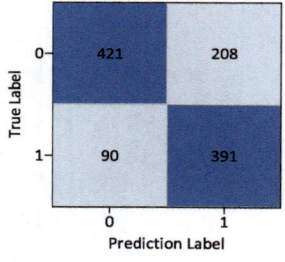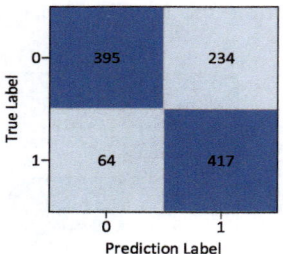

Fig. 7. Bagging and Stacking Ensemble

- ACC - Accuracy is the proportion of the correct classification, the most important parameter to evaluate a model. By comparison, we can see that the weighted SVM sacrifices some accuracy due to the control of false negatives, and its accuracy is 0.7196 compared to 0.7291 of the SVM model. MLP (0.7570) performs better among individual classifiers due to its neural network character, followed by AdaBoost (0.7371). Whether bagging (0.7574) or stacking (0.7534), the ensemble model can maintain the optimal or even better accuracy of the individual classifiers, which is the main advantage of the ensemble model.

- FNR, TPR, FPR, TNR - For our dataset, false negative is the parameter we need to pay the most attention to, besides accuracy. Due to the particularity of business data, we need to minimize false negatives to avoid missing any potential customers. From previous experiments, we verified the effectiveness of the weighted SVM in reducing false negatives, as shown in Table 2, reaching 0.1206. At the same time, we take advantage of the stacking ensemble model to significantly reduce the false negative to 0.1372 while maintaining similar accuracy. Correspondingly, the decrease in false negatives will affect the performance of false positives. It can be seen that the lower the false negative is, the higher the false positive will be, which means that many users who will not buy may be predicted as potential purchasers, which is an opposite problem that may need to be solved in other business situations.

- Precision, Recall, F-score - Precision is the ratio of correctly predicted positive observations to total predicted positive observations. The definition in this work is how many of the customers who predicted that they would buy actually bought. As Table 2 shows, both the MLP and ensemble model performs well, with the stacking ensemble achieving the best results. The recall rate is the ratio of correctly predicted purchasers to all actual purchasers. Again, the MLP and ensemble perform well, but the bagging ensemble performs best. The F-score is a harmonic average of precision and recall, so both false positives and false negatives need to be considered. F-score works best if the costs of false positives and false negatives are similar. The best F-score in our work is 0.7459 for the bagging ensemble.

- Technology Discussion - We can see that the bagging ensemble achieves the highest accuracy while stacking sacrifices some accuracy to account for false negatives. Weighted SVM achieves the lowest false negative but has the lowest accuracy, FPR, and TNR, so we need to stack other classifiers to reduce its negative impact. Collecting and maximizing the advantages of all classifiers and reducing the impact of disadvantages on the final result is necessary. Although the F-score of stacking is not the highest, the precision of stacking is the highest among all models. Compared with bagging, the precision and recall of stacking are also better, which shows the effectiveness of the weighted SVM-based stacking ensemble model.

Table 2. Performance Evaluation of Various Methods

Methods	ACC	FNR	TPR	FPR	TNR	Precision	Recall	F-score
SVM	0.7345 ± 0.0417	0.2827	0.7173	0.3529	0.6471	0.6790	0.6822	0.6765
Weighted SVM	0.7223 ± 0.0636	0.1206	0.8794	0.4722	0.5278	0.7194	0.7036	0.6780
MLP	0.7570 ± 0.0352	0.2204	0.7796	0.2814	0.7186	0.7447	0.7491	0.7438
AdaBoost	0.7371 ± 0.0374	0.2765	0.7235	0.2703	0.7297	0.7236	0.7266	0.7243
Bagging Ensemble	0.7633 ± 0.0373	0.1871	0.8129	0.3307	0.6693	0.7383	0.7411	0.7313
Stacking Ensemble	0.7561 ± 0.0360	0.1331	0.8669	0.3720	0.6280	0.7506	0.7475	0.7314

5   Conclusion

We used both structured and unstructured data from an e-commerce company to extract informative features for model training. Our proposed ensemble model effectively reduces false negatives while maintaining reasonable prediction accuracy, crucial for practical use in that e-commerce company. Moving forward, we aim to further enhance our customer behaviour prediction model by incorporating additional data sources such as phone records and social media data.

References

1. Atta-ur-Rahman, Dash, S., Luhach, A.K., et al.: A Neuro-fuzzy approach for user behaviour classification and prediction. J. Cloud Comput. **8**(1), 17 (2019). https://doi.org/10.1186/s13677-019-0144-9
2. Wong, E., Wei, Y.: Customer online shopping experience data analytics: integrated customer segmentation and customised services prediction model. Int. J. Retail Distrib. Manag. **46**(4), 406–420 (2018). https://doi.org/10.1108/IJRDM-06-2017-0130
3. Dou, X.: Online purchase behavior prediction and analysis using ensemble learning. In: 2020 IEEE 5th International Conference on Cloud Computing and Big Data Analytics (ICCCBDA), pp. 532–536 (2020). https://doi.org/10.1109/ICCCBDA49378.2020.9095554
4. Peng, C.-C., Wang, Y.-Z., Huang, C.-W.: Artificial-neural-network-based consumer behavior prediction: a survey. In: 2020 IEEE 2nd Eurasia Conference on Biomedical Engineering, Healthcare and Sustainability (ECBIOS), pp. 134–136 (2020). https://doi.org/10.1109/ECBIOS50299.2020.9203699
5. Vo, N.N.Y., Liu, S., Li, X., Xu, G.: Leveraging unstructured call log data for customer churn prediction. Knowl.-Based Syst. **212**, 106586 (2021). https://doi.org/10.1016/j.knosys.2020.106586
6. Pustokhina, I.V., et al.: Dynamic customer churn prediction strategy for business intelligence using text analytics with evolutionary optimization algorithms. Inf. Process. Manag. **58**(6), 102706 (2021). https://doi.org/10.1016/j.ipm.2021.102706

7. De Caigny, A., Coussement, K., De Bock, K.W., Lessmann, S.: Incorporating textual information in customer churn prediction models based on a convolutional neural network. Int. J. Forecast. **36**(4), 1563–1578 (2020). https://doi.org/10.1016/j.ijforecast.2019.03.029

8. Jain, H., Khunteta, A., Srivastava, S.: Telecom churn prediction and used techniques, datasets and performance measures: a review. Telecommun. Syst. **76**, 613–630 (2021). https://doi.org/10.1007/s11235-020-00727-0

9. Amin, A., Al-Obeidat, F., Shah, B., Adnan, A., Loo, J., Anwar, S.: Customer churn prediction in telecommunication industry using data certainty. J. Bus. Res. **94**, 290–301 (2019). https://doi.org/10.1016/j.jbusres.2018.03.003

10. Liu, N., Li, X., Qi, E., Xu, M., Li, L., Gao, B.: A novel ensemble learning paradigm for medical diagnosis with imbalanced data. IEEE Access **8**, 171263–171280 (2020). https://doi.org/10.1109/ACCESS.2020.3014362

11. Wang, Y., Ru, J., Jiang, Y., et al.: Adaboost-SVM-based probability algorithm for the prediction of all mature miRNA sites based on structured-sequence features. Sci. Rep. **9**, 1521 (2019). https://doi.org/10.1038/s41598-018-38048-7

12. Zhu, Z., Wang, Z., Li, D., Zhu, Y., Du, W.: Geometric structural ensemble learning for imbalanced problems. IEEE Trans. Cybern. **50**(4), 1617–1629 (2020). https://doi.org/10.1109/TCYB.2018.2877663

13. Esenogho, E., Mienye, I.D., Swart, T.G., Aruleba, K., Obaido, G.: A neural network ensemble with feature engineering for improved credit card fraud detection. IEEE Access **10**, 16400–16407 (2022). https://doi.org/10.1109/ACCESS.2022.3148298

14. Sun, J., Li, H., Fujita, H., Fu, B., Ai, W.: Class-imbalanced dynamic financial distress prediction based on Adaboost-SVM ensemble combined with SMOTE and time weighting. Inf. Fusion **54**, 128–144 (2020). https://doi.org/10.1016/j.inffus.2019.07.006

15. Lawi, A., Aziz, F.: Classification of credit card default clients using LS-SVM ensemble. In: Third International Conference on Informatics and Computing (ICIC) 2018, pp. 1–4 (2018). https://doi.org/10.1109/IAC.2018.8780427

16. Ozaki, S.T.N., Horio, K.: SVM ensemble approaches for improving texture classification performance based on complex network model with spatial information. In: International Workshop on Advanced Image Technology (IWAIT) 2018, pp. 1–3 (2018). https://doi.org/10.1109/IWAIT.2018.8369742

17. Krawczyk, B., Woźniak, M., Schaefer, G.: Cost-sensitive decision tree ensembles for effective imbalanced classification. Appl. Soft Comput. **14**(Part C), 554–562 (2014). https://doi.org/10.1016/j.asoc.2013.10.022

18. Jiang, R., Banchs, R.E., Li, H.: Evaluating and combining name entity recognition systems. In: Proceedings of the Sixth Named Entity Workshop, pp. 21–27 (2016)

19. Puth, M.T., Neuhäuser, M., Ruxton, G.D.: Effective use of Spearman's and Kendall's correlation coefficients for association between two measured traits. Anim. Behav. **102**, 77–84 (2015). https://doi.org/10.1016/j.anbehav.2015.01.004

Design and Performance Evaluation of a Fuzzy-Based System for Assessment of Emotional Trust

Shunya Higashi[1(✉)], Phudit Ampririt[1], Ermioni Qafzezi[2], Makoto Ikeda[2], Keita Matsuo[2], and Leonard Barolli[2]

[1] Graduate School of Engineering, Fukuoka Institute of Technology, 3-30-1 Wajiro-Higashi, Higashi-Ku, Fukuoka 811-0295, Japan
{mgm23108,bd21201}@bene.fit.ac.jp
[2] Department of Information and Communication Engineering, Fukuoka Institute of Technology, 3-30-1 Wajiro-Higashi, Higashi-Ku, Fukuoka 811-0295, Japan
bd20101@bene.fit.ac.jp, makoto.ikd@acm.org, {kt-matsuo,barolli}@fit.ac.jp

Abstract. Trust is used as the foundation for decision-making in many situations. Therefore, the concept of trust is attracting attention in various research fields. Emotional trust is one of the essential aspects of the trust. In this paper, we propose a fuzzy-based system for assessment of emotional trust considering three parameters: Expectation (Ex), Willingness (Wi) and Attitude (At). We evaluate the proposed system by simulations. The simulation results show that when Ex, Wi and At are increasing, the ET parameter is increased.

1 Introduction

Nowadays, human's life has to be integrated with the digital and social world. Also, the social world is greatly impacted by the information interchange in the digital world. Moreover, the linkages and connections between people and things is becoming more complicated and are many issues with the growth of digital technologies. Therefore, the security algorithms or trusts computing are needed to protect digital exchanges in the digital world [1].

Trust is used as the foundation for decision-making in many situations and is employed for interacting on an ethical level. Therefore, digital operators could identify people and determine their relationships by expressing these values of trust in the digital environment. Trust modeling is required to replicate the evaluation and decision-making processes should be carried out automatically in real life. Also, the acceptable risk of making trust should be considered to guarantee a relationship. For example, the communication between two persons needs a communication media that has a good trust value without leaking their security and privacy information (human-to-thing trust). Also, they will trust each other before starting communication (human-to-human trust) as shown in Fig. 1.

© The Author(s), under exclusive license to Springer Nature Switzerland AG 2023
L. Barolli (Ed.): CISIS 2023, LNDECT 176, pp. 116–123, 2023.
https://doi.org/10.1007/978-3-031-35734-3_12

Fig. 1. Communication with trust.

Digital environments should consider trust concept. For example, the trust concept is adopted to 5G and beyond networks to provide security guarantees and quantifies how much a system can be trusted to maintain a specific level of performance. Also, the trustworthy in Cloud, Fog and Edge computing is important for deducing the risk of a transaction and preventing data and privacy leakage [2–5].

In this paper, we propose a Fuzzy-based system for assessment of emotional trust considering three parameters: Expectation (Ex), Willingness (Wi) and Attitude (At). We evaluate the proposed system by simulations. The simulation results show that when Ex, Wi and At are increasing, the ET parameter is increased.

The rest of the paper is constructed into the following sections. In Sect. 2 is presented an overview of Trust Computing. In Sect. 3, we present Fuzzy Logic. In Sect. 4, we describe Proposed Fuzzy-based System. In Sect. 5, we describe the simulation results. In Sect. 6, we give conclusions and future work.

2   Trust Computing

The digital data and devices are increasing every day and allow sensors, smart phones and other devices to communicate with each other. When diverse devices connect with each other, the physical and social environments are integrated with the digital world [1]. The information exchange via the digital world has a strong impact on the society. Thus, digital exchanges need to be more secure and trustworthy.

Integrating of digital, social, and physical worlds bring new and more complex vulnerabilities. Thus, the authenticated access control and security protocols may be unfeasible. Also, malicious entities can cause many problems and issues in digital networks. Therefore, the trust computing is a very good approach.

Trust computing is a social concept used for interaction between people and entities. Considering the social properties of trust into the digital world, the digital entities can perceive others and choose their interactions, the same as in the real world. The trust computing is very difficult to be quantified and the perception mechanisms of physical and social world are not available to be implemented in digital environments. Therefore, trust modelling can be used for evaluation and decision-making in real life problems [1].

Trust Computing is the procedure that evaluates trust by considering many sources and the motivation for making a long-term relationship between the trustor and trustee. From Fig. 2, the trust assessment consist of Individual and Relation Trust. The Individual Trust Assessment derives trust from an individual trustee characteristics. The trust can be evaluated from logical thinking and evidence or observations (a person's emotional senses) such as feelings and propensities. On the other hand, Relational Trust Assessment derives trust from the relationships between the trustor and trustee [6].

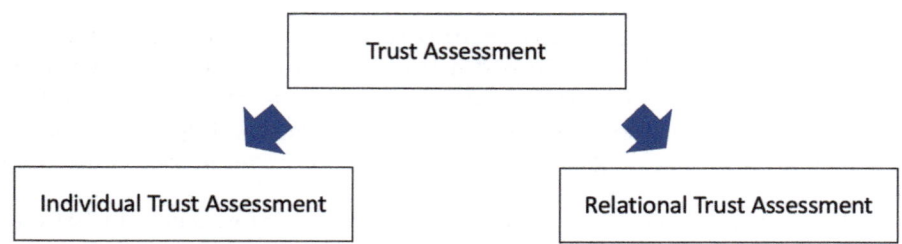

Fig. 2. The Trust Assessment process.

3   Fuzzy Logic Overview

An Fuzzy Logic (FL) system is capable of handling linguistic information and numerical data independently and can perform nonlinear mapping of the input data vector into the scalar output. In addition, it can handle opinions or demands that may be true, untrue or have a middle level of the truth value. The FL systems are implemented in several controlling applications, including TV picture correction, Sendai subway running, and airplane control (Rockwell Corp. (Sony) [7,8].

Fuzzy Logic Controller (FLC) structure consists of four components: Fuzzifier, Inference Engine, Fuzzy Rule Base and Defuzzifier as shown in Fig. 3. In the first phase, the Fuzzifier combines crisp values with linguistic variables associated with fuzzy sets to generate fuzzy inputs. Then, the Inference Engine uses fuzzified input values and fuzzy rules for inferring fuzzified output. Three different kinds of fuzzy inference techniques are typically employed: Sugeno, Mamdani, and Tsukamoto fuzzy inference. The fuzzy rules could be provided by a professional or derived from numerical data. Finally, Defuzzification convert a fuzzified output into a crisp control output [9,10].

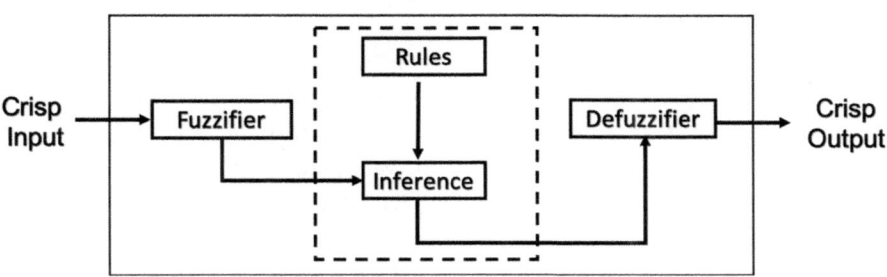

Fig. 3. FLC structure.

4   Proposed Fuzzy-Based System

In this paper, we use FL to implement the proposed system. The proposed system is called Fuzzy-based System for Assessment of Emotional Trust (FSAET). The structure of FSAET is shown in Fig. 4. For the implementation of our system, we consider three input parameters: Expectation (Ex), Willingness (Wi), Attitude (At) and the output parameter is Emotional Trust (ET).

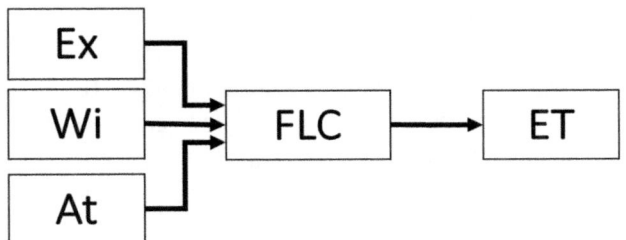

Fig. 4. Proposed system structure.

Figure 5 show the membership functions. The parameters and their term sets are shown in Table 1. Table 2 represent Fuzzy Rule Base (FRB), which has 45 rules and control rules have the form: IF "condition" THEN "control action". For example, Rule 45: "IF Ex is High, Wi is High and At is Very good, THEN ET is ET7".

Table 1. Parameter and their term sets.

Parameters	Term set
Expectation (Ex)	Low (Lo), Medium (Me), High (Hi)
Willingness (Wi)	Low (Lw), Medium (Md), High (Hg)
Attitude (At)	Very bad (Vb), Bad (Bd), Normal (No) Good (Gd), Very good (Vg)
Emotional Trust (ET)	ET1, ET2, ET3, ET4, ET5, ET6, ET7

Table 2. FRB.

Rule	Ex	Wi	At	ET	Rule	Ex	Wi	At	ET	Rule	Ex	Wi	At	ET
1	Lo	Lw	Vb	ET1	16	Me	Lw	Vb	ET1	31	Hi	Lw	Vb	ET4
2	Lo	Lw	Bd	ET1	17	Me	Lw	Bd	ET2	32	Hi	Lw	Bd	ET2
3	Lo	Lw	No	ET2	18	Me	Lw	No	ET3	33	Hi	Lw	No	ET3
4	Lo	Lw	Gd	ET3	19	Me	Lw	Gd	ET4	34	Hi	Lw	Gd	ET4
5	Lo	Lw	Vg	ET4	20	Me	Lw	Vg	ET5	35	Hi	Lw	Vg	ET5
6	Lo	Md	Vb	ET1	21	Me	Md	Vb	ET2	36	Hi	Md	Vb	ET6
7	Lo	Md	Bd	ET2	22	Me	Md	Bd	ET3	37	Hi	Md	Bd	ET3
8	Lo	Md	No	ET3	23	Me	Md	No	ET4	38	Hi	Md	No	ET4
9	Lo	Md	Gd	ET4	24	Me	Md	Gd	ET5	39	Hi	Md	Gd	ET5
10	Lo	Md	Vg	ET5	25	Me	Md	Vg	ET6	40	Hi	Md	Vg	ET6
11	Lo	Hg	Vb	ET2	26	Me	Hg	Vb	ET3	41	Hi	Hg	Vb	ET4
12	Lo	Hg	Bd	ET3	27	Me	Hg	Bd	ET4	42	Hi	Hg	Bd	ET5
13	Lo	Hg	No	ET4	28	Me	Hg	No	ET5	43	Hi	Hg	No	ET6
14	Lo	Hg	Gd	ET5	29	Me	Hg	Gd	ET6	44	Hi	Hg	Gd	ET7
15	Lo	Hg	Vg	ET6	30	Me	Hg	Vg	ET7	45	Hi	Hg	Vg	ET7

5   Simulation Results

In this section, we present the simulation result of our proposed FSAET system. The simulation results are shown in Fig. 6, Fig. 7 and Fig. 8. They show the relation of ET with At for different Wi values considering Ex as a constant parameter.

In Fig. 6, we consider the Ex value 0.1. For Wi 0.5, when At is increased from 0.2 to 0.5 and 0.5 to 0.8, the ET is increased by 17% and 19%, respectively. When At is increased, we see that ET is increased. That means when the trustor has a good attitude, it gains more emotional trust from the trustee. We compare Fig. 6 with Fig. 7 to see how Ex has affected ET. We change the Ex value from 0.1 to 0.5. The ET is increasing by 15% when the Wi value is 0.9 and the At is 0.5. In Fig. 8, we increase the value of Ex to 0.9. We see that the ET values are increased much more compared with the results of Fig. 6 and Fig. 7. For Wi value

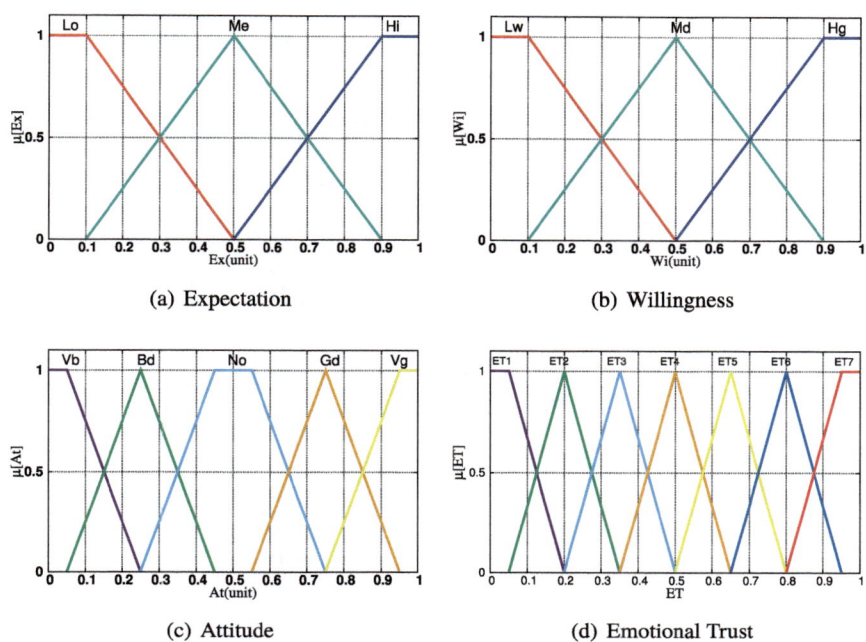

Fig. 5. Membership functions.

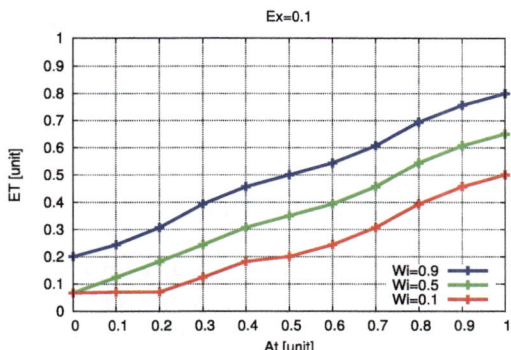

Fig. 6. Simulation results for Ex = 0.1.

0.9, all ET values are higher than 0.5, which show that the person or device is trustworthy.

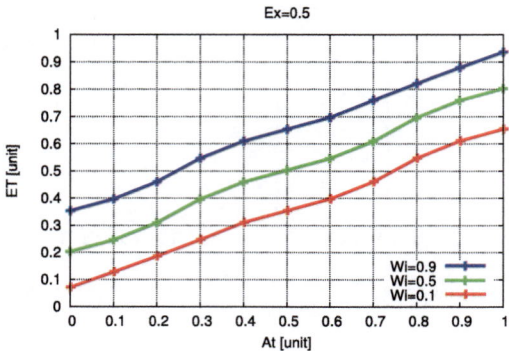

Fig. 7. Simulation results for Ex $= 0.5$.

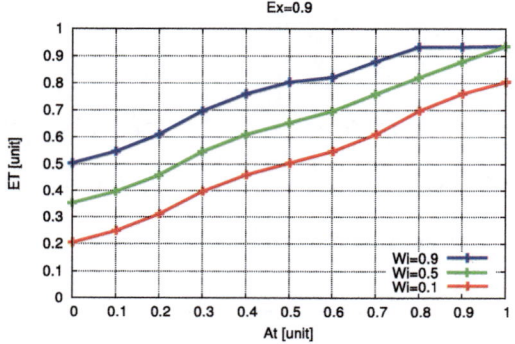

Fig. 8. Simulation results for Ex $= 0.9$.

6   Conclusions and Future Work

In this paper, we proposed and implemented a Fuzzy-based system for assessment of emotion trust. We evaluated the proposed system by simulations. From the simulation results, we found that three parameters have different effects on the ET. When Ex, Wi and At are increasing, the ET parameter is increased.

In the future, we will consider other parameters and make extensive simulations to evaluate the proposed system.

References

1. Ting, H.L.J., Kang, X., Li, T., Wang, H., Chu, C.-K.: On the trust and trust modeling for the future fully-connected digital world: a comprehensive study. IEEE Access **9**, 106743–106783 (2021). https://doi.org/10.1109/ACCESS.2021.3100767
2. Benzaïd, C., Taleb, T., Farooqi, M.Z.: Trust in 5G and beyond networks. IEEE Network **35**(3), 212–222 (2021)

3. Rahman, F.H., Au, T.-W., Newaz, S.S., Suhaili, W.S., Lee, G.M.: Find my trust-worthy fogs: a fuzzy-based trust evaluation framework. Futur. Gener. Comput. Syst. **109**, 562–572 (2020)
4. Uslu, S., Kaur, D., Durresi, M., Durresi, A.: Trustability for resilient internet of things services on 5G multiple access edge cloud computing. Sensors **22**(24), 9905 (2022)
5. Cai, H., Li, Z., Tian, J.: A new trust evaluation model based on cloud theory in e-commerce environment. In: 2011 2nd International Symposium on Intelligence Information Processing and Trusted Computing, pp. 139–142 (2011)
6. Cho, J.-H., Chan, K., Adali, S.: A survey on trust modeling. ACM Comput. Surv. (CSUR) **48**(2), 1–40 (2015)
7. Jantzen, J.: Tutorial on fuzzy logic. Technical University of Denmark, Department of Automation, Technical report (1998)
8. Zadeh, L.A.: Fuzzy logic. Computer **21**(4), 83–93 (1988)
9. Lee, C.C.: Fuzzy logic in control systems: fuzzy logic controller. I. IEEE Trans. Syst. Man Cybern. **20**(2), 404–418 (1990)
10. Mendel, J.M.: Fuzzy logic systems for engineering: a tutorial. Proc. IEEE **83**(3), 345–377 (1995)

Comparing Sampling Strategies for the Classification of Bi-objective Problems by FLACCO Features

Pavel Krömer[✉] and Vojtěch Uher

Department of Computer Science, VSB-Technical University of Ostrava,
17. listopadu 15, 70800 Ostrava, Czech Republic
{pavel.kromer,vojtech.uher}@vsb.cz

Abstract. Problem understanding and the ability to assign problems to distinct classes can improve the usability of metaheuristics. A popular problem-independent method for the characterization of optimization problems is exploratory landscape analysis (ELA). It consists of a sequence of operations that describe the hypersurfaces formed by fitness and other characteristic properties of the problem solutions on the basis of a limited number of samples. Sampling is the initial step of ELA that selects a limited number of candidate solutions for which are the characteristic properties evaluated. The solutions and the computed properties serve as the main inputs for the rest of ELA. In this work, we study the impact of different sampling strategies on machine learning-based classification of bi-objective problems on the basis of FLACCO features. A series of computational experiments demonstrates that different sampling strategies affect the value of the resulting landscape features, their suitability for problem classification, and the overhead of the sampling process. An in-depth analysis of the results also shows the relationship between classification accuracy and the structure of the training data set.

1 Introduction

A better problem understanding and the ability to classify problems into distinct classes can improve the usability of metaheuristics. Exploratory landscape analysis [12] is a popular problem characterization strategy. It describes the hypersurfaces formed by the fitness functions of a problem by identification and analysis of their characteristic properties (features).

Sampling is the initial step of ELA that selects a limited number of candidate solutions for which are the features evaluated. To perform sampling, several distinct strategies are available. Besides pseudo-random sampling, one can also use quasi-random methods, e.g., Latin Hypercube Sampling (LHS) [7,11], or low-discrepancy sequences such as the Sobol [16] and the Halton [4] ones.

Experimental studies performed on single-objective problems indicated that the values of landscape features strongly depend on the sampling strategy [14]. Moreover, the variability of feature values suggests that the features resulting

L. Barolli (Ed.): CISIS 2023, LNDECT 176, pp. 124–136, 2023.
https://doi.org/10.1007/978-3-031-35734-3_13

from various sampling strategies are of different usefulness in problem characterization. Other recent works [1,8] showed that these results extend to the domain of bi-objective problems and demonstrated that different sampling strategies lead to different accuracy of problem classification.

This work extends [8] and provides an in-depth investigation of the impact of sampling strategies on machine learning-based classification of bi-objective optimization problems. The study is performed on the set of bi-objective test problems [2] available on the COmparing Continuous Optimizers (COCO) platform [5]. The set consists of 55 problems that are divided into 15 categories based on their high-level features. Various sampling strategies are applied to collect samples from both fitness landscapes for each bi-objective problem. Then, a subset of landscape features proposed by Kerschke and Trautmann, FLACCO [7], is calculated and used by machine learning to build decision trees for predicting the problem category given its features. The experimental results confirm that the sampling strategies affect the values of landscape features, their suitability for problem classification as well as the computational cost of the sampling procedure. Further analysis of the results also reveals the relationship between classification accuracy and the structure of the training data set.

2   Exploratory Landscape Analysis and Sampling

Exploratory landscape analysis [12] is a family of techniques that analyze the objective and solution (search) spaces of a problem. They construct features that are estimated from discrete points sampled over the continuous problem landscapes. The main purpose of different *sampling strategies* is to systematically generate a set of discrete problem solutions for which the values of the fitness function(s) and other important problem properties will be evaluated. A wide range of sampling methods has been introduced to obtain points and fitness values that allow efficient and accurate characterization of the underlying problem. They are used to scatter the samples across the fitness landscape in different ways. The features computed from the samples can be different, as demonstrated for single-objective problems by Reanu et al. [13] and Škvorc et al. [15]. In this paper, we focus on bi-objective problems and investigate different sampling strategies from various points of view including, e.g., the provision of features that enable accurate problem classification and sensitivity to problem dimension and the number of samples.

The following 5 sampling strategies were considered. **Uniform random sampling (Uniform)** is the standard sampling method that uses a pseudorandom generator with a uniform probability distribution to generate the solutions. **Latin Hypercube Sampling (LHS)** generates near-random samples by dividing multi-dimensional spaces into a square grid and selecting one sample from each grid cell [11]. An optimized variant of LHS, **LHSO**, uses random permutation of coordinates to improve the space filling robustness and lower the centered discrepancy. **Sobol's sequence-based** sampling (**Sobol**) utilizes the Sobol's low-discrepancy sequence. It is a quasi-random sequence with base 2 that binarily represents the position of each dimension [16]. It uses a linear matrix scramble with digital random shifting to improve the discrepancy (i.e., the space-filling properties) of the sequence. **Halton's sequence-based** sampling **(G-Halton)** is a strategy built upon the generalized Halton's sequence with coprime integers as its bases [4].

Figure 1 illustrates the different space-filling properties of two sampling strategies, Uniform and LHSO. It clearly displays that various patterns occur among the sampled points and the density of samples generated by the methods varies in different areas of the search space. It visually confirms that different sampling strategies result in different views of the fitness landscapes due to under- or oversampling of some regions.

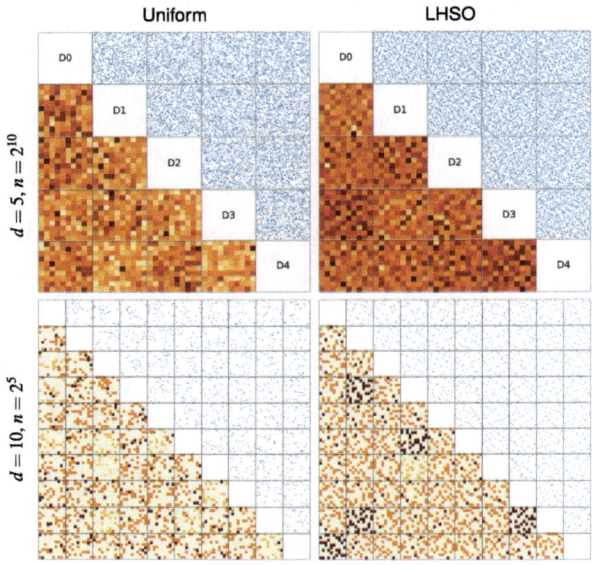

Fig. 1. Different coverage of the search space by different sampling strategies (Uniform and LHSO): the upper triangle of the scatterplot matrix shows the sampled points in all possible 2-D projections of the solution space. The lower triangle shows the densities of the points in every cell of the corresponding (10×10) grid.

3   Experimental Methodology

A series of computational experiments was conducted in the scope of this work to study the effect of different sampling strategies on the values of landscape features, the ability to classify bi-objective problems, and the potential to identify the most useful features.

The experiments used a set of 55 bi-objective test problems, bbob-biobj [2], from the COCO platform [5] representing a wide variety of bi-objective optimization problems. COCO is a popular library containing test problems and benchmarking tools designed for the analysis and performance evaluation of optimization methods. The bi-objective test problems in COCO are created from pairwise combinations of 10 of the 24 single-objective bbob functions that form COCO's single-objective benchmark [6]. The bbob-biobj problems are divided into 15 categories/groups depending on the high-level properties of the underlying single-objective functions.

The values of a subset of FLACCO features were computed for each test problem from the samples taken from both of its objective functions. This process resulted in two interim feature vectors, each corresponding to one objective function. The two vectors were concatenated into a single top-level feature vector representing the entire problem.

To analyze the differences between the investigated sampling strategies in a robust way, several randomized sets of samples were obtained for each strategy by the application of the Cranley-Patterson (CP) rotation to the original set of sampling points. This method is in this context a suitable randomization procedure because it introduces randomness, is inexpensive, and preserves the space-filling properties (e.g., the regular structure) of the rotated point sets, although not under all circumstances [9]. Altogether, 31 rotated sample sets were created for each sampling strategy, sample size, and problem dimension.

The experiments were done for several dimensions of the test problems, $d \in \{3, 5, 10\}$, and for a range of sample sizes, $n = \{32, 256, 1024, 2048, 4096, 16384, 65536\} = \{2^m \mid m \in \{5, 8, 10, 11, 12, 14, 16\}\}$. The power of two sample sizes were used because Sobol's sequence-based generator requires such size to guarantee a balanced sequence. From each sample set, a vector of FLACCO features was computed. We used the same 55 features as in [8, 13] to obtain results that would allow a broad comparison with the previous work.

3.1 Feature-Based Problem Characterization

Problem characterization is an important goal of ELA. To be able to achieve it, the feature vectors need to represent the underlying test problems well. A feature-based classification of bi-objective problems is a straightforward extension of the feature-based classification of single-objective problems [13]. It is a process that takes the feature vectors as inputs and uses a suitable classification algorithm to first learn and then detect problem types. Natural requirements for this process include high accuracy, transparency, and low computational costs. An arbitrary machine learning algorithm can be generally used for problem classification. In this work, we use a popular and efficient classifier, the Decision Tree (DT). It is a well-known universal classification model and allows an easy assessment of feature importance, although the importance is classifier-dependent.

In the problem classification exercise, the DT is first trained to divide the 55 bi-objective test problems into 15 groups. The groups were defined by the authors of COCO with respect to high-level problem properties such as separability, modality, level of conditioning, global structure, and some of their combinations. Each group contains between 3 and 4 problems. In the experiments, each problem is represented by 31 sets of feature vectors computed for each sampling strategy from 31 sets of samples randomized by the CP rotation. The trained model is later on used to classify previously unseen problems and its accuracy and robustness are evaluated.

3.2 Evaluation Methodology

The classification methodology is essential for the interpretation of its results. Previous studies have shown that the traditional train/test split to two equally sized groups of features results in high classification accuracy, often reaching 100% [1,13]. Although this is a positive result, suggesting a good ability of the feature vectors to represent the problems, such an evaluation methodology produces models prone to overfitting and reduces the ability to analyze the effects of different sampling strategies on the outcomes of problem classification. To avoid these defects, a more strict classification methodology termed leave-one-problem-out (LOPO) cross-validation was used in this work.

LOPO cross-validation takes advantage of the fine-grained classification of the feature vectors for which we know not only the group to which they belong but also the test problem from which they were computed. This allows leaving the feature vectors of one of the problems out of the training process and using them exclusively for model validation. In other words, the classifier has to learn all the information it can use to assign the test problem to the proper category from the features of other problems in the group. Classification accuracy is then computed as the average of all 55 folds of the LOPO cross-validation.

The effects of this evaluation method are visually illustrated in Fig. 2. The figure contains three plots with 2-D t-distributed stochastic neighbor embedding (t-SNE) [10] of the feature vectors obtained for three different bbob-biobj problems, F_{10}, F_{15} and F_{29}, by the Sobol's sequence-based sampler. The projection illustrates the approximate relative position of vectors that belong to the same class (brown and dark blue points) and vectors that belong to other classes (light blue points). The LOPO cross-validation is best illustrated by a look at the brown and dark blue points: the classifier has only information about the brown and light blue points. Yet, it has to assign the dark blue points to the same class as the brown ones and distinguish them from incorrect classes (all light blue points). Although only an approximate visualization, it illustrates the variance of the input data and the hardness of this classification task.

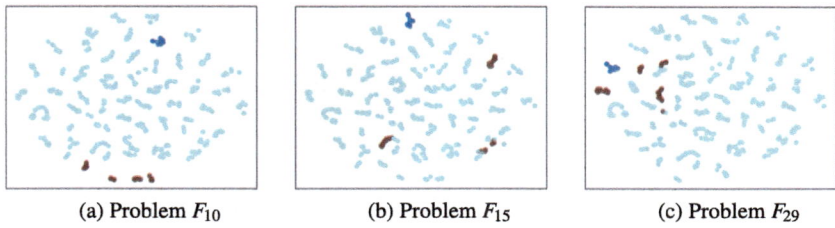

(a) Problem F_{10} (b) Problem F_{15} (c) Problem F_{29}

Fig. 2. t-SNE visualization of two classes in the LOPO cross-validation. Displayed points are 2-D projections of feature vectors computed from samples obtained by Sobol's sequence-based sampler from three different test problems with dimension $d = 3$ and the number of samples $n = 2^{10}$. The brown and dark blue points belong to the same class, and light blue points represent other classes. Dark blue points are excluded from model training and used only for validation.

4   Results and Discussion

The results of the computational experiments are summarized and discussed in this section.

4.1   Classification Accuracy

The average problem classification accuracy is summarized in Table 1 for each combination of problem dimension, sampling strategy, and the number of samples. The table clearly illustrates that with LOPO cross-validation, the average classification accuracy ranges from 0.34 (2^5 samples generated by the Halton's sequence-based method, problem dimension $d = 3$) to 0.96 (2^{16} samples generated by the Sobol's sequence-based method, problem dimension $d = 5$). Next, it shows that having 2^{10} and more samples guarantees a solid average classification accuracy of more than 0.7, no matter what sampling strategy is used. It can be also observed that when the sample size and problem dimension are the same, the classification accuracy does not change dramatically over different sampling strategies for sample sizes of 2^{10} and more.

Table 1. Average accuracy of problem classification.

n	d=3						d=5						d=10					
---	2^5	2^8	2^{10}	2^{12}	2^{14}	2^{16}	2^5	2^8	2^{10}	2^{12}	2^{14}	2^{16}	2^5	2^8	2^{10}	2^{12}	2^{14}	2^{16}
Uniform	0.39	0.74	0.77	0.77	0.78	0.79	0.74	0.90	0.90	0.92	0.91	0.90	0.76	0.92	0.94	0.91	0.93	0.95
Sobol	0.57	0.71	0.75	0.79	0.93	0.67	0.82	0.88	0.92	0.92	0.91	0.96	0.65	0.90	0.91	0.92	0.93	0.91
G-Halton	0.34	0.56	0.76	0.78	0.86	0.73	0.79	0.88	0.90	0.92	0.92	0.94	0.73	0.93	0.94	0.96	0.93	0.96
LHS	0.42	0.58	0.74	0.69	0.72	0.86	0.73	0.88	0.89	0.88	0.91	0.90	0.67	0.88	0.92	0.91	0.93	0.93
LHSO	0.38	0.68	0.71	0.78	0.79	0.74	0.71	0.91	0.89	0.93	0.93	0.94	0.79	0.90	0.92	0.93	0.94	0.92

The classification results for the Sobol and LHSO sampling strategies for one fold of the LOPO cross-validation are visually illustrated in Fig. 3. The plots use the same projection method as Fig. 2, t-SNE, but illustrate the success or failure of classification of all classes in a single picture. The points with the same color belong to the same class. Points drawn with a black edge were classified correctly, other ones were not. The figure suggests that there are some vectors that are hard to classify when using either sampling strategy (e.g., the dark blue and orange ones). However, it also shows that the use of different sampling strategies can affect the ability of the classifier to assign the correct class to feature vectors. For example, the light green points that were correctly classified when using Sobol's sequence-based sampling strategy (Fig. 3a) were misclassified with the LHSO-based sampling (Fig. 3b).

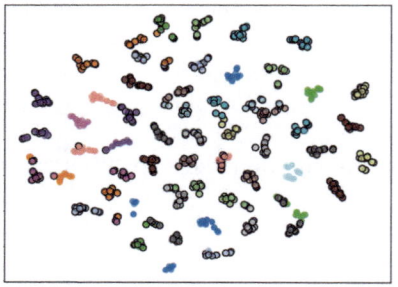
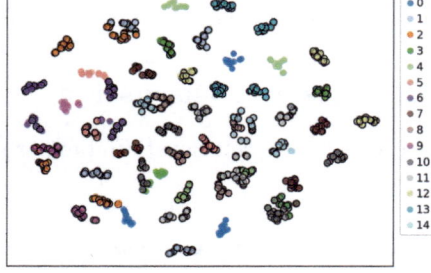

(a) Sobol's sequence-based sampling, average accuracy = 0.75

(b) LHSO-based sampling, average accuracy = 0.71

Fig. 3. t-SNE visualization of feature vectors computed from samples generated by the Sobol sequence-based sampler (a) and the LHSO-based sampler (b). The test problems have dimension $d = 3$ and the number of samples is $n = 2^{10}$. The colors represent the 15 COCO classes. Points with black edges are classified correctly, the points without black edges are misclassified. The accuracies are averaged over all 55 LOPO folds.

The results of the classification are further investigated in more detail. Confusion matrices computed for each fold of the LOPO cross-validation can show if and how the use of different sampling strategies affects the classification outcomes. An example of the confusion matrices is shown in Fig. 4. They confirm the observations made with the t-SNE visualizations. Although there are clearly vectors that are hard to classify correctly, e.g., 15 to 32 members of class 2 are labeled incorrectly as class 0 by all sampling strategies, the use of certain strategies can be associated with specific types of classification errors. For example, the incorrect classification of class 12 vectors as class 3 can be observed only with LHS and LHSO-based sampling strategies. This further supports the original assumption that the sampling strategy has a significant impact on ELA.

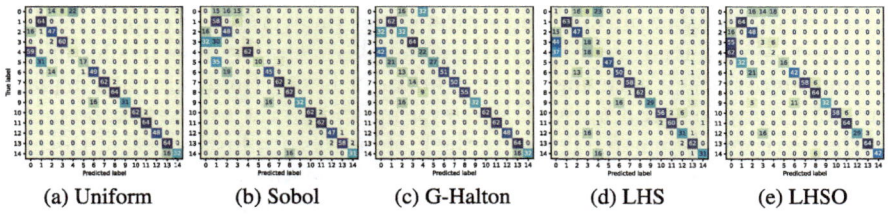

| (a) Uniform | (b) Sobol | (c) G-Halton | (d) LHS | (e) LHSO |

Fig. 4. Aggregate confusion matrices with classification results within all 15 COCO groups for all five investigated sampling strategies in one fold of the LOPO cross-validation. Dimension of the test problems is $d = 3$, number of samples is set to $n = 2^{10}$.

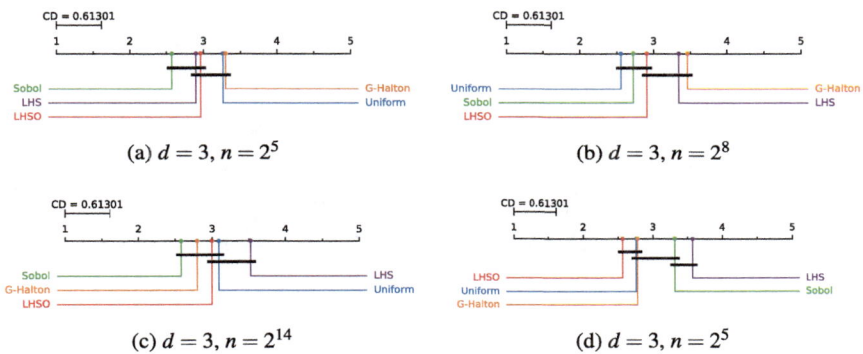

Fig. 5. Average ranks of classifiers obtained for different sample sizes and problem dimensions. The difference between ranks connected by a horizontal bar is not statistically significant at $\alpha = 0.05$.

4.2 Statistical Evaluation

To obtain robust results, the classification accuracy was assessed using appropriate statistical methods. The non-parametric Friedman test and the post hoc Nemenyi test at significance level $\alpha = 0.05$ [3] were applied to determine if there are significant differences between the accuracy of problem classification when different sampling strategies are used and to find out which differences are significant. The significance level, α, was set to 0.05 so that the significance of each pairwise test was thanks to the Bonferroni correction kept at 0.01.

The results of the analysis are for all cases with statistical significance between sampling strategies' ranks illustrated by critical distance plots in Fig. 5. The plots show the average rank of classification accuracy obtained by different sampling strategies across all folds of the LOPO cross-validation. No significant differences in average ranks were observed for any other combinations than those shown in Fig. 5.

Finally, the classification accuracies were summarized in the form of an Olympic-like ranking in Table 2. It shows for each sampling strategy and possible rank (1–5) the number of times the strategy was, on average, ranked at the position. The table is sorted according to the principles of the Olympic medal ranking: the sampling strategy with the higher number of better (higher) ranks is placed on top of a sampling strategy with the lower number of the same rank, no matter how the two compare when it comes to the lower ranks. It illustrates that the highest number of the top ranks, 6, was obtained with the sampling strategy based on Sobol's sequence. The second-best top ranking was accomplished with the LHSO-based sampling strategy.

Table 2. Olympic medal ranking of the sampling strategies.

	Ranking				
	1st	2nd	3rd	4th	5th
Sobol	6	1	4	1	6
LHSO	5	2	5	3	3
G-Halton	3	7	3	3	2
Uniform	3	6	4	4	1
LHS	1	2	1	7	7

Table 3. Olympic medal ranking, significant differences only.

	Ranking				
	1st	2nd	3rd	4th	5th
LHSO	17	1	0	0	0
Sobol	17	1	0	0	0
Uniform	17	1	0	0	0
G-Halton	16	2	0	0	0
LHS	15	2	1	0	0

However, the Olympic-like ranking in Table 2 does not reflect the results of the statistical analysis. A revised ranking, shown in Table 3, takes that into account and assigns each strategy to a rank when it is ranked on this position or the difference between the rank of the strategy and the position is not statistically significant at $\alpha = 0.05$. This demonstrates that the effect of the use of different sampling strategies is, in fact, weaker. When the statistical significance is considered, LHSO-based, Sobol's sequence-based, and uniform random sampling strategies perform equivalently. This is an interesting observation because out of the three methods, uniform random sampling is the only one that makes no attempt to cover the solution space evenly. LHSO is the most computationally expensive one and the Sobol's sequence provides a good ratio of space-filling properties and computational complexity.

Table 4. The average number of features used by the trained DT.

	Uniform			Sobol			G-Halton			LHS			LHSO		
n	d=3	d=5	d=10	d=3	d=5	d=10	d=3	d=5	d=10	d=3	d=5	d=10	d=3	d=5	d=10
2^5	25.3	25.0	20.3	24.3	20.1	25.6	21.5	20.5	23.5	18.7	22.5	26.6	21.3	16.2	21.0
2^8	26.7	15.8	12.1	24.0	14.1	10.3	28.5	13.7	10.2	25.8	16.6	12.4	27.2	14.4	12.8
2^{10}	17.3	13.4	12.6	17.6	13.9	9.5	17.9	10.2	9.5	21.0	16.2	10.0	19.3	13.9	9.9
2^{12}	22.5	13.7	12.3	23.0	14.3	10.4	23.7	15.2	10.5	16.9	16.4	10.5	15.9	13.8	10.7
2^{14}	15.5	10.5	10.8	15.9	13.4	11.2	16.5	13.2	11.1	17.7	14.4	10.5	15.0	11.3	10.6
2^{16}	16.0	14.9	11.0	21.0	10.2	11.4	16.6	14.1	11.4	16.3	11.4	10.9	15.4	14.3	10.9

Table 5. Average time needed to generate the sampling points in milliseconds (Uniform) or as a multiplication of the time of the uniform sampler (all other algorithms).

n	Uniform			Sobol			G-Halton			LHS			LHSO		
	d=3	d=5	d=10	d=3	d=5	d=10	d=3	d=5	d=10	d=3	d=5	d=10	d=3	d=5	d=10
2^5	0.0033	0.0029	0.0036	162.3	124.2	166.0	106.1	160.0	217.0	23.3	27.2	22.5	31743	60703	95309
2^8	0.0056	0.0079	0.0143	46.6	45.9	41.9	96.1	93.5	81.8	13.9	11.4	8.0	83776	109195	95051
2^{10}	0.0185	0.0274	0.0509	14.1	13.2	11.8	68.4	62.2	48.8	6.3	5.2	4.2	81014	80758	53210
2^{12}	0.0589	0.0994	0.1914	4.7	3.9	3.4	71.9	55.6	41.3	4.1	3.5	3.3	92699	60730	39966
2^{14}	0.2335	0.3844	0.7653	1.5	1.2	1.0	68.1	54.0	38.5	3.2	3.1	2.9	83545	56911	38044
2^{16}	0.9161	1.5446	3.0544	0.7	0.5	0.9	68.8	53.3	39.2	3.1	2.9	3.8	95670	70145	70269

4.3 Identification of Important Features

The use of DT makes it possible to analyze the features that were used by the trained model. This has a significant practical impact on ELA because the evaluation of some landscape features is computationally very expensive, in particular for large sample sizes such as $2^{12} = 4096$ and more. The average number of features used by the trained DTs is shown in Table 4 for all combinations of sampling strategies, problem dimensions, and sample sizes. A closer look at the table shows several interesting observations. First, the highest average number of FLACCO features used by the DT is 28.5. That is less than 55% of the total number of features used in this experiment (55). The lowest average number of features is 9.5, i.e., less than 18% of all considered features. This shows the potential to reduce the computational costs associated with the classification of new problems by computing only the features that are indeed used by the trained models, no matter what sampling strategy, sample sizes, and problem dimensions are used. Second, it is possible to observe the trends. The average number of features generally decreases with the growing problem dimension and the number of samples. This means that although computing landscape features from large sample sizes is a problem, the time costs can be reduced by eliminating the features that are not used by the classifiers. This fact is especially interesting because the classification accuracy generally grows with the number of samples, as already demonstrated in Table 1.

4.4 Sampling Performance

The computational costs associated with the generation of sampling points can have a large impact on the ELA process. Some sampling strategies, for example, LHSO, use internal optimization to improve the quality of the samples. In particular, LHSO randomly permutes the coordinates of the sampling points to decrease the centered discrepancy of the sample and, therefore, achieve better space-filling properties.

The performance of the sampling strategies used in this work was measured in the last batch of computational experiments. All strategies were implemented in Python with the use of standard libraries, in particular scipy.stats.qmc, and the time they needed to produce the sets of sampling points of a given size was measured and averaged across 31 executions of the algorithm. Although the program can have a different execution time on other configurations, the relative performance will remain the same. The results of the experiments are shown in Table 5. The table contains two types of information. For the baseline uniform random sampling strategy (first three columns), it shows the average time (in milliseconds) needed to generate the set of sampling points of each size for every problem dimension. For all other sampling strategies, it shows the time needed to generate the sampling points relative to the time of the uniform algorithm.

The table well illustrates the trends in computational costs associated with the generation of sampling points for problems with different dimensionality. The first three columns clearly show how the computational costs exponentially increase with the size of the sample. The factor by which the processing time increases grows higher with the dimensionality of the problem. The rest of Table 5 allows a detailed comparison of the performance of the investigated sampling strategies. It can be immediately seen that they do perform differently. The Sobol, G-Halton, and LHS-based sampling strategies follow a similar pattern: they are much slower than the uniform random strategy for smaller sample sizes. However, with the increasing sample size, the difference becomes smaller. G-Halton is clearly the slowest of the three. LHS starts as the fastest one but Sobol's sequence-based sampling becomes faster than LHS for large sample sizes ($n = 2^{14}$ and $n = 2^{16}$) and even faster than the baseline uniform random sampling for $n = 2^{16}$. The LHSO-based sampling shows a different behavior. It is slower than the other sampling strategies by two (small sample sizes) to four (large sample sizes) orders of magnitude, which makes it clearly the slowest sampling strategy. This can be attributed to the internal optimization performed by the sampler to decrease the discrepancy of the generated sequence. Latin hypercube-based sampling without this optimization (LHS) is significantly faster but the produced sampling points have been shown less useful for the classification of bi-objective problems (see Table 3).

5   Conclusions

This work studied the effect of different sampling strategies on the ability to characterize black-box bi-objective problems for metaheuristic optimization. It analyzed and thoroughly evaluated five distinct methods that can be used to generate sampling points across multidimensional search spaces. The investigated sampling strategies were based on pseudo-random as well as quasi-random techniques. They were used to obtain samples of fitness landscapes from which landscape characteristics such as the FLACCO features can be computed. The features were subsequently used to train a machine learning-based model, the decision tree, to classify test problems into expert-defined categories.

A series of computational experiments that followed a challenging leave-one-problem-out cross-validation methodology empirically showed that the sampling strategies differ at several levels. It was demonstrated that they have in certain cases significantly different ability to detect the class of previously unknown problems, are able to do so under different conditions, e.g., with a different minimum number of samples taken, and have different computational costs. LHSO, Sobol's sequence, and the uniform random sampling strategies were the most accurate in problem classification in the majority of test cases. On the other hand, performance analysis of the computational costs showed that LHSO is by a wide margin the worst-performing sampling method. It is more than 5 orders of magnitude slower than uniform random sampling and more than 3 orders of magnitude slower than Sobol's sequence for any combination of problem dimension and sample size. Sobol's sequence, on the other hand, performed for larger sample sizes on par (for $n = 2^{14}$) or better ($n = 2^{16}$) than uniform random sampling.

Last but not least, the analysis of selected features showed that all the methods required less than 55% of the total number of landscape features for classification, showing clear potential for efficient characterization of unknown problems.

Acknowledgments. This work was supported by the project "Constrained multi-objective Optimization Based on Problem Landscape Analysis" funded by the Czech Science Foundation (grant no. GF22-34873K) and in part by the grant of the Student Grant System no. SP2023/12, VSB - Technical University of Ostrava.

References

1. Andova, A., Vodopija, A., Krömer, P., Uher, V., Tušar, T., Filipič, B.: Initial results in predicting high-level features of constrained multi-objective optimization problems. In: Slovenian Conference on Artificial Intelligence, Proceedings of the 25th International Multiconference Information Society, vol. A, pp. 7–10 (2022)
2. Brockhoff, D., Auger, A., Hansen, N., Tušar, T.: Using well-understood single-objective functions in multiobjective black-box optimization test suites. Evol. Comput. **30**(2), 165–193 (2022)
3. Demšar, J.: Statistical comparisons of classifiers over multiple data sets. J. Mach. Learn. Res. **7**, 1–30 (2006)
4. Halton, J.H.: Algorithm 247: radical-inverse quasi-random point sequence. Commun. ACM **7**(12), 701–702 (1964)
5. Hansen, N., Auger, A., Ros, R., Mersmann, O., Tušar, T., Brockhoff, D.: COCO: a platform for comparing continuous optimizers in a black-box setting. Optim. Methods Softw. **36**(1), 114–144 (2021)
6. Hansen, N., Finck, S., Ros, R., Auger, A.: Real-parameter black-box optimization benchmarking 2009: noiseless functions definitions. Research report RR-6829, INRIA (2009)
7. Kerschke, P., Trautmann, H.: Automated algorithm selection on continuous black-box problems by combining exploratory landscape analysis and machine learning. Evol. Comput. **27**(1), 99–127 (2019)
8. Krömer, P., Uher, V., Andova, A., Tušar, T., Filipič, B.: Sampling strategies for exploratory landscape analysis of bi-objective problems. In: Proceedings of the 2022 International Conference on Computational Science and Computational Intelligence (CSCI) (2022, in print)

9. Lemieux, C.: Monte Carlo and Quasi-Monte Carlo Sampling. Springer, New York (2009). https://doi.org/10.1007/978-0-387-78165-5
10. van der Maaten, L., Hinton, G.: Visualizing data using t-SNE. J. Mach. Learn. Res. **9**(86), 2579–2605 (2008)
11. McKay, M.D., Beckman, R.J., Conover, W.J.: A comparison of three methods for selecting values of input variables in the analysis of output from a computer code. Technometrics **42**(1), 55–61 (2000)
12. Mersmann, O., Bischl, B., Trautmann, H., Preuss, M., Weihs, C., Rudolph, G.: Exploratory landscape analysis. In: Proceedings of the 13th Annual Genetic and Evolutionary Computation Conference (GECCO), pp. 829–836. ACM (2011)
13. Renau, Q., Doerr, C., Dreo, J., Doerr, B.: Exploratory landscape analysis is strongly sensitive to the sampling strategy. In: Bäck, T., et al. (eds.) PPSN 2020. LNCS, vol. 12270, pp. 139–153. Springer, Cham (2020). https://doi.org/10.1007/978-3-030-58115-2_10
14. Renau, Q., Dreo, J., Doerr, C., Doerr, B.: Towards explainable exploratory landscape analysis: extreme feature selection for classifying BBOB functions. In: Castillo, P.A., Jiménez Laredo, J.L. (eds.) EvoApplications 2021. LNCS, vol. 12694, pp. 17–33. Springer, Cham (2021). https://doi.org/10.1007/978-3-030-72699-7_2
15. Skvorc, U., Eftimov, T., Korosec, P.: The effect of sampling methods on the invariance to function transformations when using exploratory landscape analysis. In: 2021 IEEE Congress on Evolutionary Computation (CEC), Kraków, Poland, pp. 1139–1146. IEEE (2021)
16. Sobol, I.M.: On the distribution of points in a cube and the approximate evaluation of integrals. Zhurnal Vychislitel'noi Matematiki i Matematicheskoi Fiziki **7**(4), 784–802 (1967)

Efficient FPGA Implementation of a Convolutional Neural Network for Surgical Image Segmentation Focusing on Recursive Structure

Takehiro Miura, Shuto Abe, Taito Manabe, Yuichiro Shibata^(✉),
Taiichiro Kosaka, and Tomohiko Adachi

Nagasaki University, Nagasaki, Japan
`yuichiro@nagasaki-u.ac.jp`

Abstract. This paper discusses FPGA implementation of a convolutional neural network (CNN) for surgical image segmentation, which is part of a project to develop an automatic endoscope manipulation robot for laparoscopic surgery. From a viewpoint of hardware design, the major challenge to be addressed is that simple parallel implementation with spatial expansion requires a huge amount of FPGA resources. To cope with this problem, we propose a highly efficient implementation approach focusing on the recursive structure of the proposed network. Experimental results showed that the dominant computing resources could be reduced by about half in exchange for a 6% increase in memory resources and a 0.01% increase in latency. It was also observed that the operations performed on the network itself did not change, keeping the same inference results and throughput.

1 Introduction

In an effort to meet the growing demand for laparoscopic surgery, we are developing an automatic endoscope operation robot for cholecystectomy. For the purpose of automatically adjusting the position and orientation of the endoscope by a robot so that a specific area around the gallbladder fits into the image, we have proposed a convolutional neural network (CNN), which identifies the area around the gallbladder in an endoscopic image [8].

However, the huge parameter space of modern large CNNs is likely to cause overfitting for medical application domains where only small training data sets are typically available. Thus, we have proposed a convolutional neural network that can handle small data sets while incorporating the U-Net's characteristic skip connections [8]. Key features are to construct a network by combining several small networks that are used recursively to suppress overfitting by intentionally reducing the number of parameters. In addition, recursive use of the network for data with different resolutions can be expected to enable more universal feature extraction independent of size, which also leads to suppression of overfitting.

L. Barolli (Ed.): CISIS 2023, LNDECT 176, pp. 137–148, 2023.
https://doi.org/10.1007/978-3-031-35734-3_14

In general, image processing is highly compatible with FPGA implementation, and especially when stream processing is possible, it is advantageous in terms of low power consumption and real-time performance [2]. However, since general neural networks require a large number of sum-of-products operations, the increase in hardware resources becomes a problem when parallelized with simple spatial expansion [1]. Therefore, in this work, we propose a resource reduction method that shares arithmetic units, focusing on idle time in the pipelined recursive networks.

2   Related Work

U-Net [10] is a representative neural network for medical image segmentation, which won both the Dental X-Ray Image Segmentation Challenge and the Cell Tracking Challenge at the 2015 ISBI. Like other methods, U-Net combines image downsampling and upsampling in convolutional processing. What characterizes U-Net is that the data before downsampling is concatenated during upsampling. This connection is called a skip connection, which has the effect of preventing detailed information from being lost in data propagation and segmentation results from becoming coarse. However, its network size is still large for applications in which the available number of training data images is limited.

A typical approach to reduce resource usage for hardware-based CNNs is a quantization of arithmetic [4,11]. On the other hand, Kim et al. have proposed a CNN with a recursive structure called the deeply-recursive convolutional network (DRCN) and have revealed that the introduction of a recursive structure into CNNs reduces the number of parameters [5]. Although the idea of DRCN should also be effective for hardware resource reduction, it is not the case when the network is deployed on a simple pipeline structure. Even the technique to dynamically decide the number of iterations of recursive blocks, which was addressed in the network called the dynamic recursive neural network (DRNN) [3], does not solve this problem.

3   Proposed Network

Figure 1 shows the overall diagram of the proposed network [8]. First, the input image is fed to a small network called Ext, which generates a feature map. Ext has 3 RGB input channels and n output channels, where n can be set arbitrarily by parameters. Next, the feature map is reduced by 2×2 average pooling and then passed to a small network called Rdc. Rdc has n channels for both input and output. Recursively repeating the reduction by pooling and application of the Rdc network, feature maps are aggregated to lower dimensions and the global information can be used. After reaching the bottom layer, it is expanded by unpooling and passed to a small network called Mrg with feature maps of the same size passed directly from the previous stage. Therefore, Mrg has $2n$ input channels and n output channels. The expansion by unpooling and the integration of the feature map in the previous stage aim to recover the location information lost in the pooling layer. This expansion and integration by Mrg are performed the same number of times as the reduction and application of Rdc. Finally, a 1×1 convolution operation is performed on the feature map that has the same size as the original image to generate a likelihood map for the specified four classes: gallbladder, cystic duct, common bile duct, and background.

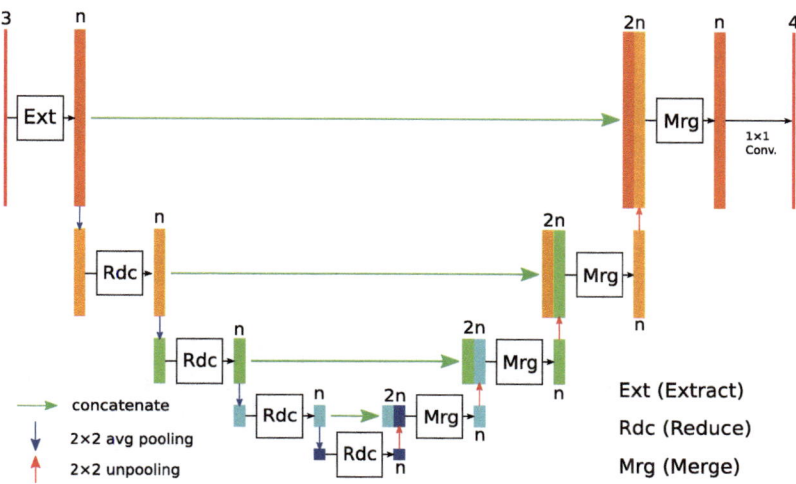

Fig. 1. Proposed network diagram

3.1 Convolution Layer

The network mainly consists of three small networks. Figure 2 shows each shape. The filter sizes are all 3×3, and padding is performed in each layer to keep the input and output sizes the same. The activation function is the Leaky ReLU [6]: $f(x) = \max(ax, x)$, where $a = 0.25$. ReLU, which is often used as the activation function for general neural networks, outputs 0 for any negative input, and negative numbers with large absolute values may appear in the middle of operations as learning progresses [7]. Since this implementation uses fixed-point numbers for hardware, this phenomenon leads to an increase in the bit width required for arithmetic operations and a decrease in arithmetic precision. Leaky ReLU, on the other hand, has a non-zero weight even in the negative region, which mitigates this problem. For the value of a, we adopted $a = 0.25 = 2^{-2}$, which can be implemented simply by a bit shift operation.

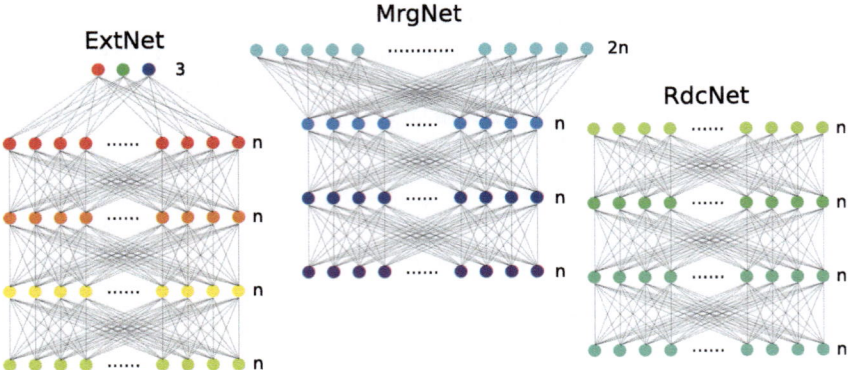

Fig. 2. 3 types of small networks

3.2 Pooling and Unpooling Layers

In the pooling layer, a 2×2 average pooling is applied as in a typical CNN. The input is divided into regions with the size of 2×2, and each region is replaced by an average of 4 pixels. Therefore, the application of the pooling layer reduces the height and width of the image by half. On the other hand, the process performed in the unpooling layer is just outputting each pixel 2×2 times, in other words, it is the same as the expansion using the nearest neighbor method. The operation of the unpooling layer is shown in Fig. 3.

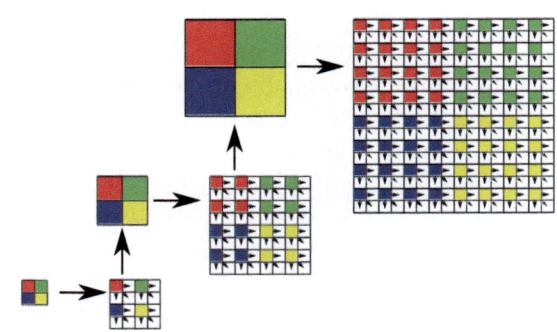

Fig. 3. Expansion of feature maps with unpooling layers

4 Design and Implementation

For the hardware implementation of the system, the input image is given as a stream of pixel values (in_pixel) with horizontal and vertical coordinates (hcnt, vcnt) in a raster scan. The output gives a class label corresponding to one of the four classes for each pixel along with the coordinates. Implementation in FPGA was done by RTL description using Verilog HDL. The target FPGA was Xilinx Virtex UltraScale xcvu095-ffva2104-2-e and Vivado 2018.3 was used for logic synthesis.

4.1 Design Focused on Recursive Structure

Without considering the recursive use of small networks, a simple spatial expansion of the network in Fig. 1 would look like Fig. 4. For pipeline processing, multiple modules corresponding to Rdc and Mrg are generated, and thus resource usage increases with the number of network stages.

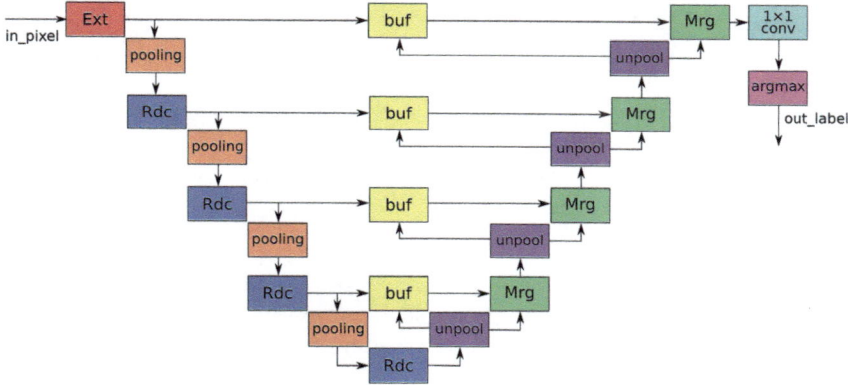

Fig. 4. Configuration diagram with simple spatial expansion

On the other hand, it is worth noting that the pipeline for the pooling process partially stalls. Since the pooling module outputs a valid value only once for the 4 pixels in each region, there are idle cycles when the network should stop operating after pooling is applied. Therefore, we took advantage of this idle time and reduced resource usage by consolidating multiple modules of Rdc or Mrg into a single module that operates on multiple inputs of different stages. Its configuration is shown in Fig. 5. In the following, the modules used for implementation are explained together with the control method.

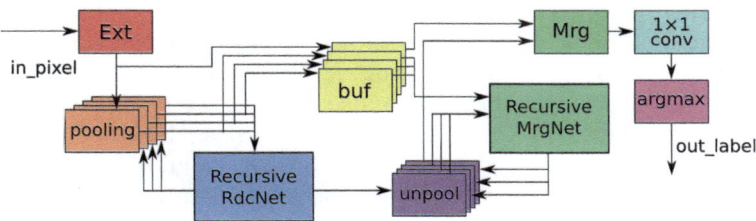

Fig. 5. Configuration diagram focusing on recursive structure

4.2 Convolution Hardware

Figure 6 shows the structure of the stream_patch module, which is used to cut out an image (patch) of the same size as the filter each time a valid value is input. The line buffer memory (FIFO) is implemented as a simple dual-port memory generated by the Vivado IP catalog.

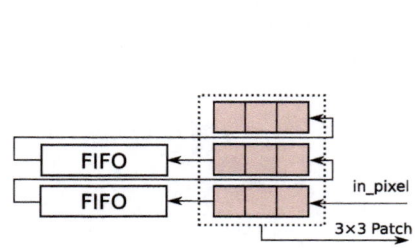

Fig. 6. Patch cutout with stream_patch module

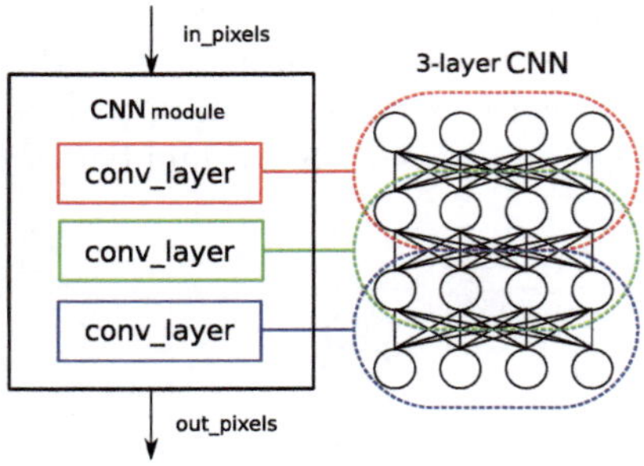

Fig. 7. Structure of conv_layer modules

The proposed network consists of three types of small networks (Ext, Rdc, Mrg), which have many common parts, such as sum-of-products operations and activation functions. Therefore, we aim to simplify the design by implementing the computation of neurons between layers as a common module (conv_layer module). Figure 7 shows an overview of network construction with the conv_layer modules. The module is designed so that parameters such as bit width, filter size, and the number of neurons can be specified independently for each layer, allowing arbitrary networks to be constructed by combining multiple modules.

The conv_layer module is responsible for operations in each layer of CNN. Pixel values input to this module are cut out as patches of the same size as the filter set by the stream_patch module and multiplied by the weights for each channel. The result of the multiplication is passed to the adder tree to obtain the sum for each channel. The biases are then added and the activation function is applied. The activation function is the Leaky ReLU with slope $a = 0.25$, as described in the previous section. When the input value is a negative number, the result is rounded to the nearest integer, i.e., $f(x) = \lfloor \frac{x}{4} + 0.5 \rfloor$.

As aforementioned, the recursive_conv_layer module is a modification of the conv_layer module for sharing computational resources among recursively used modules as described The configuration is shown in Fig. 8. When operating on multiple inputs, delay buffers are added outside the module to avoid conflicts between the inputs. Each input pixel value is passed to the corresponding stream_patch module and is cut out as a patch corresponding to the filter. Since the input pixels of different patches are exclusively given, the output timings for each patch are not overlapped, so that the pipeline operations can be naturally interleaved without any additional arbitration mechanisms.

4.3 Pooling and Unpooling Hardware

The pooling module is responsible for operations equivalent to average pooling. Figure 9 shows the structure of this module. Pixel values input to this module are cut into 2×2 size patches by the stream_patch module. The sum of the four pixels for each channel is then obtained, and the average value of the area is obtained by a 2-bit right arithmetic shift operation. The result is rounded to the nearest integer by adding 2 in advance.

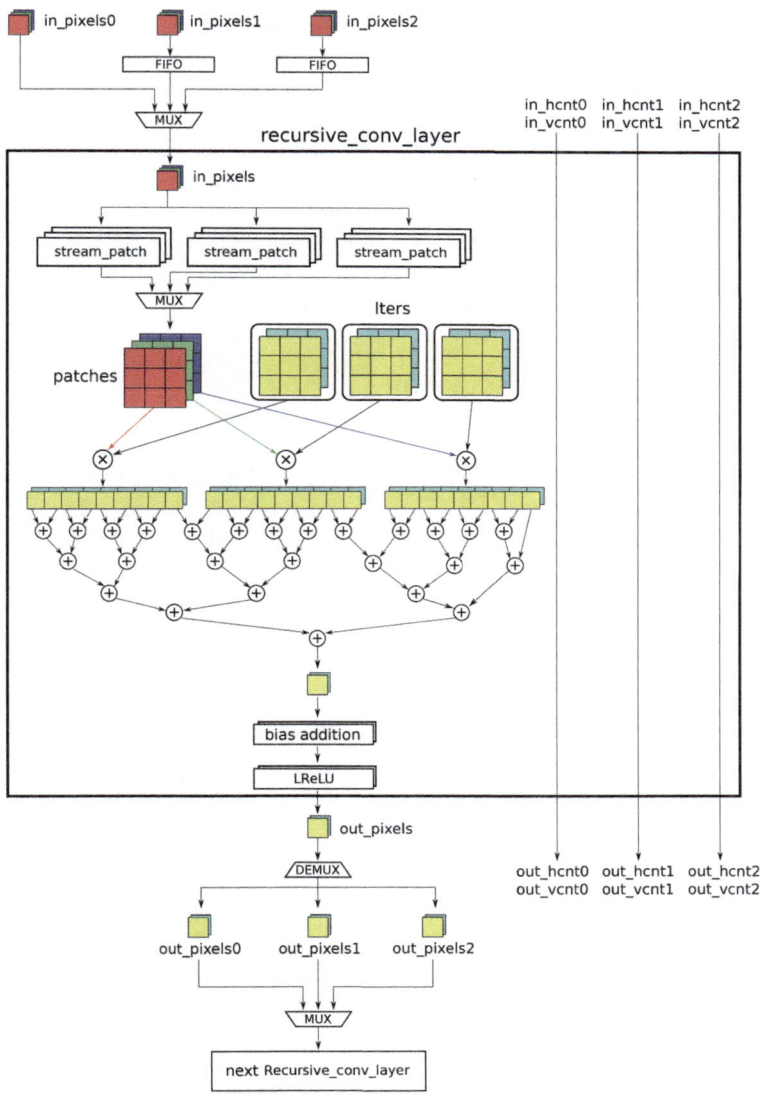

Fig. 8. Structure of recursive_conv_layer module

The unpooling module is responsible for pixel expansion operations equivalent to unpooling. Figure 10 shows the configuration of this module. As shown in the image diagram shown in Fig. 11, by adding an appropriate delay and repeatedly outputting the input pixels, an operation equivalent to transferring the value to the surrounding pixels is realized.

4.4   Training

The weights and biases used for inference are parameters obtained from training with PyTorch [9]. The result of software learning is given in 32-bit floating-point number type, but the use of floating-point number type in hardware implementations is undesirable in terms of resource capacity. Therefore, the parameters obtained from the training are converted to 17-bit fixed-point number type (10-bit integer part and 7-bit decimal part) and implemented. The data set consisting of image data and label data used for training the neural network contains 183 pairs, of which 138 pairs are for training and 45 pairs are for evaluation. The image data are actual surgical images, and the size is 640×512. These images were visually consulted and manually annotated by medical doctors to create the label data. Examples of these images are shown in Fig. 12 and Fig. 13.

5   Evaluation and Discussion

For evaluation, the network configuration shown in Fig. 1, where $n = 16$, was synthesized and implemented with Xilinx Vivado. The size of the input/output images is 640×512, as in Sect. 4.4. The inference results obtained by the Verilog simulator for the naive design, where each network module is implemented as an

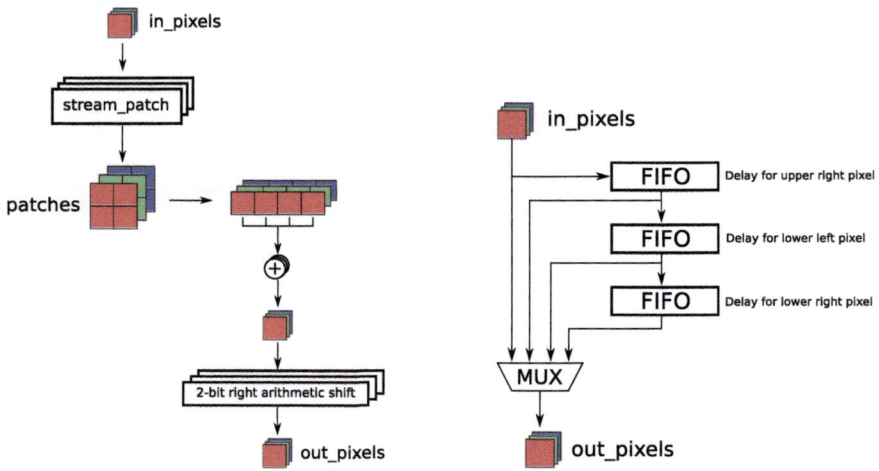

Fig. 9. Structure of pooling module   **Fig. 10.** Structure of unpooling module

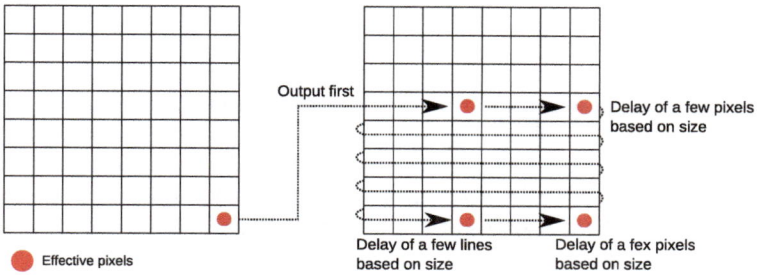

Fig. 11. Behavior of unpooling module

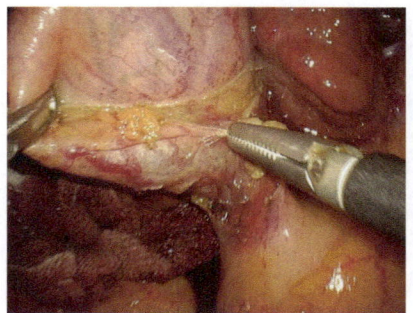

Fig. 12. Actual surgical image

Fig. 13. Labeled image

Fig. 14. Inference result (naive)

Fig. 15. Inference result (proposed)

independent instance, and the proposed recursive design are shown in Fig. 14 and Fig. 15. The two implementations produced identical label images as expected, suggesting that the proposed hardware reduction method was implemented correctly.

5.1   Resource Usage

The resource usage is summarized in Fig. 16. The results demonstrate that the application of the proposed method reduced the amount of dominant computational resources (LUT, FF, and CARRY8) to about half. Although there is an increase of about 6% in memory resources due to the additional delay buffers, their utilization is small and acceptable compared to the other resources. The required number of FPGA chips are also reduced from 7 to 3, which is beneficial in terms of cost and portability.

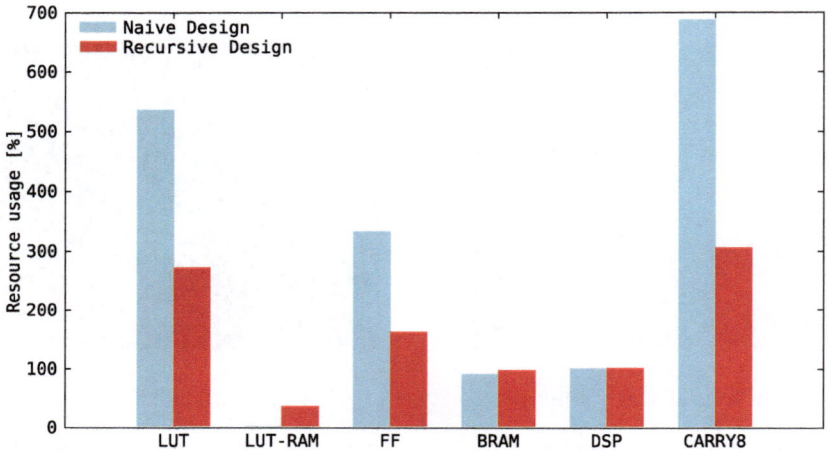

Fig. 16. Resource usage comparison

5.2   Operating Frequency and Latency

The pixel clock frequency for a 640×512 pixel image input at a frame rate of 60 fps is 19.7 MHz. Logic synthesis was performed with a clock frequency constraint of 100 MHz, and WNS (Worst Negative Slack) was obtained by Vivado static timing analysis of the results. Based on these WNS values, the maximum operating frequency and maximum frame rate were calculated and the results are shown in Table 1. While these are estimated values, the results did not change with or without module sharing, giving the prospect of achieving an operating frequency sufficient to sustain the required frame rate. Examination of the results of the timing analysis confirmed that the critical path in this system is on the access wires to the BRAM and that the additional multiplexers for input switching resulting from the application of the proposed method did not affect the critical path. The maximum frame rate is estimated to be 305.18 fps when operating at 100 MHz, which is the clock constraint for logic synthesis.

Table 1. Estimated operating frequency and frame rate

	Naive	Proposed
WNS (ns)	4.836	4.836
Maximum operating frequency (MHz)	193.65	193.65
Maximum frame rate (fps)	590.97	590.97

Table 2. Estimated operating frequency and frame rate

	Naive	Proposed
Processing time (μs)	991.65	991.75
Clock cycles	99, 165	991, 75
Breakdown of cycles	(154lines + 604)	(154lines + 614)

Based on the Verilog simulation results, latency is evaluated. Table 2 shows the latency for each implementation, assuming a 10 ns clock period. The latency increase was 10 clock cycles, which was caused by the delay added to resolve input contention. The increase in latency is 0.01%, and even after the increase, the system is expected to operate with a low latency of approximately 991.75 μs.

6 Conclusion

In this paper, we discussed the FPGA implementation of a semantic segmentation system for surgical images using convolutional neural networks. A naive implementation with simple spatial expansion requires a lot of resources equivalent to five Virtex UltraScale xcvu095 chips. Therefore, we focused on the recursive use of the network, which is a feature of the proposed network, and proposed and examined a method for reducing the resources used by the computing modules. It was found that when this system operates at 100 MHz, it can operate at a frame rate of about 300 fps and a low latency of about 1 ms. Further resource reduction methods such as time-sharing of arithmetic units, are to be considered in the future.

References

1. Araki, Y., Matsuda, M., Manabe, T., Ishizuka, Y., Shibata: FPGA implementation of convolutional neural networks. In: Proceedings of International Conference on Complex, Intelligent, and Software Intensive Systems (CISIS), pp. 223–233 (2020)
2. Bailey, D.G. (ed.): Image Processing Using FPGAs. MDPI AG (2019)
3. Guo, Q., Yu, Z., Wu, Y., Liang, D., Qin, H., Yan, J.: Dynamic recursive neural network. In: Proceedings of IEEE/CVF Conference on Computer Vision and Pattern Recognition (CVPR), pp. 5142–5151 (2019)

4. Gupta, S., Agrawal, A., Gopalakrishnan, K., Narayanan, P.: Deep learning with limited numerical precision. In: Proceedings of International Conference on International Conference on Machine Learning, vol. 37, pp. 1737–1746 (2015)
5. Kim, J., Lee, J.K., Lee, K.M.: Deeply-recursive convolutional network for image super-resolution. In: Proceedings of IEEE Conference on Computer Vision and Pattern Recognition (CVPR), pp. 1637–1645 (2016)
6. Maas, A., Hannun, A., Ng, A.: Rectifier nonlinearities improve neural network acoustic models. In: Proceedings of International Conference on Machine Learning (ICML) Workshop on Deep Learning for Audio, Speech and Language Processing (2013)
7. Manabe, T., Shibata, Y., Oguri, K.: FPGA implementation of a real-time super-resolution system using flips and an RNS-based CNN. IEICE Trans. Fundam. Electron. Commun. Comput. Sci. **E101-A**(12), 2280–2289 (2018)
8. Manabe, T., Tomonaga, K., Shibata, Y.: CNN architecture for surgical image segmentation systems with recursive network structure to mitigate overfitting. In: Proceedings of International Symposium on Computing and Networking (CANDAR), pp. 171–177 (2019)
9. Paszke, A., et al.: Pytorch: an imperative style, high-performance deep learning library. Adv. Neural. Inf. Process. Syst. **32**, 8024–8035 (2019)
10. Ronneberger, O., Fischer, P., Brox, T.: U-Net: convolutional networks for biomedical image segmentation. In: Proceedings of International Conference on Medical Image Computing and Computer-Assisted Intervention, pp. 234–241 (2015)
11. Zhang, D., Yang, J., Ye, D., Hua, G.: LQ-Nets: learned quantization for highly accurate and compact deep neural networks. In: Proceedings of European Conference on Computer Vision (ECCV), pp. 373–390 (2018)

A Mobile-Oriented GPU Implementation of a Convolutional Neural Network for Object Detection

Yasutoshi Araki, Takuho Kawazu, Taito Manabe, Yoichi Ishizuka,
and Yuichiro Shibata[✉]

Nagasaki University, Nagasaki, Japan
{yaraki,kawazu}@pca.cis.nagasaki-u.ac.jp,
{tmanabe,isy2,yuichiro}@nagasaki-u.ac.jp

Abstract. Realtime image detection and recognition on low-power edge devices are key to implementing many social applications such as automated traffic guard systems. This paper proposes mobile-oriented convolutional neural network implementation for object detection, focusing on computational reduction techniques such as depthwise separable convolution and grouped convolution on an edge GPU device. Through empirical evaluation and analysis, it is shown that the use of grouped convolution and half-precision floating point arithmetic is effective for the calculation of 3×3 convolution, and the grouped convolution with 8 groups for YOLOv3-tiny achieves 16 FPS with 86% detection accuracy in a 5 W low power mode on Jetson Nano. We also discuss how the processing time and recognition accuracy are affected by the floating arithmetic types and power consumption modes that the GPU device offers.

1 Introduction

Advanced image recognition technology based on the rapid development of deep learning methods is expected to have many social applications such as autopilot UAVs and automated traffic guard systems. While convolutional neural networks (CNNs) enable highly accurate object detection and recognition tasks, their computational costs are becoming increasingly expensive. Thus, realtime image inference on low-power edge devices is key to implementing such applications, and has motivated various research attempts on mobile-oriented neural network optimization techniques. Mobile GPUs are promising as a computing acceleration platform for CNNs for edge tasks. However, the tradeoff relationship between processing performance, accuracy, and power consumption has not been fully revealed, which makes it difficult for application developers to make the right design decisions.

Aiming at addressing this problem, we propose modified CNN structures for object detection, where computational reduction techniques for convolution are applied to a well-known YOLOv3-tiny network. We also discussed practical design alternatives for CNNs on an NVIDIA Jetson Nano mobile GPU platform.

© The Author(s), under exclusive license to Springer Nature Switzerland AG 2023
L. Barolli (Ed.): CISIS 2023, LNDECT 176, pp. 149–160, 2023.
https://doi.org/10.1007/978-3-031-35734-3_15

The architecture offers two floating point arithmetic types: half precision (FP16) and single precision (FP32), and two power consumption modes: 10 W and 5 W. Through empirical evaluation and analysis, we will reveal how the processing time and recognition accuracy are affected by the floating arithmetic types and power modes.

The contributions of the paper are as follows:

- Mobile-oriented variants of the YOLOv3-tiny object detection CNN are proposed and evaluated, introducing depthwise separable convolution and grouped convolution techniques.
- Impacts of selection of floating point arithmetic types are evaluated in terms of inference time and accuracy.
- Tradeoffs between the frame rates and the power consumption on Jetson Nano are empirically evaluated.

The rest of the paper is organized as follows. After Sect. 2 shows related work, Sect. 3 explains efficient convolution techniques. The proposed models are shown in Sect. 4 and evaluation results are discussed in Sect. 5. Finally, the paper is concluded in Sect. 6.

2   Related Work

Extensive research has been actively conducted on CNNs for object detection and recognition [5]. YOLO [9] is a CNN-based object detection neural network that simultaneously performs the identification and classification of object bounding boxes (BBoxes) on input images. Conventional object detection neural networks have a two-stage structure in which BBox candidates are first identified by the extraction process, then a CNN is applied to each of them for classification. On the other hand, YOLO performs both object candidate region extraction and classification in a single network, making it faster than conventional neural networks and suitable for applications that require realtime performance. YOLOv3 [10] has a Darknet-53 structure as its backbone, which requires a relatively large amount of computing resources and is not suitable for embedded platforms. YOLOv3-tiny [8] reduces the number of trainable parameters and improves the model for embedded applications. Although some improved versions of YOLO have also been announced [1], they are computationally more expensive than YOLOv3-tiny and are not oriented for low-power applications.

Literature [2] presents a benchmark analysis of YOLO performance on mobile GPU devices and reveals that YOLOv3-tiny only achieves 6.8 FPS on NVIDIA Jetson Nano. Alireza et al. implemented a vehicle detection system with YOLOv3-tiny on Jetson Nano for traffic monitoring applications [11]. Based on the assumption that cameras are installed in fixed locations and the background is fixed, the model was lightened by pruning batch normalization layers and other less important layers in the accuracy of the YOLOv3-tiny model. For performance improvement, 16-bit half-precision floating point arithmetic (FP16) was used. The evaluation demonstrated the resulting frame rate was 17 FPS and the mAP degradation was only 1.82%. However, it is not discussed how the selection of the power mode of Jetson Nano affects the performance and accuracy.

3 Computational Reduction Techniques for Convolution

Since the dominant operation in CNN inference is convolution, several computational reduction techniques for convolution have been proposed.

3.1 Depthwise Separable Convolution

Depthwise separable convolution is a parameter reduction technique for CNNs, which was originally proposed for MobileNet [4]. The depthwise separable convolution is a sort of approximation for a normal convolution, in which the computation is split into spatial (depthwise) and channel (pointwise) directions. The number of parameters required for a normal convolution method is expressed as

$$N \cdot M \cdot K^2 \tag{1}$$

where N, K, and M are the number of input channels, filter size, and the number of output channels, respectively. On the other hand, the numbers of parameters for depthwise and pointwise convolution are $N \cdot K^2$ and $N \cdot M$, respectively, making a total of

$$N \cdot (K^2 + M). \tag{2}$$

Therefore, the parameter reduction ratio brought about by depthwise separable convolution is expressed as

$$\frac{N \cdot (K^2 + M)}{N \cdot M \cdot K^2} = \frac{1}{M} + \frac{1}{K^2}. \tag{3}$$

Since the filter size is fixed, the effect of parameter reduction increases as the number of output channels M increases.

3.2 Grouped Convolution

Grouped convolution is a method proposed in AlexNet [7] that performs convolution by sparsely connecting input and output channels to reduce computational cost. Grouped convolution divides input channels and filters into G groups. When the number of input channels is N and the number of output channels is M, each group performs convolution with $\frac{N}{G}$ input channels and $\frac{M}{G}$ filters. While this technique reduces the computational cost, it has the disadvantage that each output channel can only depend on a limited number of input channels. The required number of parameters for grouped convolution is

$$G \cdot \left(\frac{N}{G} \cdot \frac{M}{G} \cdot K^2 \right) = \frac{1}{G} \cdot N \cdot M \cdot K^2, \tag{4}$$

which means a reduction of $\frac{1}{G}$.

4   Proposed Network Models

Based on the YOLOv3-tiny, we designed the following variant models introducing depthwise separable convolution and grouped convolution.

- Tiny-sep model
  In this model, depthwise separable convolution was applied to all 3×3 convolution layers except the first layer in the YOLOv3-tiny.
- Tiny-sep-grouped model
  In this model, both depthwise separable convolution and grouped convolution were applied, aiming at reducing the computational costs of pointwise (1×1) convolution in the tiny-sep model. To avoid loss of accuracy, grouped convolution was applied to the 3rd, 5th, 7th, 9th, 11th, and 12th convolution layers rather than all pointwise convolution layers. Three numbers of groups (G) were compared: 2, 4, and 8.
- Tiny-grouped model
  For comparison, only grouped convolution was applied to the original YOLOv3-tiny in this model. The target layers are the same as the abovementioned tiny-sep-grouped model. Also, three numbers of G were compared: 2, 4, and 8.

Hereafter, the original YOLOv3-tiny is referred to simply as the tiny model. The evaluated models and their sizes are listed in Table 1.

Table 1. Compared models and their sizes

Model Name	Model Size [MB]
tiny	34.0
tiny-grouped $(G = 2)$	20.0
tiny-grouped $(G = 4)$	13.0
tiny-grouped $(G = 8)$	9.4
tiny-sep	4.9
tiny-sep-grouped $(G = 2)$	3.4
tiny-sep-grouped $(G = 4)$	2.6
tiny-sep-grouped $(G = 8)$	2.2

5   Evaluation and Discussion

The network models were implemented on an NVIDIA Jetwon Nano platform. The software tools we used for the evaluation are Pytorch 1.11.0, ONNX opset 11, and TensorRT 8.2.1. Data labeled "Car" from Google Open Images Dataset [3] were used as the training dataset for the model. Of these, $62,856$ were used as the training dataset while $17,697$ were used as the validation dataset for adjusting hyperparameters during training. As the test dataset for evaluation, we used 244 vehicle images, which were taken at a construction site with alternating traffic on one side. A portion of the test data set is shown in Fig. 1.

As a learning algorithm, Adam [6] was used with the hyperparameters shown in Table 2. The training was performed on a Linux PC equipped with an NVIDIA GeForce GTX 1080 Ti GPU.

Table 2. Hyperparameters used for training the model

Parameter Name	Set Value
Number of epochs	150
Batch size	32
Learning rate	0.0001
Weight decay	0.0005

Fig. 1. Example images and BBoxes in test dataset

5.1 Recognition Accuracy

Using the above mentioned environment, we evaluated the mean average precision (mAP) on the test dataset. The results are summarized in Table 3. The mAP achieved with FP32 was 87.01% for the tiny model and 89.87% for the tiny-sep model, showing no decrease in accuracy with the introduction of the depthwise separable convolution. For grouped convolution, on the other hand, mAP decreases as the number of groups increases, except for the tiny-grouped ($G = 8$) model. An mAP drop of approximately 2 to 4 points is occurring when the number of groups is doubled. It is also shown that the use of FP16 degrades mAP by 2 to 5 points compared to FP32.

Table 3. Resulting mAP for test data set

Model	FP32 [%]	FP16 [%]
tiny	87.01	81.59
tiny-grouped $(G = 2)$	87.97	84.12
tiny-grouped $(G = 4)$	84.46	80.55
tiny-grouped $(G = 8)$	88.39	86.41
tiny-sep	89.87	87.19
tiny-sep-grouped $(G = 2)$	87.77	85.16
tiny-sep-grouped $(G = 4)$	85.58	80.97
tiny-sep-grouped $(G = 8)$	81.63	77.18

5.2 Inference Processing Time

During the build phase on TensorRT, multiple CUDA kernels are exhaustively executed, and the implementation strategy with the lowest latency is selected. Here, we evaluated the inference time by extracting the CUDA kernel processing time from the TensorRT log output. Table 4 shows the results.

Table 4. Processing time required for inference [ms]

Model	FP32 (10 W)	FP32 (5 W)	FP16 (10 W)	FP16 (5 W)
tiny	33.18	46.31	21.21	29.83
tiny-grouped $(G = 2)$	25.62	35.28	17.11	24.07
tiny-grouped $(G = 4)$	22.87	31.11	15.45	21.96
tiny-grouped $(G = 8)$	21.80	29.56	15.20	21.73
tiny-sep	18.71	25.08	16.23	22.68
tiny-sep-grouped $(G = 2)$	17.55	22.94	15.71	21.86
tiny-sep-grouped $(G = 4)$	17.18	22.39	15.72	21.70
tiny-sep-grouped $(G = 8)$	17.50	23.09	16.26	22.25

Regarding the effects of depthwise separable convolution, the processing time in FP32 mode (10 W) is 33.18 ms for the tiny model and 18.71 ms for the tiny-sep model, revealing that the use of depthwise separable convolution achieves a 1.77 times speedup. Figure 2 compares the processing times of the tiny and tiny-sep models in FP32 mode (10 W) for each convolution layer. Comparing the early-stage layers with a small number of channels to the late-stage layers with a large number of channels, it is clear that the larger the number of channels, the greater the effect of reducing processing time by introducing the depthwise separable convolution.

Figure 3 shows the ratio of the processing time for depthwise convolution and pointwise convolution of the tiny-sep model in the FP32 mode (10 W). In depthwise separable convolution, the computation of pointwise convolution is found to be more dominant than that of depthwise convolution, and this trend is more pronounced for layers with a large number of channels. The suppressive effect of the gropued convolution on tiny-sep models suggests that TensorRT CUDA kernels allow 3×3 filters to be more efficiently computed, probably due to the SIMD characteristics of the GPU architecture.

Regarding the grouped convolution, as shown in Fig. 4, the tiny model shows a greater reduction in processing time than the tiny-sep model thanks to the introduction of grouped convolution. The tiny-sep model did not benefit from the increase in the number of groups. The grouped convolution is performed for 3×3 filters in the tiny model, while the grouped convolution is applied to the pointwise convolution of 1×1 filters in the tiny-sep model.

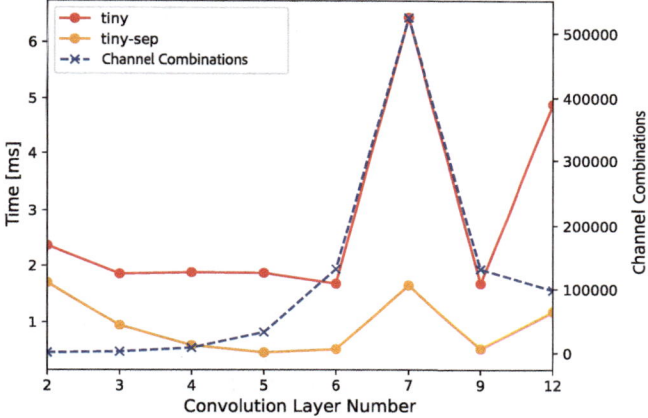

Fig. 2. Comparison of processing times by convolution layer: tiny and tiny-sep models

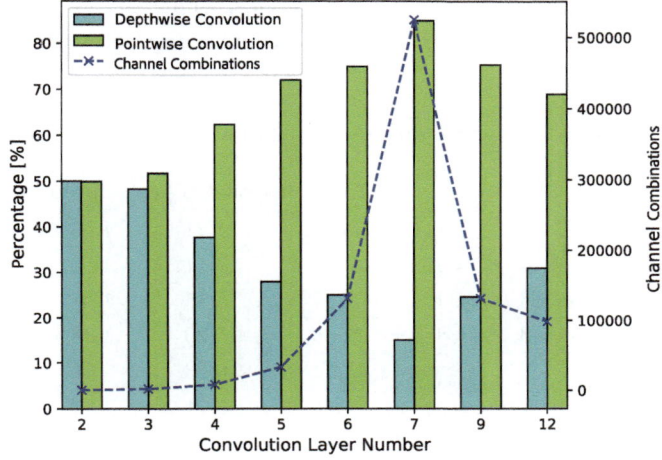

Fig. 3. Comparison of processing time between depthwise convolution and pointwise convolution

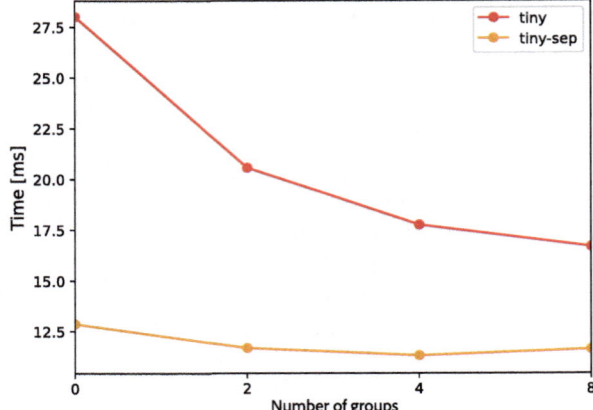

Fig. 4. Comparison of processing time and the number of groups in grouped convolution

5.3   Floating Point Arithmetic Types

Table 5 compares the processing times of FP32 and FP16 in 10 W mode. By utilizing FP16, the processing time of the tiny model is reduced by about 35%, while that of the tiny-sep model is reduced by only about 10%. Table 6 shows detailed breakdowns of processing times for the tiny and tiny-sep models in 10 W mode. For tiny-sep, FP16 reduced the processing time for convolution by approximately 4 ms, while the processing time for Reformatting-copy-node, an operation to change the shape of the tensor, increased by approximately 3 ms. The reduced processing time of convolution is offset by the increased processing time of Reformatting-copy-node, where two FP16 data are packed into one word for SIMD operations. In the FP16 depthwise separable convolution, this tensor shape change seems to become a bottleneck since the 1×1 convolution operation does not benefit from the SIMD acceleration compared to the usual 3×3 convolution.

Table 5. Comparison of processing time: FP32 and FP16

Model	FP32 [ms]	FP16 [ms]	Speedup
tiny	33.18	21.21	1.56
tiny-grouped ($G = 2$)	25.62	17.11	1.50
tiny-grouped ($G = 4$)	22.87	15.45	1.48
tiny-grouped ($G = 8$)	21.80	15.20	1.43
tiny-sep	18.71	16.23	1.15
tiny-sep-grouped ($G = 2$)	17.55	15.71	1.12
tiny-sep-grouped ($G = 4$)	17.18	15.72	1.09
tiny-sep-grouped ($G = 8$)	17.50	16.26	1.08

Table 6. Breakdowns of processing time [ms]

Layer Type	tiny (FP32)	tiny (FP16)	tiny-sep (FP32)	tiny-sep (FP16)
Convolution	27.97	17.21	12.82	9.02
Reformatting-copy-node	0.00	0.68	0.00	3.16
Maxpooling	2.07	1.04	2.01	1.06
LeakyReLU	2.56	1.77	3.29	2.31
Up-samling	0.13	0.16	0.14	0.11
Zero-padding	0.09	0.05	0.09	0.05
Tensor-copy	0.36	0.31	0.35	0.52

5.4   Power Consumption Mode

Tables 7 and 8 show how the selection of power mode affects the inference processing times for FP32 and FP16, respectively. Inference time in 5 W mode is about 40% longer than in 10W mode. We also evaluated the transfer time between CPU host memory and GPU device memory. FP16 is expected to reduce the transfer time compared to FP32 since the data size is shrunk. While Table 9 shows the transfer time from the host memory to the device memory, Table 10 shows the transfer time from the device memory to the host memory. The transfer time from the host memory to the device memory of FP16 is approximately 50% shorter than that of FP32, and the transfer time from the device memory to the host memory of FP16 is approximately 30% shorter than that of FP32.

Table 7. Comparison of 10 W and 5 W mode for FP32

Model	10 W [ms]	5 W [ms]	Increase rate [%]
tiny	33.18	46.31	39
tiny-grouped $(G = 2)$	25.62	35.28	37
tiny-grouped $(G = 4)$	22.87	31.11	36
tiny-grouped $(G = 8)$	21.80	29.56	35
tiny-sep	18.71	25.08	34
tiny-sep-grouped $(G = 2)$	17.55	22.94	30
tiny-sep-grouped $(G = 4)$	17.18	22.39	30
tiny-sep-grouped $(G = 8)$	17.50	23.09	31

Table 8. Comparison of 10 W and 5 W mode for FP16

Model	10 W [ms]	5 W [ms]	Increase rate [%]
tiny	21.21	29.83	40
tiny-grouped ($G = 2$)	17.11	24.07	40
tiny-grouped ($G = 4$)	15.45	21.96	42
tiny-grouped ($G = 8$)	15.20	21.73	42
tiny-sep	16.23	22.68	39
tiny-sep-grouped ($G = 2$)	15.71	21.86	39
tiny-sep-grouped ($G = 4$)	15.72	21.70	38
tiny-sep-grouped ($G = 8$)	16.26	22.25	36

Table 9. Data transfer time from host to device [ms]

Model	FP32 (10 W)	FP32 (5 W)	FP16 (10 W)	FP16 (5 W)
tiny	0.209	0.246	0.104	0.123
tiny-grouped ($G = 2$)	0.210	0.247	0.104	0.123
tiny-grouped ($G = 4$)	0.207	0.246	0.105	0.123
tiny-grouped ($G = 8$)	0.207	0.253	0.104	0.129
tiny-sep	0.206	0.248	0.104	0.124
tiny-sep-grouped ($G = 2$)	0.206	0.249	0.103	0.123
tiny-sep-grouped ($G = 4$)	0.206	0.248	0.103	0.123
tiny-sep-grouped ($G = 8$)	0.211	0.248	0.106	0.124

Table 10. Data transfer time from device to host [ms]

Model	FP32 (10 W)	FP32 (5 W)	FP16 (10 W)	FP16 (5 W)
tiny	0.0071	0.0116	0.0054	0.0073
tiny-grouped ($G = 2$)	0.0086	0.0114	0.0055	0.0077
tiny-grouped ($G = 4$)	0.0081	0.0110	0.0055	0.0074
tiny-grouped ($G = 8$)	0.0083	0.0110	0.0057	0.0076
tiny-sep	0.0085	0.0114	0.0054	0.0070
tiny-sep-grouped ($G = 2$)	0.0082	0.0114	0.0054	0.0075
tiny-sep-grouped ($G = 4$)	0.0082	0.0109	0.0051	0.0075
tiny-sep-grouped ($G = 8$)	0.0082	0.0112	0.0055	0.0072

5.5   Resulting Frame Rates

Finally, the frame rate was empirically measured by connecting a camera device, a Logitech C270n USB camera. We measured processing time for 100 frames, including the acquisition of camera images, image preprocessing, inference pro-

Table 11. Resulting frame rates [FPS]

Model	FP32 (10 W)	FP32 (5 W)	FP16 (10 W)	FP16 (5 W)
tiny	15.7	11.2	20.8	12.8
tiny-grouped (G = 2)	17.3	12.8	20.2	15.2
tiny-grouped (G = 4)	20.8	12.6	22.5	13.3
tiny-grouped (G = 8)	20.5	13.8	22.8	16.0
tiny-sep	20.1	14.4	20.8	15.5
tiny-sep-grouped (G = 2)	20.6	15.6	22.9	15.9
tiny-sep-grouped (G = 4)	21.3	15.5	22.3	16.0
tiny-sep-grouped (G = 8)	20.3	14.5	22.2	15.6

cessing, post-processing of inference results (NMS), and rendering of detected BBoxes, and then averaged the results. The frame rates for each model are shown in Table 11. Compared to 6.8 FPS in [2] and 17 FPS in [11], the effectiveness of the proposed approach is highlighted even in the low-power mode of 5 W.

6 Conclusion

In this paper, mobile-oriented optimization of YOLOv3-tiny on Jetson Nano was discussed, focusing on the use of depthwise separable convolution and grouped convolution. The effects of using FP16 arithmetic were quantitatively evaluated in terms of the processing speed and detection accuracy, for both 10 W and 5 W power consumption modes. The evaluation results revealed that the use of grouped convolution and FP16 arithmetic was effective for the calculation of 3×3 convolution. The implementation experiments demonstrated that the introduction of grouped convolution with 8 groups achieved 16 FPS with 86% detection accuracy under the 5 W power constraint. Our future work includes a more detailed analysis of efficiency relating to SIMD instructions for further performance improvement.

References

1. Bochkovskiy, A., Wang, C., Liao, H.M.: YOLOv4: optimal speed and accuracy of object detection. arXiv preprint arXiv:2004.10934 (2020)
2. Feng, H., Mu, G., Zhong, S., Zhang, P., Yuan, T.: Benchmark analysis of YOLO performance on edge intelligence devices. Cryptography **6**(2), 1–16 (2022)
3. Google: Open images dataset. https://storage.googleapis.com/openimages/web/index.html. Accessed 30 Mar 2023
4. Howard, A.G., et al.: MobileNets: efficient convolutional neural networks for mobile vision applications. arXiv preprint arXiv:1704.04861 (2017)
5. Jiao, L., et al.: A survey of deep learning-based object detection. IEEE Access **7**, 128837–128868 (2019)

6. Kingma, D.P., Ba, J.: Adam: a method for stochastic optimization. In: Proceedings of the International Conference on Learning Representations (ICLR), pp. 1–15 (2015)
7. Krizhevsky, A., Sutskever, I., Hinton, G.E.: ImageNet classification with deep convolutional neural networks. In: Proceedings International Conference on Neural Information Processing Systems, pp. 1097–1105 (2012)
8. Redmon, J.: YOLO: real-time object detection. https://pjreddie.com/darknet/yolo/. Accessed 30 Mar 2023
9. Redmon, J., Divvala, S.K., Girshick, R.B., Farhadi, A.: You only look once: unified, real-time object detection. In: Proceedings of the IEEE Conference on Computer Vision and Pattern Recognition (CVPR), pp. 779–788 (2016)
10. Redmon, J., Farhadi, A.: YOLOv3: an incremental improvement. arXiv preprint arXiv:1804.02767 (2018)
11. Tajar, A.T., Ramazani, A., Mansoorizadeh, M.: A lightweight tiny-YOLOv3 vehicle detection approach. J. Real-Time Image Proc. **18**, 2389–2401 (2021)

A Fuzzy-Based Error Driving System for Improving Driving Performance in VANETs

Ermioni Qafzezi[1]([✉]), Kevin Bylykbashi[1], Shunya Higashi[2], Phudit Ampririt[2], Keita Matsuo[1], and Leonard Barolli[1]

[1] Department of Information and Communication Engineering, Fukuoka Institute of Technology (FIT), 3-30-1 Wajiro-Higashi, Higashi-Ku, Fukuoka 811-0295, Japan
{qafzezi,kevin}@bene.fit.ac.jp, {kt-matsuo,barolli}@fit.ac.jp
[2] Graduate School of Engineering, Fukuoka Institute of Technology (FIT), 3-30-1 Wajiro-Higashi, Higashi-Ku, Fukuoka 811-0295, Japan
bd21201@bene.fit.ac.jp

Abstract. The primary reason for car accidents is caused by driver's error. In this work, we consider some critical factors related to driver's behavior that can lead to a car accident such as recognition error, decision error and performance error. We propose a fuzzy logic system to determine the Driver's Error Value (DEV) for evaluating the driver's decision in real-time. We show through simulations the effect of each considered parameter and demonstrate some actions to prevent a possible crash when the driver shows a high error value. The fuzzy logic system is implemented in vehicles in a Vehicular Ad hoc Network (VANET) environment which are able to exchange information with each-other in real time and based on the DEV output the drivers are informed when they have approached a potential dangerous situation.

1 Introduction

According to National Highway Traffic Safety Administration 94% of car accidents are because of bad choices made by drivers [6]. The drivers should be held accountable for their actions, even though some of these situations are unintended. However, it is possible to prevent fatal crashes through technology. Therefore, the industry, government and academic institutions are conducting substantial research to implement proper systems and infrastructure for car accident prevention. The initiatives of many governments for a collaboration of such researchers have concluded to the establishment of Intelligent Transport Systems (ITSs).

ITS makes use of different networking techniques and data communication technologies to create applications that serve environmental conservation, the efficiency, comfort, cost, safety and management of transportation systems. This is achievable by integrating the system of people, roads and vehicles into a single coordination manageable system. As a result, drivers are better informed and can

L. Barolli (Ed.): CISIS 2023, LNDECT 176, pp. 161–169, 2023.
https://doi.org/10.1007/978-3-031-35734-3_16

take better decisions. Some of these technologies include vehicular navigation, camera and image processing to monitor vehicle speed, vehicle detection, security and accident detection, smart toll payments, weather information, detection of traffic signal control and rules violation, and so on.

Vehicular Ad hoc Network (VANET) is a relevant component of ITS, which enhance the safety and efficiency in transportation systems. In VANETs, the vehicles are able to interact with the environment by communicating with the infrastructure and other nearby vehicles. Moreover, different artificial intelligent approaches including Fuzzy Logic (FL) are used to make better predictions in these systems.

In a previous work [1], we proposed an intelligent system based on FL that evaluated the driver's stress as one the most important factors that leads to accident. The driver's stress was determined based on driver's impatience signs, the dangerous behavior of other nearby drivers and traffic condition. In this work, we take into consideration internal factors that can make drivers take a wrong decision. Such factors include recognition error caused by internal and external distractions, decision error which happens by a wrong decision taken consciously, and performance error which is the improper physical response done unintentionally, as the most important ones. We evaluate the proposed system by computer simulations. Based on the output value, the proposed system can determine whether the driver poses a risk for himself and other road users.

The structure of the paper is as follows. In Sect. 2, we present an overview of VANETs and its evolution from traditional to cloud-fog-edge SDN-VANETs. Section 2 we describe in detail the combination of two emerging technologies that are SDN and VANETs, and the advantages this combination brings. In Sect. 4, we describe the proposed fuzzy-based system and its implementation. In Sect. 5, we discuss the simulation results. Finally, conclusions and future work are given in Sect. 6.

2   Overview of VANETs

VANETs is the network of vehicles able to interact with each other and the infrastructure. These vehicles are equipped with an On Board Unit (OBU) which enable the communication with other entities, and sensors which gather information about the vehicle, the driver and the surrounding environment. The information is then shared via Vehicle to Vehicle (V2V) and Vehicle to Infrastructure (V2I) communication links.

VANETs are a subset of Mobile Ad hoc Networks (MANETs) and as such they share some similar characteristics. For instance, they are self configuring networks and rely on peer-to-peer and multihop for data transmission. On the other side, VANETs have their own unique characteristics such as frequent and dynamic topology change, vehicles do not depend on limited power resources and they have more storage and processing capability, regular movement of vehicles in designated areas and so on.

However, because of these unique characteristics, they encounter many problems. Some issues are intermittent connectivity, short contact duration and the exponential growth of data generated in the network. The integration of cloud-fog-edge computing with VANETs is the solution to handle complex computation, provide mobility support, low latency and high bandwidth. While the integration of SDN is considered necessary to provide programmability, flexibility, management and a better intelligent use of VANETs.

The most critical applications in VANETs are real-time traffic and safety information for traffic control management and accident avoidance. For example, if a vehicle notices congestion in the road, it informs the other vehicles coming from all directions to avoid the crowded segments. In another situation where a driver is unable to see an obstacle on the road, VANETs can use V2I communication to alert the driver in advance and prevent a collision. This information should be provided to all vehicles in real time. By providing real-time information about traffic conditions, road hazards, and other vehicles' location and speed, VANETs can help drivers make more informed decisions and avoid accidents.

VANETs can also enable new applications, such as cooperative driving, where vehicles can cooperate with each other to form platoons and reduce fuel consumption. Finally, VANETs can also support new business models, such as car-sharing services and mobility-as-a-service (MaaS) platforms, by enabling seamless connectivity between different vehicles and infrastructure. Overall, VANETs have the potential to transform the way we use vehicles and interact with the transportation infrastructure, paving the way for a safer, more efficient, and more connected future. While there are still some technical and regulatory challenges to be overcome, the potential benefits of VANETs make them a technology to watch in the coming years.

3   SDN-VANETs

Software-Defined Networking (SDN) is a promising networking paradigm that allows network administrators to dynamically manage network resources, adapt to changing traffic demands, and improve network performance and security. When combined with VANET, SDN can significantly enhance the performance and efficiency of communication between vehicles, infrastructure, and other devices. SDN-VANETs can enable efficient data transmission, reduce network latency, and improve network scalability, which are essential for the deployment of advanced Intelligent Transportation Systems (ITS) applications such as traffic management, collision avoidance, and cooperative driving. The use of SDN in VANETs can also enable flexible and centralized management of the network, facilitate network virtualization, and improve network resilience against cyber-attacks and network failures. SDN-VANETs can also enable new applications, such as vehicle-to-vehicle communication and cooperative driving. Overall, SDN-VANETs hold great promise for the development of intelligent and connected transportation systems that can improve road safety, reduce congestion, and enhance the overall driving experience.

4   Proposed Fuzzy-Based System

Despite technological advancements, human errors remain one of the main causes of traffic accidents. However, VANETs have the potential to reduce the likelihood of driver errors and improve road safety. By providing real-time information about traffic conditions, road hazards, and other vehicles' location and speed, VANETs can help drivers make more informed decisions and avoid accidents. With V2I communication, VANETs can warn drivers of potential hazards or obstacles, such as accidents or debris on the road, and provide them with alternative routes to avoid traffic congestion. V2V communication allows vehicles to communicate with each other, exchange information, and adjust their speed and position to maintain a safe distance from each other. By providing these real-time information and assistance, VANETs can help reduce the likelihood of human driver errors, such as sudden braking or incorrect lane changes. With the increasing prevalence of connected vehicles and the growing maturity of VANET technology, we can expect to see significant improvements in road safety and overall driving experience in the near future.

Our research work focuses on developing an intelligent non-complex driving support system that determines the driving risk level in real time based on different types of parameters. In this work, we propose a system, called Fuzzy System for Assessment of Driver's Error Value (FS-ADEV) that calculates Driver's Error Value (DEV) in order to avoid a potential accident from happening. The proposed system aims to keep the driver alerted when he is performing poorly or when other drivers in its vicinity take wrong decisions. For this, we take into consideration the recognition error which happens when the driver is inattentive, the decision error which occurs when the driver responds in the wrong way because of underestimating a critical situation and performance error which is the faulty physical execution of the driver even though being conscious of the dangerous situation. The input parameters of FS-ADEV are described in following.

Recognition Error (RE): occurs when the driver fails to identify and comprehend a certain danger or changes in the road or is not properly paying attention while driving. This can happen when the driver is distracted because of external (everything outside the car that is distracting the driver) or internal factors (when the driver is using the cellphone, adjusting controls or talking with passengers).

Decision Error (DE): occurs when a driver responds improperly because of underestimating the critical situation. The driver recognizes the dangerous situation, however does not properly assess the threat and decides to take a wrong decision deliberately. Typical examples are when taking a curve with high speed, driving fast in bad weather conditions, and so on.

Table 1. FS-ADEV parameters and their term sets.

Parameters	Term Sets
Recognition Error (RE)	Small (S), Medium (M), Big (B)
Decision Error (DE)	Small (Sm), Medium (Me), Big (Bi)
Performance Error (PE)	Small (Sma), Medium (Med), Big (Bg)
Driver's Error Value (DEV)	DEV1, DEV2, DEV3, DEV4, DEV5, DEV6, DEV7

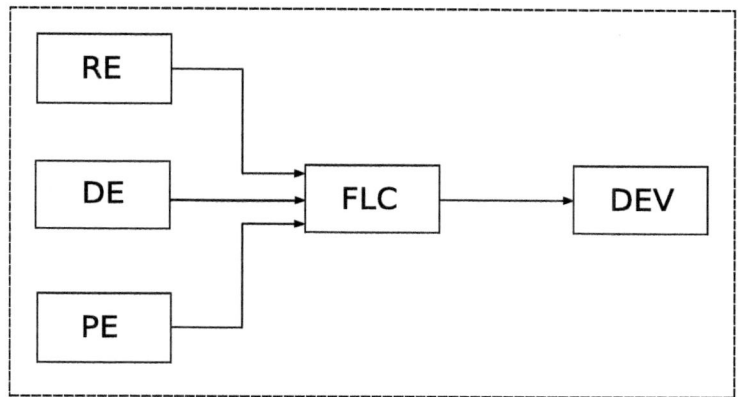

Fig. 1. Proposed system structure.

Performance Error (PE): is the faulty physical execution of the driver done unintentionally, even though being conscious of the dangerous situation. This can happen when the driver panics and does not take any preventive action, overreacts, or fails to keep the proper direction in its lane.

To implement the proposed system we use Fuzzy Logic (FL), as it is able to take real-time decisions, despite the lack of precision of the given input data [2–5,7,8]. The structure of the proposed system is shown in Fig. 1, whereas the term sets of the input and output parameters are given in Table 1. The membership functions of input and output parameters are shown in Fig. 2 and the Fuzzy Rule Base (FRB) is given in Table 2.

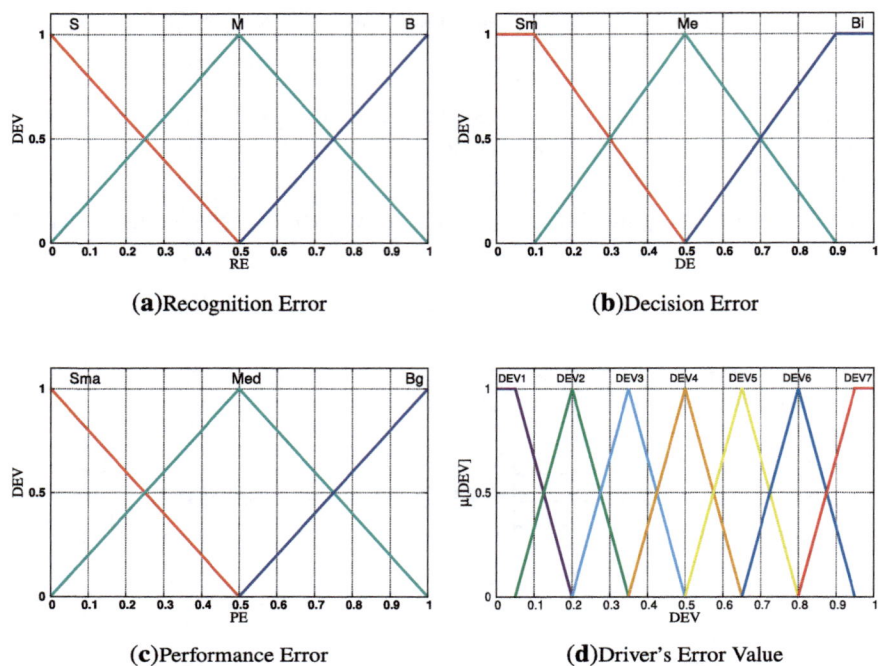

(**a**)Recognition Error

(**b**)Decision Error

(**c**)Performance Error

(**d**)Driver's Error Value

Fig. 2. Membership functions of FS-ADEV.

5   Simulation Results

The simulations were conducted using FuzzyC and the results are shown for three scenarios. Figure 3 shows the results for small, moderate and big recognition error. We show the relation between DEV and PE for different DE values.

Figure 3(a) shows the scenario when the driver makes a low number of recognition errors. For this reason, DEV values are lower when compared with other scenarios. In this case the probability of having an accident is low, therefore there is no need to alert the driver. However, when the value of RE increases, the probability of having an accident also increases as seen for moderate values of RE in Fig. 3(b). The increase of DE indicates that the driver is underestimating a critical situation, therefore, for Big DE, the situation is considered dangerous, regardless the PE value. The situation becomes more dangerous when RE = 0.9. As seen in Fig. 3(c), in most of the cases the driver is put in danger even when his performance ability are good (PE is Sma). In this situation the driver is informed by the system for his poor behavior and to be more vigilant in order to avoid a dangerous situation.

Table 2. FRB of FS-ADEV.

No	RE	DE	PE	DEV
1	S	Sm	Sma	DEV1
2	S	Sm	Med	DEV2
3	S	Sm	Bg	DEV3
4	S	Me	Sma	DEV2
5	S	Me	Med	DEV3
6	S	Me	Bg	DEV4
7	S	Bi	Sma	DEV3
8	S	Bi	Med	DEV4
9	S	Bi	Bg	DEV5
10	M	Sm	Sma	DEV2
11	M	Sm	Med	DEV3
12	M	Sm	Bg	DEV4
13	M	Me	Sma	DEV3
14	M	Me	Med	DEV4
15	M	Me	Bg	DEV5
16	M	Bi	Sma	DEV4
17	M	Bi	Med	DEV5
18	M	Bi	Bg	DEV6
19	B	Sm	Sma	DEV4
20	B	Sm	Med	DEV5
21	B	Sm	Bg	DEV6
22	B	Me	Sma	DEV5
23	B	Me	Med	DEV6
24	B	Me	Bg	DEV7
25	B	Bi	Sma	DEV6
26	B	Bi	Med	DEV6
27	B	Bi	Bg	DEV7

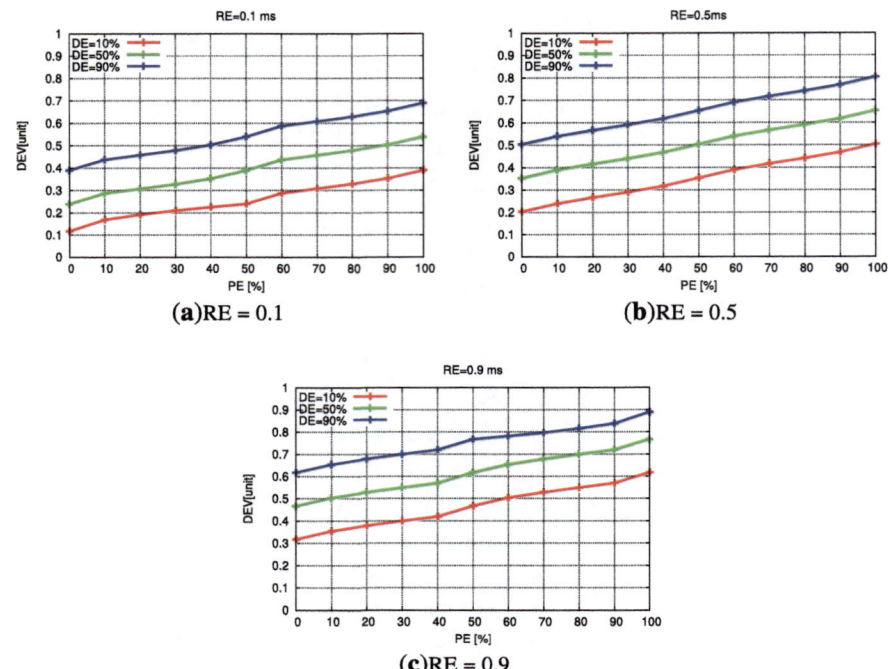

Fig. 3. Simulation results for DEV.

6   Conclusions

Despite technological advancements, human errors remain one of the main causes of traffic accidents. However, VANETs have the potential to reduce the likelihood of driver errors and improve road safety. In this work, we presented the implementation of an FL approach that assess the driver's error value in real-time by considering the their recognition error, decision error and performance error. We showed through simulations the effect of the considered parameters for improving driving performance and avoiding a potential accident from happening in VANETs. The simulations show that drivers are exposed to accidents when they fail to properly assess a situation. The worst condition happens when drivers underestimate a risky situation or panics. In this case, the system immediately informs the driver for his poor behavior and other nearby drivers for being attentive. In order to avoid danger, the system can even trigger some preventive actions like decreasing the vehicle speed.

References

1. Bylykbashi, K., Qafzezi, E., Ampririt, P., Ikeda, M., Matsuo, K., Barolli, L.: A fuzzy-based system for safe driving in VANETs considering impact of driver impatience on stress feeling level. In: Barolli, L., Kulla, E., Ikeda, M. (eds.) Advances

in Internet, Data & Web Technologies. EIDWT 2022. Lecture Notes on Data Engineering and Communications Technologies, vol. 118, pp. 236–244. Springer, Cham (2022). https://doi.org/10.1007/978-3-030-95903-6_25

2. Kandel, A.: Fuzzy Expert Systems. CRC Press Inc., Boca Raton (1992)
3. Klir, G.J., Folger, T.A.: Fuzzy Sets, Uncertainty, and Information. Prentice Hall, Upper Saddle River (1988)
4. McNeill, F.M., Thro, E.: Fuzzy Logic: A Practical Approach. Academic Press Professional Inc., San Diego (1994)
5. Munakata, T., Jani, Y.: Fuzzy systems: an overview. Commun. ACM **37**(3), 69–77 (1994)
6. Singh, S.: Critical reasons for crashes investigated in the national motor vehicle crash causation survey (2015)
7. Zadeh, L.A., Kacprzyk, J.: Fuzzy Logic for the Management of Uncertainty. Wiley, New York (1992)
8. Zimmermann, H.J.: Fuzzy control. In: Fuzzy Set Theory and Its Applications, pp. 203–240. Springer, Dordrecht (1996). https://doi.org/10.1007/978-94-015-8702-0_11

Performance Evaluation of FC-RDVM and RIWM Methods for WMNs by WMN-PSOHCDGA System Considering Different Instances and Subway Distribution

Admir Barolli[1], Shinji Sakamoto[2], Leonard Barolli[3(✉)], and Makoto Takizawa[4]

[1] Department of Information Technology, Aleksander Moisiu University of Durres, L.1, Rruga e Currilave, Durres, Albania
admirbarolli@uamd.edu.al
[2] Department of Information and Computer Science, Kanazawa Institute of Technology, 7-1 Ohgigaoka, Nonoichi, Ishikawa 921-8501, Japan
shinji.sakamoto@ieee.org
[3] Department of Information and Communication Engineering, Fukuoka Institute of Technology, 3-30-1 Wajiro-Higashi, Higashi-Ku, Fukuoka 811-0295, Japan
barolli@fit.ac.jp
[4] Department of Advanced Sciences, Faculty of Science and Engineering, Hosei University, 3-7-2, Kajino-machi, Koganei-shi, Tokyo 184-8584, Japan
makoto.takizawa@computer.org

Abstract. In this paper, we consider three intelligent algorithms: Particle Swarm Optimization (PSO), Hill Climbing (HC) and Distributed Genetic Algorithm (DGA), and implement a new hybrid intelligent system for Wireless Mesh Networks (WMNs) called WMN-PSOHCDGA. For distribution of mesh clients, we consider Subway distribution and we carry out simulations for different instances (different scales) of WMNs. We compare the simulation results of a Fast Convergence Rational Decrement of Vmax Method (FC-RDVM) with Random Inertia Weight Method (RIWM). By simulation results, we found that FC-RDVM performs better than RIWM for load balancing and middle scale WMNs.

1 Introduction

In the designing and engineering process of Wireless Mesh Networks (WMNs) should be considered various parameters such as network connectivity, user coverage, Quality of Service (QoS), network cost and so on. For the mesh router node placement problem in WMNs, the mesh router nodes should have an optimal allocation in order to achieve good network connectivity and client coverage while also balancing the load of mesh routers. For the optimization process, we consider three parameters: Size of Giant Component (SGC), Number of Covered Mesh Clients (NCMC) and Number of Covered Mesh Clients per Router (NCMCpR).

There are different research works for mesh node placement in WMNs [2,4–6,13,14]. In previous work, some intelligent algorithms have been proposed for the

© The Author(s), under exclusive license to Springer Nature Switzerland AG 2023
L. Barolli (Ed.): CISIS 2023, LNDECT 176, pp. 170–178, 2023.
https://doi.org/10.1007/978-3-031-35734-3_17

node placement problem [1,3,7,8]. In [9,10], we implemented intelligent simulations systems for WMNs considering simple heruistic algorithms.

In this paper, we present a new hybrid intelligent system for Wireless Mesh Networks (WMNs) called WMN-PSOHCDGA, which integrates three intelligent algorithms: Particle Swarm Optimization (PSO), Hill Climbing (HC) and Distributed Genetic Algorithm (DGA). We compare the results of a Fast Convergence Rational Decrement of Vmax Method (FC-RDVM) with Random Inertia Weight Method (RIWM) considering Subway distribution of mesh clients and different instances (different scales of WMNs). The simulation results show that FC-RDVM has better performance and load balancing than RIWM for middle scale WMNs.

The paper is organized as follows. In Sect. 2, we introduce intelligent algorithms. In Sect. 3, we present the implemented hybrid simulation system. The simulation results are given in Sect. 4. Finally, we give conclusions and future work in Sect. 5.

2   Intelligent Algorithms

2.1   Particle Swarm Optimization

The PSO is a local search algorithm, which considers particles swarm for optimization process. The particles in the swarm move according to some rules considering the previous and the best known position. The PSO algorithm generates the initial positions of particles in random way and after every iteration the position and velocity of each particle is changed in order to move towards the desired location. The acceleration coefficients are weighting in order to improve the efficiency of local search and convergence to the global optimum solution.

2.2   Hill Climbing Algorithm

The HC algorithm is also a local search algorithm, which can find the best solution from a set of possible solutions. The HC algorithm is easily implemented and is used for different optimization problems. The HC algorithm can find quickly the local optima, but sometime it can get stuck in local optima and may not find the global optimum. The performance of HC algorithm depends on initial solution, so a good initial solution is needed.

2.3   Genetic Algorithm

Genetic Algorithm (GA) searches through a population of individuals and can operate on different representations. The main GA operators are Selection, Crossover and Mutation. The GA is successfully applied to different optimization problems, but it is computationally expensive and time-consuming. In our research work, we use Distributed GA (DGA), which has an additional mechanism to escape from local optima by considering multiple islands.

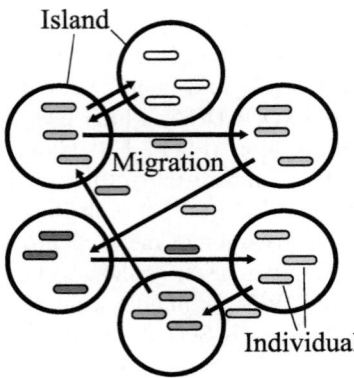

Fig. 1. Model of Migration in DGA.

3   Implemented WMN-PSOHCDGA System

In this section, we present the proposed and implemented WMN-PSOHCDGA hybrid intelligent simulation system. We consider the integration of three intelligent algorithms: PSO, HC and DGA in order to improve the convergence and solution of proposed system.

The migration process in DGA is shown in Fig. 1 and the proposed system flowchart is shown in Fig. 2. The system includes these components: initialization, particle-pattern, fitness function, distribution of mesh clients and replacement methods.

The system generates the initial solution randomly by *ad hoc* methods [15] and the velocity of particles is determined by a random process considering the area size.

We consider a particle as a mesh router. The fitness value of a particle-pattern is computed by considering the position of mesh routers and mesh clients. The solution for each particle-pattern is shown in Fig. 3. Eeach individual in the population is a combination of mesh routers and a WMN is represented by a gene.

In WMN-PSOHCDGA, we use the following fitness function:

$$Fitness = \alpha \times SGC(x_{ij}, y_{ij}) + \beta \times NCMC(x_{ij}, y_{ij}) + \gamma \times NCMCpR(x_{ij}, y_{ij}).$$

In fitness function, the SGC is the maximum number of connected routers, NCMC is the number of covered mesh clients by mesh routers and NCMCpR is the number of clients covered by each router, which is used for load balancing.

In this work, we consider Subway distribution of mesh clients as shown in Fig. 4. In this distribution, the mesh clients are positioned the same as in a city subway.

There are many router replacement methods. In this paper, we consider RIWM and FC-RDVM.

In RIWM, the ω parameter is changing randomly from 0.5 to 1.0 and C_1 and C_2 are considered 2.0. The ω can be estimated by the week stable region and the average value of ω is 0.75 [12].

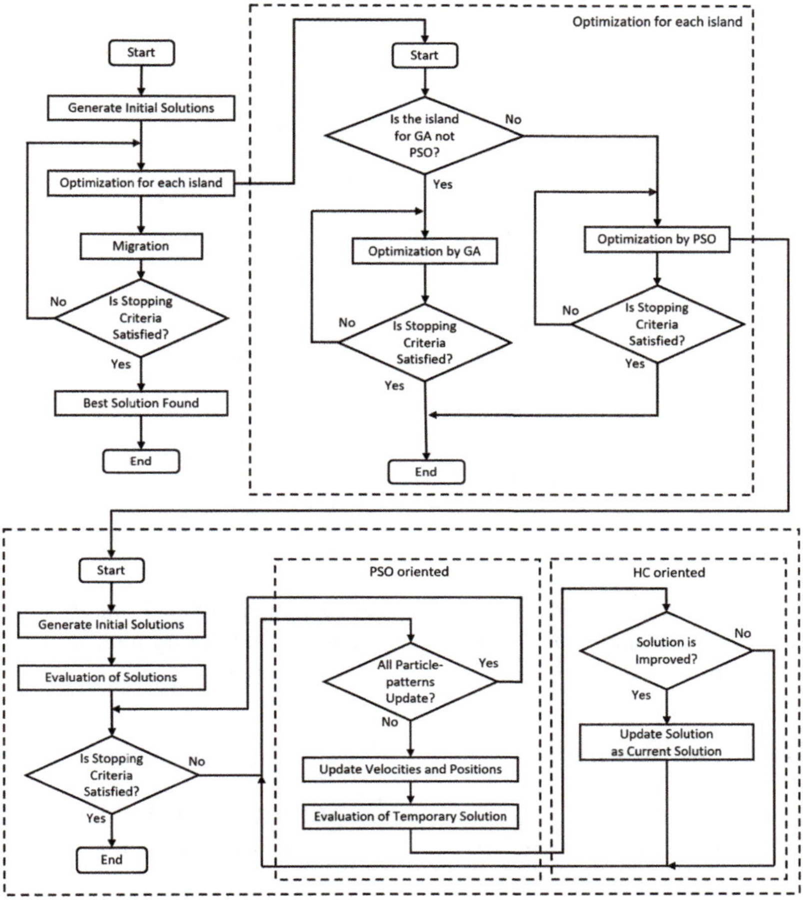

Fig. 2. Flowchart of WMN-PSOHCDGA system.

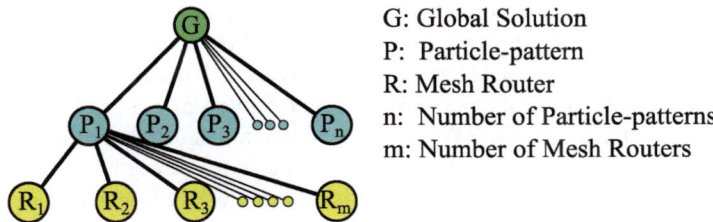

G: Global Solution
P: Particle-pattern
R: Mesh Router
n: Number of Particle-patterns
m: Number of Mesh Routers

Fig. 3. Relationship among global solution, particle-patterns, and mesh routers in PSO part.

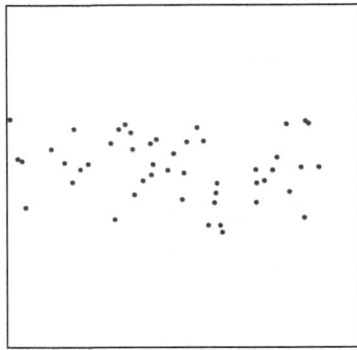

Fig. 4. Subway distribution of mesh clients.

In FC-RDVM [11], V_{max} is the maximum velocity of particles, which is decreased with increasing of iterations as shown in Eq. (1).

$$V_{max}(k) = \sqrt{W^2 + H^2} \times \frac{T - k}{T + \delta k} \tag{1}$$

where W and H are width and height of the considered area, while T and k are the total number and current number of iterations, respectively. The k is varying from 1 to T and δ is the curvature parameter.

4   Simulation Results

In this work, we present the simulation results of RIWM and FC-RDVM. The fitness function coefficients are set as $\alpha = 0.1$, $\beta = 0.8$, $\gamma = 0.1$ and other parameters used for simulations are shown in Table 1.

The visualization results after optimization for small scale WMN and middle scale WMN are shown in Fig. 5 and Fig. 6, respectively. In Fig. 5, we can see that for both RIWM and FC-RDVM router replacement methods, all mesh routers are connected so the SGC is maximized. However, for FC-RDVM there is a concentration of mesh routers on the right side area. So, we need less mesh routers to cover all mesh clients for FC-RDVM. We will deal with this issue in our future work.

In Fig. 7 and Fig. 8 are shown the NCMC of RIWM and FC-RDVM methods for small scale WMN and middle scale WMN, respectively. For small scale WMN, all mesh clients are covered for both methods. However, for middle scale WMN, FC-RDVM cover all mesh clients, while for RIWM one mesh client is not covered.

In Fig. 9 and Fig. 10 is shown the relation of standard deviation with the number of updates of RIWM and FC-RDVM methods for small scale WMN and middle scale WMN, respectively. These figures shows the data, regression line and r, which is the

Table 1. Simulation parameters.

Parameters	Values	
	Small Scale WMN	Middle Scale WMN
$\alpha : \beta : \gamma$	1 : 8 : 1	
Number of GA Islands	16	
Evolution Steps	9	
Number of Migrations	200	
Number of Mesh Routers	16	32
Number of Mesh Clients	48	96
Mesh Client Distribution	Subway Distribution	
Selection Method	Rulette Selection Method	
Corssover Method	SPX	
Mutation Method	Uniform Mutation	

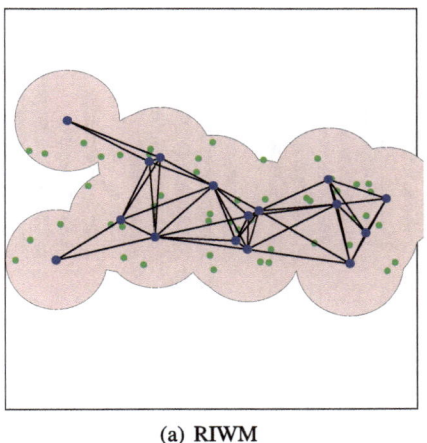 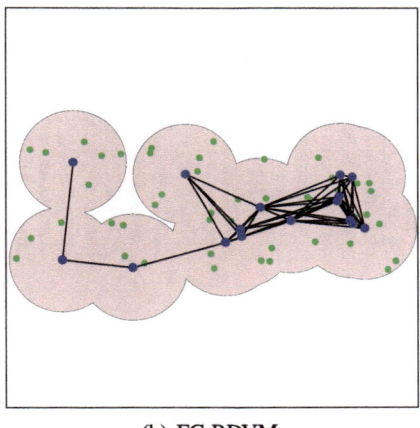

(a) RIWM

(b) FC-RDVM

Fig. 5. Visualization results after optimization (Small Scale WMN).

correlation coefficient. When the standard deviation is a decreasing line, the load balancing among routers is better. In Fig. 9, the load balancing of RIWM is slighty better than FC-RDVM. However, in Fig. 10, we can see that the standard deviation for RIWM is an increasing line, while FC-RDVM is a decreasing line. Thus, the FC-RDVM performs better than RIWM for load balancing and middle scale WMN.

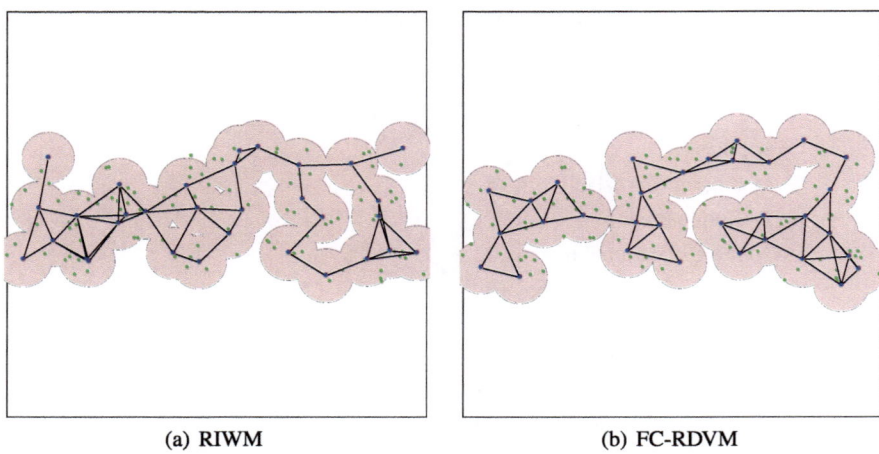

(a) RIWM   (b) FC-RDVM

Fig. 6. Visualization results after optimization (Middle Scale WMN).

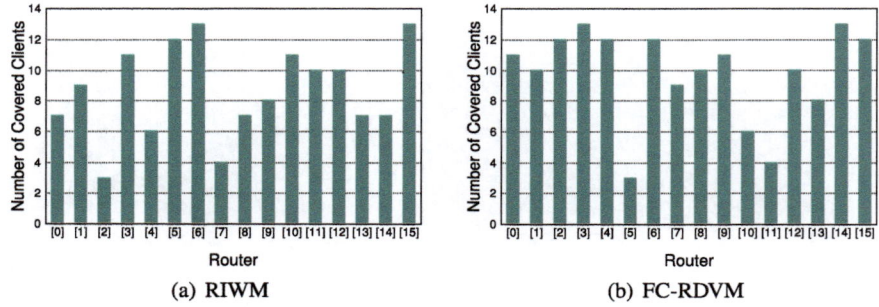

(a) RIWM   (b) FC-RDVM

Fig. 7. Number of covered mesh clients (Small Scale WMN).

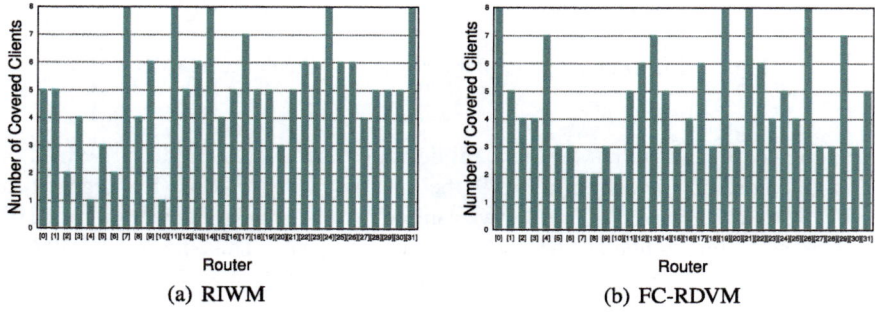

(a) RIWM   (b) FC-RDVM

Fig. 8. Number of covered mesh clients (Middle Scale WMN).

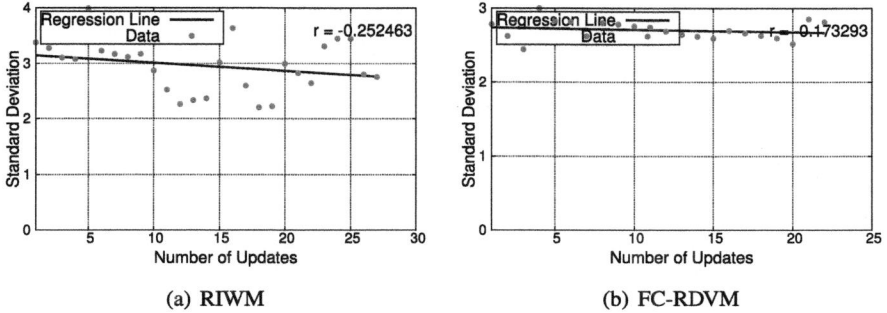

(a) RIWM   (b) FC-RDVM

Fig. 9. Standard deviation, regression line and correlation coefficient (Small Scale WMN).

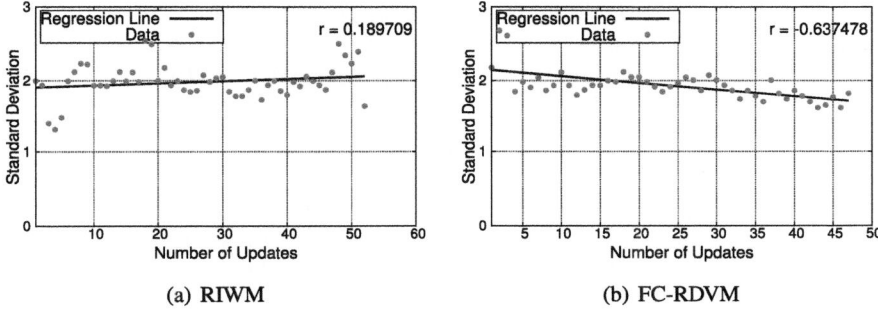

(a) RIWM   (b) FC-RDVM

Fig. 10. Standard deviation, regression line and correlation coefficient (Middle Scale WMN).

5   Conclusions

In this work, we presented the implementation of WMN-PSOHCDGA system and carried out a comparison study between RIWM and FC-RDVM methods for WMNs. For simulations, we considered Subway distribution of mesh clients and different scales of WMNs. By simulation results, we found that both methods achieve high connectivity, but they don not have the same performance for client coverage and load balancing, with FC-RDVM outperforming RIWM for load balancing and middle scale WMNs.

In future work, we will consider less number of mesh routers and carry out simulations for small scale WMNs when using FC-RDVM.

References

1. Barolli, A., Sakamoto, S., Ozera, K., Barolli, L., Kulla, E., Takizawa, M.: Design and implementation of a hybrid intelligent system based on particle swarm optimization and distributed genetic algorithm. In: Barolli, L., Xhafa, F., Javaid, N., Spaho, E., Kolici, V. (eds.) EIDWT 2018. LNDECT, vol. 17, pp. 79–93. Springer, Cham (2018). https://doi.org/10.1007/978-3-319-75928-9_7

2. Franklin, A.A., Murthy, C.S.R.: Node placement algorithm for deployment of two-tier wireless mesh networks. In: Proceedings of the Global Telecommunications Conference, pp. 4823–4827 (2007)
3. Girgis, M.R., Mahmoud, T.M., Abdullatif, B.A., Rabie, A.M.: Solving the wireless mesh network design problem using genetic algorithm and simulated annealing optimization methods. Int. J. Comput. Appl. **96**(11), 1–10 (2014)
4. Lim, A., Rodrigues, B., Wang, F., Xu, Z.: k-center problems with minimum coverage. Theor. Comput. Sci. **332**(1–3), 1–17 (2005)
5. Maolin, T., et al.: Gateways placement in backbone wireless mesh networks. Int. J. Commun. Netw. Syst. Sci. **2**(1), 44–50 (2009)
6. Muthaiah, S.N., Rosenberg, C.P.: Single gateway placement in wireless mesh networks. In: Proceedings of the 8th International IEEE Symposium on Computer Networks, pp. 4754–4759 (2008)
7. Naka, S., Genji, T., Yura, T., Fukuyama, Y.: A hybrid particle swarm optimization for distribution state estimation. IEEE Trans. Power Syst. **18**(1), 60–68 (2003)
8. Sakamoto, S., Kulla, E., Oda, T., Ikeda, M., Barolli, L., Xhafa, F.: A comparison study of simulated annealing and genetic algorithm for node placement problem in wireless mesh networks. J. Mob. Multimedia **9**(1–2), 101–110 (2013)
9. Sakamoto, S., Kulla, E., Oda, T., Ikeda, M., Barolli, L., Xhafa, F.: A comparison study of hill climbing, simulated annealing and genetic algorithm for node placement problem in WMNs. J. High Speed Netw. **20**(1), 55–66 (2014)
10. Sakamoto, S., Oda, T., Ikeda, M., Barolli, L., Xhafa, F.: Implementation and evaluation of a simulation system based on particle swarm optimisation for node placement problem in wireless mesh networks. Int. J. Commun. Netw. Distrib. Syst. **17**(1), 1–13 (2016)
11. Sakamoto, S., Barolli, A., Liu, Y., Kulla, E., Barolli, L., Takizawa, M.: A fast convergence RDVM for router placement in WMNs: performance comparison of FC-RDVM with RDVM by WMN-PSOHC hybrid intelligent system. In: Barolli, L. (ed.) CISIS 2022. LNNS, vol. 497, pp. 17–25. Springer, Cham (2022). https://doi.org/10.1007/978-3-031-08812-4_3
12. Shi, Y.: Particle swarm optimization. IEEE Connect. **2**(1), 8–13 (2004)
13. Vanhatupa, T., Hannikainen, M., Hamalainen, T.: Genetic algorithm to optimize node placement and configuration for WLAN planning. In: Proceedings of the 4th IEEE International Symposium on Wireless Communication Systems, pp. 612–616 (2007)
14. Wang, J., Xie, B., Cai, K., Agrawal, DP.: Efficient mesh router placement in wireless mesh networks. In: Proceedings of the IEEE International Conference on Mobile Adhoc and Sensor Systems (MASS-2007), pp. 1–9 (2007)
15. Xhafa, F., Sanchez, C., Barolli, L.: Ad hoc and neighborhood search methods for placement of mesh routers in wireless mesh networks. In: Proceedings of 29th IEEE International Conference on Distributed Computing Systems Workshops (ICDCS-2009), pp. 400–405 (2009)

A Cuckoo Search Based Simulation System for Node Placement Problem in Wireless Mesh Networks

Kaho Asakura and Shinji Sakamoto[✉]

Department of Information and Computer Science, Kanazawa Institute of Technology,
7-1 Ohgigaoka, Nonoichi, Ishikawa 921-8501, Japan
b1906534@planet.kanazawa-it.ac.jp, shinji.sakamoto@ieee.org

Abstract. Wireless mesh networks (WMNs) are a popular technology due to their numerous advantages, but they have some issues that are related with wireless communication. Node placement is a critical factor in addressing these issues, but optimizing the placement of mesh routers is a complex problem that falls under the category of NP-hard problems. To tackle this problem, we propose a system for node placement problem in WMNs based on a metaheuristic algorithm called Cuckoo Search (CS). We call this system WMN-CS. We evaluated the proposed system by computer simulations considering Normal Distribution and Uniform distribution of mesh clients. The simulation results show that the proposed system can achieve good results for Normal distribution of mesh clients. For Uniform distribution, we need to add more mesh routers to cover all mesh clients. But this increases the cost of WMN.

1 Introduction

Wireless Mesh Networks (WMNs) are a cost-effective, easy-to-deploy technology that offers high robustness. However, the performance of WMNs can be severely impacted by the wrong placement of mesh routers [12], resulting in interference, congestion, low throughput, high packet loss, and long delays.

Mesh routers must be located effectively to overcome the challenges associated with the placement of WMN nodes. However, WMNs nodes placement optimization is known to be NP-hard [2], which means that computing optimal solutions with a polynomial-time algorithm becomes intractable for complex real-size instances [3,4,7]. As a result, approximate optimization algorithms such as heuristic, meta-heuristic, and hybrid algorithms have been proposed to obtain good solutions.

In this work, we focus on the node placement problem in WMNs. To solve this problem, we propose a system based on Cuckoo Search (CS) called WMN-CS. We evaluated the proposed system by computer simulations considering Normal Distribution and Uniform distribution of mesh clients. The simulation results show that the proposed system can achieve good results for Normal distribution of mesh clients. For Uniform distribution, we need to add more mesh routers to cover all mesh clients. But this increases the cost of WMN.

© The Author(s), under exclusive license to Springer Nature Switzerland AG 2023
L. Barolli (Ed.): CISIS 2023, LNDECT 176, pp. 179–187, 2023.
https://doi.org/10.1007/978-3-031-35734-3_18

The rest of this paper is organised as follows. In Sect. 2, we brifly introduce the related works. The mesh router nodes placement problem is defined in Sect. 3. In Sect. 4, we present CS algorithm. The proposed simulation system is discussed in Sect. 5. Simulation results are presented in Sect. 6, followed by our conclusions and suggestions for future work in Sect. 7.

2   Related Works

Many researchers have investigated to solve the node placement problem, resource allocation, node selection, and routing problems which are essential for the future of WMNs [1, 14]. There rae many solutions to solve these problems, including mixed-integer linear programming (MILP)-based optimization models and heuristic-based methods such as Hill Climbing (HC), Simulated Annealing (SA), Genetic Algorithm (GA), and Particle Swarm Optimization (PSO) [5, 6, 8–11, 13]. Heuristic-based methods are commonly used to solve NP-hard problems and are more time-efficient than MILP-based methods [15].

Many proposed solutions for WMNs are evaluated based on simulation results. Heuristic-based methods are promising approaches for solving the node placement problem in WMNs and other related issues [16].

3   Node Placement Problem in WMNs

For this problem, we consider a continuous area where we aim to distribute a fixed number of mesh router nodes and mesh client nodes that are located at arbitrary positions. The objective is to find the optimal location assignment for mesh routers within the area to maximize network connectivity and client coverage. The network connectivity is measured using the Size of Giant Component (SGC) and the user coverage is defined as the Number of Covered Mesh Client (NCMC) that fall within the radio coverage of at least one mesh router node and is measured using the NCMC metric. It should be noted that network connectivity and user coverage directly affect the network's performance, so SGC and NCMC are the essential metrics in WMNs.

To formalize the problem, we use an adjacency matrix of the WMN graph. Where the nodes correspond to the router and client nodes. Also, the edges represent the links between nodes in the mesh network. Each mesh node in the graph is represented by a triple $v = <x, y, r>$, which denotes the 2D location point and the radius of the transmission range. An arc exists between two nodes u and v if v is within the circular transmission range of u.

4   Cuckoo Search

4.1   Summary of Cuckoo Search

The CS is one of the nature-inspired search algorithms, which considers the brood parasitism of some cuckoo species.

Algorithm 1. Pseudo Code of CS Algorithm.

1: Initialize Parameters:
2:     Computation Time $t = 0$ and T_{max}
3:     Number of Nests $n(n > 0)$
4:     Host Bird Recognition Rate $p_a(0 < p_a < 1)$
5:     Lévy Distribution Scale Parameter $\gamma(\gamma > 0)$
6: Fitness Function to get Fitness Value as f
7: Generate Initial n Solutions S_0
8: **while** $t < T_{max}$ **do**
9:     **while** $i < n$ **do**
10:         $j := i \% \text{len}(S_t)$ // j is the remainder of dividing i by number of solutions.
11:         Generate a new solution S_{t+1}^i from S_t^j by Lévy Flights.
12:         **if** $(f(S_{t+1}^i) < f(S_t^j))$ and $(\text{rand}() < p_a)$ **then**
13:             Discard Solution S_{t+1}^i
14:         **end if**
15:         $i = i + 1$
16:     **end while**
17:     $t = t + 1$
18: **end while**
19: **return** Best solution

To apply CS as an optimization tool, there are three rules [17]:

1. Each cuckoo can lay one egg at a time and randomly place it in a chosen nest.
2. The nests with high-quality eggs are considered the best and are carried over to the next generations.
3. The number of available host nests remains fixed and the host bird recognizes the egg laid by a cuckoo with a probability of $p_a \in (0, 1)$. If the host bird recognizes the egg, it can either get rid of it or abandon the nest and build a new one elsewhere.

We show the pseudo code of CS in Algorithm 1. In CS algorithm, there are three hyper parameters: number of nests, host bird recognition rate and scale parameter of Lévy distribution.

The CS algorithm begins with initializing the parameters. Then, the fitness of each nest is evaluated and the best nest among all nests is identified. The algorithm continues until a stopping criterion ($t \geq T_{max}$) is met.

For each iteration, a new solution is generated by Lévy flight, which is a random walk that generates steps with a heavy-tailed distribution. The fitness of the new solution is also evaluated, and if it is better than the current nest, the current nest is replaced with the new solution. In addition, a nest is randomly selected, and if it is worse than the current nest, it is abandoned. The best nest found so far is updated after each iteration.

After all nests have been evaluated, the best nests are kept and the others are discarded. New solutions are generated to replace the discarded ones. The fitness of each nest is evaluated again and the best nest among all nests is identified. This process is repeated until the stopping criterion is met.

4.2 Lévy Flight

In the CS algorithm, the Lévy flight generates new solutions through a stochastic process in the solution space. The Lévy flight, which is based on the Lévy distribution, is known for its long-tailed nature, enabling it to traverse long distances occasionally. This feature allows CS to explore the search space and potentially discover superior solutions compared to other search algorithms.

The Lévy flight is a type of random walk characterized by some unique features. It is a movement model that primarily covers short distances but occasionally takes long leaps. The Lévy flight is found in nature. For example, honeybees use it to search for flower gardens. They move short distances within a garden and then take long leaps to find another one. Humans also exhibit the Lévy flight behaviour.

The Lévy distribution is a continuous probability distribution widely used in probability theory and statistics to model non-negative random variables. It is a specific case of the inverse-gamma distribution and is classified as stable.

The probability density function of Lévy distribution is:

$$P(x; \mu, \gamma) = \sqrt{\frac{\gamma}{2\pi}} \frac{e^{-\gamma/2(x-\mu)}}{(x-\mu)^{3/2}}, \tag{1}$$

where μ is a local parameter ($\mu \leq x$) and γ is a scale parameter.

The cumulative distribution function of Lévy distribution is:

$$F(x; \mu, \gamma) = \text{erfc}\left(\sqrt{\frac{\gamma}{2(x-\mu)}}\right), \tag{2}$$

where erfc is the complementary error function.

We can generate values which follows Lévy distribution, by using the inverse transformation method.

$$F^{-1}(x; \mu, \gamma) = \frac{\gamma}{2(\text{erfc}^{-1}(x))^2} + \mu \tag{3}$$

5   WMN-CS Simulation System

We show the flow-chart of WMN-CS in Fig. 1. In following, we present the initialization, nests and eggs and fitness function.

WMN-CS generates the initial solution randomly, where solutions are considered nests and selected solutions are considered eggs. When the cuckoo discovers a good nest to lay its egg, it replaces the solution routers using a value that follows the Lévy distribution. This process imitates the cuckoo's behaviour, which follows a Lévy flight to find better nests to lay the eggs.

WMN-CS then evaluates the new solutions quantitatively using the fitness function, which compares the nests found by cuckoo search. The cuckoo must decide whether to lay an egg, but sometimes the host bird recognizes that the egg has been swapped, resulting in the worse solution being discatded. This process continues until a termination condition is satisfied. The next generation then searches for new solutions.

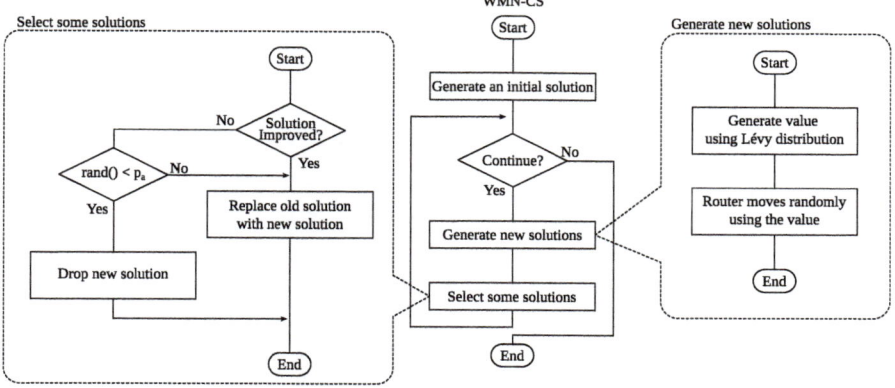

Fig. 1. WMN-CS flowchart.

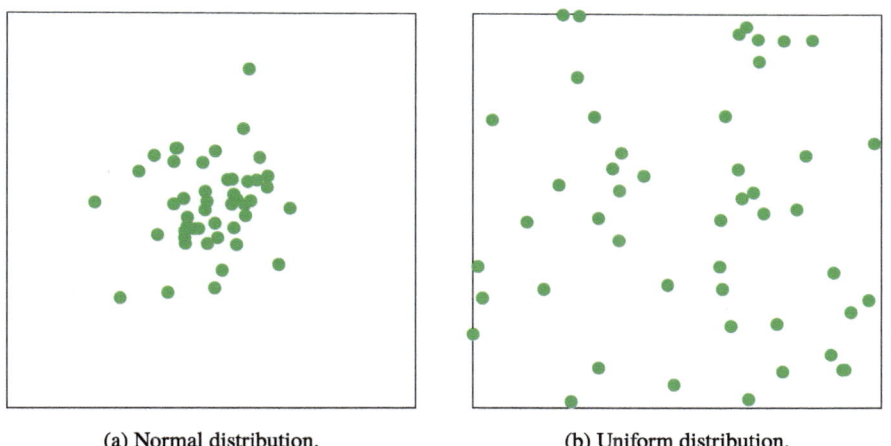

(a) Normal distribution. (b) Uniform distribution.

Fig. 2. Examples of clients distributions.

We define the fitness function as:

$$\text{Fitness} = \alpha \times \text{SGC}(x, y, r) + \beta \times \text{NCMC}(x, y, r).$$

The fitness function in WMN-CS is a combination of two metrics: the Size of the Giant Component (SGC) and the Number of Covered Mesh Clients (NCMC). Both of which are calculated from WMN graph using vectors x, y, and r representing the positions and communication distances of mesh nodes. The SGC measures the connectivity of mesh routers, while NCMC measures the user coverage. The fitness function includes weight coefficients α and β to balance the contributions of SGC and NCMC, respectively.

WMN-CS can generate different mesh clients distributions. In this paper, we consider Normal and Uniform distribution of mesh clients as shown in Fig. 2. The Normal distribution is a distribution where mash clients tends to be distributed in the center place in of the considered area. While, in Uniform distribution, mesh clients are spread out in the considered area. Thus, it is more difficult to cover all mesh clients compared with Normal distribution.

6   Simulation Results

In this section, we show simulation results using WMN-CS. The area size is considered 32×32. The simulation parameters for WMN-CS are shown in Table 1. We conducted simulations 10 times.

We present the simulation results in Fig. 3 and Fig. 4. Figure 3(a) and Fig. 3(b) shows that for both metrics the values are maximized. This shows that all mesh routers are connected and all mesh clients are covered as shown in Fig. 3(c).

For Uniform distribution of mesh clients, the SGC is 100% as shown in Fig. 4(a). However, the is about 50% as shown in Fig. 4(b). This is because the number of mesh routers is not enough to cover all mesh clients as shown in Fig. 4(c).

From the simulation results, we conclude that the proposed system achieved good results for Normal distribution of mesh clients. For Uniform distribution, we need to add more mesh routers to cover all mesh clients. But this increases the cost of WMN.

Table 1. Parameter settings.

Parameters	Values
Clients Distribution	Normal, Uniform
Area Size	32×32
Number of Mesh Routers	16
Number of Mesh Clients	48
Total Iterations	2000
Iteration per Phase	10
Number of Nests	50
Radius of a Mesh Router	From 2.0 to 3.0
Fitness Function Weight-coefficients (α, β)	0.7, 0.3
Scale Parameter (γ)	0.1
Host Bird Recognition Rate (p_a)	0.05

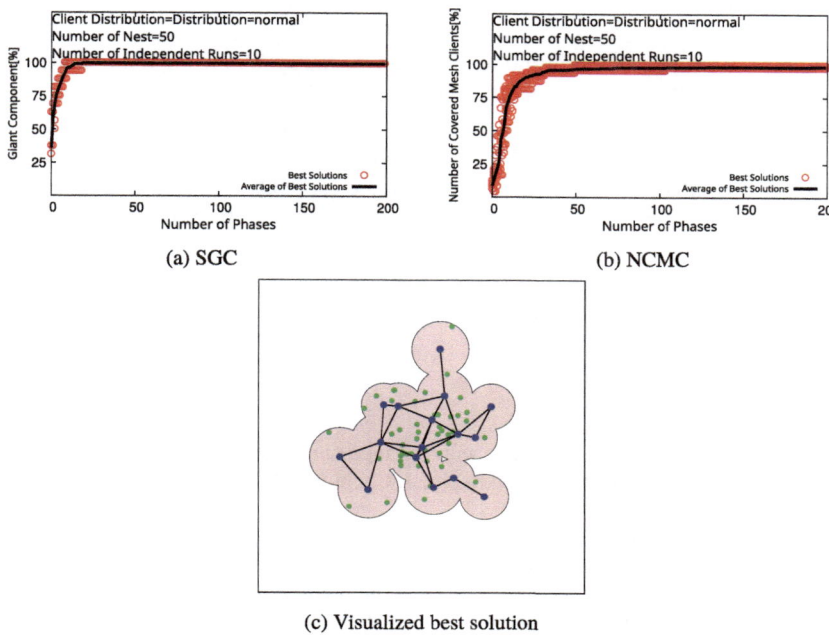

(a) SGC

(b) NCMC

(c) Visualized best solution

Fig. 3. Simulation results of WMN-CS for Normal distribution of mesh clients.

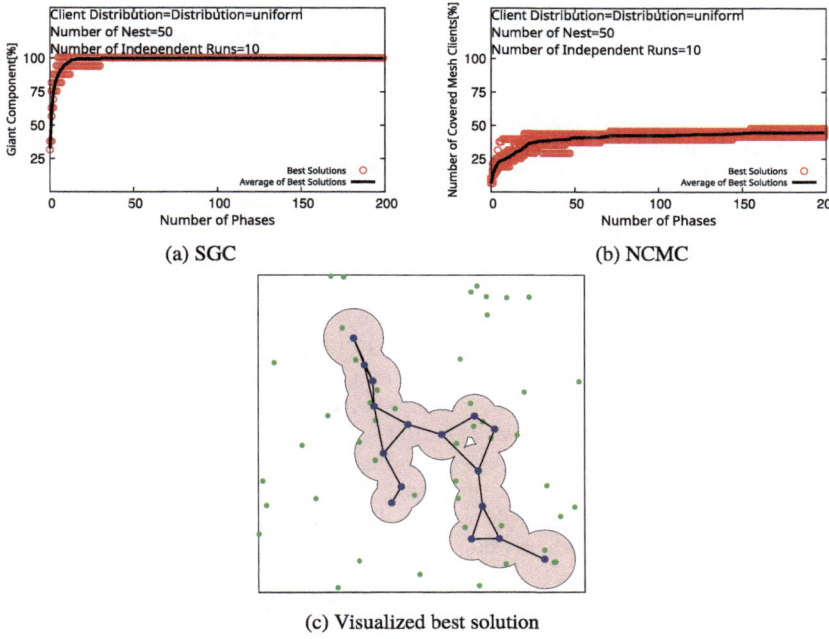

(a) SGC

(b) NCMC

(c) Visualized best solution

Fig. 4. Simulation results of WMN-CS for Uniform distribution of mesh clients.

7 Conclusions

In this work, we proposed and implemented a CS based system to solve the node placement problem in WMNs. We called the system WMN-CS. We conducted simulations to evaluate the performance of WMN-CS considering Normal distribution and Uniform distribution of mesh clients. The simulation results show that the proposed system achieved good results for Normal distribution of mesh clients. For Uniform distribution, we need to add more mesh routers to cover all mesh clients. But this increases the cost of WMN.

In our future work, we would like to evaluate the performance of the proposed system for different parameters and scenarios.

References

1. Ahmed, A.M., Hashim, A.H.A.: Metaheuristic approaches for gateway placement optimization in wireless mesh networks: a survey. Int. J. Comput. Sci. Netw. Secur. (IJCSNS) **14**(12), 1 (2014)
2. Amaldi, E., Capone, A., Cesana, M., Filippini, I., Malucelli, F.: Optimization models and methods for planning wireless mesh networks. Comput. Netw. **52**(11), 2159–2171 (2008)
3. Azzoug, Y., Boukra, A.: Bio-inspired VANET routing optimization: an overview: a taxonomy of notable VANET routing problems, overview, advancement state, and future perspective under the bio-inspired optimization approaches. Artif. Intell. Rev. **54**, 1005–1062 (2021)
4. Basirati, M., Akbari Jokar, M.R., Hassannayebi, E.: Bi-objective optimization approaches to many-to-many hub location routing with distance balancing and hard time window. Neural Comput. Appl. **32**, 13267–13288 (2020)
5. Coelho, P.H.G., do Amaral, J.F., Guimaraes, K., Bentes, M.C.: Layout of routers in mesh networks with evolutionary techniques. In: The 21st International Conference on Enterprise Information System (ICEIS-2019), pp. 438–445 (2019)
6. Elmazi, D., Oda, T., Sakamoto, S., Spaho, E., Barolli, L., Xhafa, F.: Friedman test for analysing WMNs: a comparison study for genetic algorithms and simulated annealing. In: 2015 9th International Conference on Innovative Mobile and Internet Services in Ubiquitous Computing, pp. 171–178. IEEE (2015)
7. Gharehchopogh, F.S., Shayanfar, H., Gholizadeh, H.: A comprehensive survey on symbiotic organisms search algorithms. Artif. Intell. Rev. **53**, 2265–2312 (2020)
8. Hirata, A., Oda, T., Saito, N., Hirota, M., Katayama, K.: A coverage construction method based hill climbing approach for mesh router placement optimization. In: Barolli, L., Takizawa, M., Enokido, T., Chen, H.-C., Matsuo, K. (eds.) BWCCA 2020. LNNS, vol. 159, pp. 355–364. Springer, Cham (2021). https://doi.org/10.1007/978-3-030-61108-8_35
9. Lee, S.C., Tan, S.W., Wong, E., Lee, K.L., Lim, C.: Survivability evaluation of optimum network node placement in a hybrid fiber-wireless access network. In: IEEE Photonic Society 24th Annual Meeting, IEEE, pp. 298–299 (2011)
10. Lin, C.C.: Dynamic router node placement in wireless mesh networks: a PSO approach with constriction coefficient and its convergence analysis. Inf. Sci. **232**, 294–308 (2013)
11. Oda, T., Elmazi, D., Barolli, A., Sakamoto, S., Barolli, L., Xhafa, F.: A genetic algorithm-based system for wireless mesh networks: analysis of system data considering different routing protocols and architectures. Soft. Comput. **20**, 2627–2640 (2016)
12. Qiu, L., Bahl, P., Rao, A., Zhou, L.: Troubleshooting wireless mesh networks. ACM SIGCOMM Comput. Commun. Rev. **36**(5), 17–28 (2006)

13. Sakamoto, S., Oda, T., Ikeda, M., Barolli, L.: Design and implementation of a simulation system based on particle swarm optimization for node placement problem in wireless mesh networks. In: 2015 International Conference on Intelligent Networking and Collaborative Systems, pp. 164–168. IEEE (2015)

14. Sanni, M.L., Hashim, A.H.A., Anwar, F., Naji, A.W., Ahmed, G.S.: Gateway placement optimisation problem for mobile multicast design in wireless mesh networks. In: 2012 International Conference on Computer and Communication Engineering (ICCCE), pp. 446–451. IEEE (2012)

15. Seetha, S., Anand John Francis, S., Grace Mary Kanaga, E.: Optimal placement techniques of mesh router nodes in wireless mesh networks. In: Haldorai, A., Ramu, A., Mohanram, S., Chen, M.-Y. (eds.) 2nd EAI International Conference on Big Data Innovation for Sustainable Cognitive Computing. EICC, pp. 217–226. Springer, Cham (2021). https://doi.org/10.1007/978-3-030-47560-4_17

16. Taleb, S.M., Meraihi, Y., Gabis, A.B., Mirjalili, S., Ramdane-Cherif, A.: Nodes placement in wireless mesh networks using optimization approaches: a survey. Neural Comput. Appl. **34**(7), 5283–5319 (2022)

17. Yang, X.S.: Nature-Inspired Metaheuristic Algorithms. Luniver Press, UK (2010)

A Lightweight Botnet Exploiting HTTP for Control Flow Denial on Open-Source Medical Systems

Wei Lu[✉]

Department of Computer Science, Keene State College, The University System of New Hampshire, Keene, NH 03431, USA
wlu@usnh.edu

Abstract. The recent emergence of open-source medical cyber-physical systems has rapidly transformed the healthcare industry. This can be attributed to advancements in 3D printing technology and the growing popularity of open-source microcomputer systems like Arduino and Raspberry Pi. However, the increased use of these systems in hospitals has also raised cybersecurity concerns. In particular, new technologies, such as IoT devices and other mobile devices, have posed new challenges in exploiting modern botnets and determining their effectiveness with limited resources. In this paper, we propose a lightweight and full-encrypted cross-platform botnet system that provides a proof-of-concept demonstration of how a botnet attack can block control flow from the syringe pump in a testbed of an IoT medical network. The emphasis is placed on minimal deployment time and resource usage, making this lightweight botnet different from most traditional botnets, thus furthering cybersecurity research in intrusion detection for open-source medical systems.

1 Introduction

The IoMT (Internet of Medical Things) refers to using connected devices and technologies in the healthcare field to collect, transmit, and analyze medical data for various purposes, such as patient monitoring, diagnostics, treatment, and health management. Medical devices in a typical IoMT range in size from tiny implantable medical devices to massive objects such as MRI scanners, and open-source medical devices and clinic laboratory instruments using desktop 3D printers and open-source electronic microcomputer systems, including such fluorescence imaging devices [1], micro-dispensers [2] and syringe pumps [3, 4].

The IoMT has facilitated the development of innovative healthcare applications but has also given rise to new security and privacy concerns that could impede its progress. For example, in 2011, there were reports of malicious attacks on insulin pumps [5], while in 2018, Halperin et al. revealed the potential for wireless attacks on FDA-approved Implantable Cardiac Defibrillator (ICD) devices [6].

Although there are many good ideas for security mechanisms in the medical device domain, they still need to develop fully. Most existing security solutions focus on prevention, which employs authentication, encryption, and trust-based security management to

© The Author(s), under exclusive license to Springer Nature Switzerland AG 2023
L. Barolli (Ed.): CISIS 2023, LNDECT 176, pp. 188–199, 2023.
https://doi.org/10.1007/978-3-031-35734-3_19

protect commercial wearable, implantable, and portal medical devices [7, 8]. However, open-source medical devices are being overlooked. In addition, poorly implemented security mechanisms make it easy for potential attackers to gain remote control of smart medical devices using malware or botnets [9, 10]. They can then manipulate sensitive data by injecting false health data or cause malfunctions by flooding the IoMT network with many illegitimate requests.

This paper aims to address the security challenges posed by open-source medical devices. To this end, we manufacture an open-source medical syringe pump prototype using simple 3D printed hardware parts, a Raspberry Pi system, an Arduino microcomputer, and an open-source software program. We then propose and develop a lightweight and full-encrypted cross-platform botnet system that provides a proof-of-concept demonstration of how a botnet attack can block the control flow command sent from the Raspberry Pi to the syringe pump. This botnet can completely block the communication between the microcomputer system and the mechanical pump. It can also cause the pump to dispense an unexpectedly large amount of fluid after the attack is terminated, creating a potential risk of overdose when used at the bedside. Such a zero-day botnet attack may disturb the network traffic pattern of the connected medical devices. Thus, it helps further the cybersecurity research in network traffic analysis and intrusion detection/prevention in the domain where network flows collected from this botnet are publicly available for feature engineering and adversarial machine learning.

The remainder of this paper is structured as follows. Section 2 discusses the concept of botnet attacks. Section 3 describes the prototype of the open-source medical syringe pump, including a step-by-step guide to its manufacturing process. In Sect. 4, we introduce the centralized botnet system that is based on secure HTTP protocol and is fully encrypted. Section 5 presents the installation process of this botnet system; then, the Distributed Denial of Service (DDoS) attacks using this botnet system are conducted against the syringe pump in a controlled testbed IoT network where relevant network traffic using packet capture tools is captured. Finally, in Sect. 6, we offer concluding remarks and discuss future work.

2   Concept of Botnet Attacks

A bot is a self-operating software program controlled by a remote operator known as the botmaster for performing malicious activities, often without the knowledge or consent of the victim whose computer it has been installed on. The bot allows the remote operator to take control of the victim's system and instruct it to carry out malicious tasks, including but not limited to mass spamming, distributed denial of service attacks, click fraud, and distributed computing for password cracking or other types of cybercrime.

There are various methods that the bot uses to establish this network structure. Command and control channels must efficiently deliver orders from the botmaster to individual bots while evading detection by security measures [11, 12]. The IRC-based channels are very efficient mainly because of their ease of implementation and the capability to form large networks, thanks to their simple network architecture. However, network traffic monitoring can quickly reveal the messages being exchanged between the server and individual clients, making detecting botnets based on message content analysis easy. As an alternative, botnets use HTTP traffic for command and control (CC) schemes, as it can provide stealth by using a legitimate communication channel and evading traditional firewall-based security. To avoid detection based on deep packet analysis, packets are frequently encrypted. The communication channel between bots and botmasters can be protected and kept from being identified using robust encryption methods.

According to Feily et al. [13], there are five phases in the life cycle of a botnet. The first phase is called an initial infection; it involves an attacker exploiting a known vulnerability in a target system to infect it with malware, providing the attacker with additional capabilities on the victim's machine. A malicious binary, called secondary injection, will be fetched during the second phase by executing additional scripts or programs. Once the binary is installed, the victim's computer becomes a bot. Then in the third phase of the connection, the bot attempts to connect to the C&C server using various methods, officially joining the botnet once the connection is established. The final phase is to maintain the bots for updating their binaries to defend against new attacks. Furthermore, a simplified way of categorizing the life cycle of a botnet has been described into four phases: formation, command, and control (C&C), attack, and post-attack [14]. The attack phase is when a bot executes malicious actions in response to orders received from the botmaster, while the post-attack phase is akin to the maintenance phase.

Existing botnet detection techniques mainly focus on detecting bot activity during the attack, initial infection, and secondary injection phases. These techniques often use traditional intrusion detection methods, which identify botnets by analyzing the behavior of underlying malicious activities and comparing them to known signatures of attacks. In our study, we introduce and implement a lightweight centralized botnet attack exploiting secure HTTP protocol. Unlike traditional botnet attacks, such as Sinit [15], Phatbot (which utilized WASTE command) [16], Nugache [17], and Peacomm (Storm worm) [18], the proposed botnet system has a highly secure communication protocol between the bots and the botmaster, making it challenging to detect during the command and control phase.

3   Manufacturing an Open-Source Medical Syringe Pump

This section provides a step-by-step guide on creating an open-source medical syringe pump using simple hardware components, a Raspberry Pi system, an Arduino micro-computer, and open-source software. A 3D printer with Cura software manufactures the pump's physical parts. The Arduino system controls the pump through a CNC shield, while the Raspberry Pi is the control center to send data commands to the Arduino. Additionally, an open-source program monitors the syringe pump's working process when ejecting fluids from the syringe. This study uses a regular Creality Ender 3 printer with a fully open-source resume printing function [19]. Manufacturing one syringe pump set costs approximately $410, cheaper than similar commercial products. Details on the raw materials and their prices can be found in Table 1.

Throughout the manufacturing process, the most demanding aspect of 3D printing is guaranteeing that the filament is correctly positioned and that the bed is leveled. This

Table 1. Raw materials for manufacturing the open-source medical syringe pump.

Materials	Price
Creality CR Touch Auto Bed Leveling Sensor [20]	$39.0
HATCHBOX 1.75 mm Cool Gray PLA 3D Printer Filament [21]	$24.99
uxcell® M5 × 14 mm 316 Stainless Steel Metric Fully Thread Hex Socket Cap [22]	$9.49
SanDisk SDSDQM-016G-B35A 16 GB Memory Card [23]	$6.75
Raspberry Pi 7″ Touch Screen Display [24]	$69.99
5 mm to 5 mm Aluminum Flexible Shaft Coupling [25]	$14.09
CNC Shield V3.0 & Keyestudio R3 Board & Nema 17 Stepper Motor [26]	$36.17
Linear Ball Bearings, Linear Motion Ball Bearing Bushing [27]	$10.62
uxcell 2pcs 6 mm × 200 mm Metal Machine Turning Tool Rod Bar [28]	$7.49
uxcell a16071500ux0127 M5 × 170 mm 304 Stainless Steel Fully Threaded Rod [29]	$11.51
Raspberry Pi 3 Model B + Board [30]	$169.99
uxcell M3 × 10 mm Thread 304 Stainless Steel Hex Socket Head Cap Screw Bolt [31]	$9.99

can require several attempts with trial and error to obtain an accurate print. A heuristic approach to determine if we have achieved a successful print is to observe if the filament adheres to the bed firmly and does not detach easily.

Figure 1 showcases the produced medical pump, which is capable of dispensing fluids from the syringe.

Fig. 1. The syringe pump

The medical pump acquires data from the Arduino and the CNC (Computer Numerical Control) shield, which allows the motors to rotate clockwise or counterclockwise [32]. In addition, the Raspberry Pi system is linked to a touchscreen pad, serving as the control center for the medical pump [33]. All the directives to the Arduino are executed through this interface. The Arduino consists of two components, the lower part contains the Arduino, which receives data from the Raspberry Pi, and the upper part comprises the CNC shield, which assists the Arduino in managing the medical pump's motors.

To create the 3D-printed parts, we utilized Cura as our program of choice [34]. Numerous public tutorials can aid beginners in setting up this program on their 3D printer before using it. We employed this program to generate the 3D prints by obtaining premade 3D printed files [33]. We then drag these files into the program and configure the settings to print the file accurately, which is relatively easy to accomplish. Typically, we used Dynamic Quality with 80% infill, and the prints were highly successful. Next, Arduino programming is applied to upload data to the physical Arduino device. Finally, we can download and install a third-party library called *AccelStepper* to transmit data from the Arduino to the CNC shield. While transferring data to the Arduino, it is crucial to ensure that the data-sharing cord is compatible with the serial communication ports of the device. Afterward, the open-source python controlling program is utilized to commence running the motors.

4   Database Schema, Command Format, and Monitoring

The proposed botnet system in the paper is completely encrypted and serves as a proof-of-concept for further research in network traffic analysis and intrusion detection and prevention by monitoring its network flow behaviors. It comprises three key components: an HTTPS web server, an SSL bot server, and an execution shell. The web server is built on Apache2 with *mod_wsgi*, and is responsible for calling a Python script that utilizes the *Flask* module to provide web services. Meanwhile, the execution shell, also built using Python, utilizes a PostgreSQL database shared with the web service. Finally, the bot server connects to the web service to provide updates on its status and to the execution shell to respond to commands.

The PostgreSQL database utilized by this botnet system is specifically designed for monitoring bot activity. It includes a concise set of tables for storing relevant data such as command logs, bot notifications, and account information for the web service such as IP addresses, port numbers, and command id.

A command language was created for network communication, and the syntax used for these commands slightly differs from the syntax used when entering them through the command shell. This is because the command shell applies automatic transformations to the commands.

The syntax of the commands below represents how they are transmitted between the execution node and the bot server. These commands transmit and execute Python scripts and monitor and identify active bots.

The *put* statement is utilized to carry out simple file transfer operations. First, the unprocessed file data is transmitted with a filename to associate with the file. Upon receiving the command, the bot generates a write handle for the file referenced by the given filename and truncates it according to the associated flag. Subsequently, the file data is written into the file, and the stream is terminated. It is presumed that the file data represents a Python script.

The *execute* command is intended to be executed after a put command. When a *put* command is executed, a file is stored in the execution directory of the bot server and identified by a specific name. The name of the file is determined by the execution node when it issues the *put* command. This file can then be retrieved and executed by name using Python's eval statement.

The *ping* statement confirms that the bot server is still operational. Despite its name, this statement does not generate an ICMP echo request. Instead, it establishes a connection over the same socket for all other bot communication. When the bot receives this command, it dispatches a ping notification to the web server.

It is important to keep track of the number of active bots at any given time. In addition, activity monitoring tools can assist the bot manager in identifying network connectivity problems or software issues. Currently, activity monitoring is carried out on-demand, meaning the execution node must issue a *ping* command. Furthermore, we need to ensure that malicious sources cannot commandeer our bots, so we require a method for our bots to authenticate the commands they receive.

Each bot updates the control server whenever it starts up, performs an activity, or shuts down. For example, an activity is storing a file (in response to a *put* command), executing a file (in response to an *exec* command), or responding to a *ping*. The bot will transmit these notifications to the address specified in *bot.conf*. In addition, each notification includes the port number on which the bot server operates. When the control server receives these notifications, it will update the bot_status table to reflect the appropriate notification time, message, and port.

When a bot receives communication through its listening socket, the first step is verifying the command. To verify the command, MD5 is used to hash the command, which is then transmitted through a GET request to the configured *validate_addr*. The MD5 command hash is sent as this request's 'command' parameter. The web service checks the command_log table when the validated request is received. It computes the MD5 of all commands transmitted within the last 10 s. If the supplied hash is found among them, the value *True* will be transmitted in response to the bot. If not, *False* will be sent instead. This mechanism verifies both the origin of the command (i.e., the execution node) and the timeliness of the command.

5   Experimental Evaluation of Botnet Effectiveness

The prerequisites for our experimental evaluation of the effectiveness of the proposed botnet system include the python program for the web server, execution shell, and bot with several additional modules, including *flask*, *flask-login*, and *psycopg2*. The control database for the backend runs on PostgreSQL. After installing the server process, an account for the "medibot" user must be created by adding the following line *local all medibot md5* to the postgres configuration.

The *medibot* database schema can be created by logging in as a Postgres user, which can be done on UNIX platforms. After the *medibot* user has been created, the database schema can be created using the included schema file called "schema.sql". This is done by executing the command *$ psql schema.sql --username = medibot*.

To install the *medibot* package, we invoke the setup script in the *medibot* directory using the following syntax. The execution shell is a basic Python script that wraps the core functions of the *medibot* package, i.e. *$ python setup.py install.*

After installing the package, we edit the *medibot.conf* file to reflect the database configuration and then test the installation by launching the executor script *$ python executor.py../medibot.conf.*

The web service installation primarily involves installing and configuring Apache2 (i.e., the apache2 package on Ubuntu). In addition to the base Apache installation, the *mod_wsgi* and *mod_ssl* packages must also be installed. OpenSSL is then applied to generate an SSL private key and certificate. Once the SSL private key and certificate have been generated, we copy the contents of the *Medibot-Web* folder to a web folder managed by Apache, such as /var/www/medibot. After this, we modified the Apache configuration file located at */etc./apache2/apache2.conf* on Ubuntu Linux. The values indicated in angle brackets depend on the actions taken in previous steps and the specific system configuration. To create a link to the *medibot.conf* file created during the "execution shell" setup within the web directory, we need to execute the following command from the web directory where the *medibot* services were copied on UNIX:

$ ln < path to medibot.conf > medibot.conf

In addition to the base configuration, the bot code has several prerequisites, including the OpenSSL server and the Python bindings to OpenSSL, *pyopenssl*. The bot packages are installed as part of the *medibot* package installation so that we can run the command *python setup.py install* for the package installation. The next step is to update the bot configuration file to point to the correct web server. Moreover, the botnet system can use either a statically defined port which may or may not be available when the bot server starts up, or a dynamically assigned port which is guaranteed to be available. After completing the configuration settings, we can run it using a command *$./bot server.*

During the experimental testing of the botnet system, we discovered that it could successfully disable the control host of a medical syringe pump. However, as shown in Fig. 2, before the attack, the communication between the syringe pump and the control host was functioning normally, with commands to set fluid parameters being sent and received without issue.

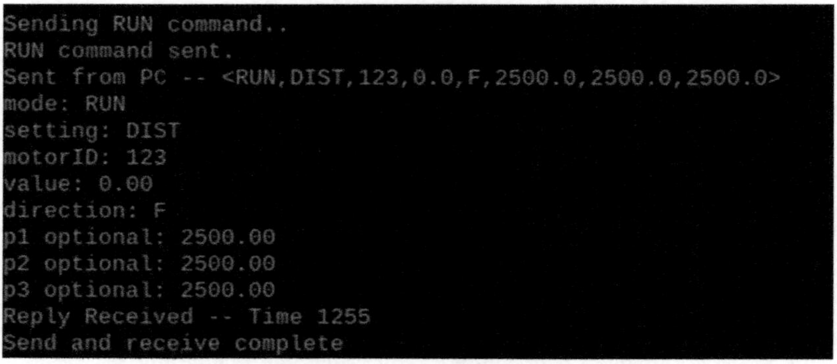

```
Sending RUN command..
RUN command sent.
Sent from PC -- <RUN,DIST,123,0.0,F,2500.0,2500.0,2500.0>
mode: RUN
setting: DIST
motorID: 123
value: 0.00
direction: F
p1 optional: 2500.00
p2 optional: 2500.00
p3 optional: 2500.00
Reply Received -- Time 1255
Send and receive complete
```

Fig. 2. Command sent/received between syringe pump and central control system.

However, after approximately two minutes of the DDoS attack, the central control system's graphical user interface experienced some responsiveness delays. This created a potential safety hazard as a medical operator may accidentally click the "run" button multiple times due to the slow response time, resulting in an overdose of fluid injected into patients. For example, in our simulation, we observed that if the operator clicked the "run" button three times, it would cause three times the amount of fluid to be injected, increasing the dose from 5 mm to 15 mm. In addition, with a continuous DoS attack, the syringe pump completely froze after about five minutes, leading to potential undersupply issues.

The Open Argus network monitoring system [35] collected 211,364 instances, each with 13 features described in Table 2. Table 3 visually represents the descriptive statistics for these 13 features.

Table 2. Feature description.

Feature	Description
SrcBytes/DstBytes	Number of bytes from source to destination (or from destination to source)
SrcLoad/DstLoad	Source to destination bits per second (Destination to source bits per second)
SrcPkts	Number of packets from source to destination
DstPkts	Number of packets from destination to source
SrcRate	Number of packets per second from source to destination
DstRate	Number of packets per second from destination to source
Dur	Transaction record total duration
TotPkts	Total transaction packets count
TotBytes	Total transaction bytes
Load	Total transaction bits per second
Rate	Number of packets per second

Table 3. Descriptive statistics of features for a total of 211,364 data instances.

Feature	Mean	Std	Min	25%	50%	75%	Max
SrcBytes	286.255	2822.596	0.0	71.0	120.0	180.0	435324
SrcPkts	2.439	22.385	0.0	1.0	1.0	3.0	4890
DstPkts	0.938	10.365	0.0	0.0	0.0	1.0	2411
DstBytes	1052.62	57168.03	0.0	0.0	0.0	142.0	12858140
Dur	0.442	0.801	0.0	0.0	0.0	0.598	4.999
TotPkts	3.377	30.879	1.0	1.0	2.0	3.0	5979
SrcLoad	4765462	55782630	0.0	0.0	0.0	683.77	15168000000
DstLoad	12886.2	292332	0.0	0.0	0.0	0.0	74981780
SrcRate	1551.3	15807.7	0.0	0.0	0.0	1.4	4000000
DstRate	1.983	17.58	0.0	0.0	0.0	0.0	1851.57
TotBytes	1338.9	58882.5	54	74	180	243	13230630
Load	4778348	55782350	0.0	0.0	0.0	684.4	15168000000
Rate	2535.2	17395.6	0.0	0.0	1.25	250	4000000

6   Conclusions and Future Work

This paper investigates a denial of control flow attack on an open-source medical syringe pump system. We first create a functioning system prototype using a 3D printer and open-source microcomputer systems like Arduino, Raspberry Pi, and CNC shields to do this. We then explore the vulnerabilities of the open-source software monitoring system that controls the medical syringe pump. Our proof-of-concept approach shows that the pump could dispense too much fluid, not enough fluid, or completely stop injecting fluid into patients in the event of a DoS attack launched by the proposed lightweight centralized botnet system.

The future work for this research primarily involves (1) differentiating such attacks based on the collected network traffic payloads from both malicious and malware-free environments [36]. This can be achieved using feature selection using clustering [37] or co-clustering [38] and advanced machine learning models for transfer learning with a focus on deep transfer learning (DTL) models to enable the detection of such DoS attacks against medical syringe pumps across various IoMT networking environments; and (2) enhance the performance of the botnet system by automating tasks such as database structure creation, SSL key pair generation, and then integrating the web service with the command shell and execution node.

Acknowledgments. This research is supported by New Hampshire – INBRE through an Institutional Development Award (IDeA), P20GM103506, from the National Institute of General Medical Sciences of the NIH.

References

1. Nuñez, I., et al.: Low cost and open source multi-fluorescence imaging system for teaching and research in biology and bioengineering. PLoS ONE **12**(11), e0187163 (2017). https://doi.org/10.1371/journal.pone.0187163.PMID:29140977;PMCID:PMC5687719
2. Forman, C.J., Tomes, H., Mbobo, B., et al.: Openspritzer: an open hardware pressure ejection system for reliably delivering picolitre volumes. Sci. Rep. **7**, 2188 (2017). https://doi.org/10.1038/s41598-017-02301-2
3. Wijnen, B., Hunt, E.J., Anzalone, G.C., Pearce, J.M.: Open-source Syringe Pump Library. Plos One **9**, e107216 (2014). https://doi.org/10.1371/journal.pone.0107216
4. Croatt Group DIY Flow Chemistry Setup – UNC-Greensboro. https://chem.uncg.edu/croatt/flow-chemistry/. Retrieved on 27 Mar 2023
5. Li, C.X., Raghunathan, A., Jha, N.K.: Hijacking an insulin pump: security attacks and defenses for a diabetes therapy system. In: 2011 IEEE 13th International Conference on e-Health Networking, Applications and Services, pp. 150–156 (2011). https://doi.org/10.1109/HEALTH.2011.6026732
6. Halperin, D., et al.: Pacemakers and implantable cardiac defibrillators: software radio attacks and zero-power defenses. In: 2008 IEEE Symposium on Security and Privacy, pp. 129–142 (2008).https://doi.org/10.1109/SP.2008.31
7. Yanambaka, V.P., Mohanty, S.P., Kougianos, E., Puthal, D.: PMsec: physical unclonable function-based robust and lightweight authentication in the internet of medical things. IEEE Trans. Consum. Electron. **65**(3), 388–397 (2019). https://doi.org/10.1109/TCE.2019.2926192
8. Su, J., Vasconcellos, D. V., Prasad, S., Sgandurra, D., Feng, Y., Sakurai, K.: Lightweight classification of IoT malware based on image recognition. In: 2018 IEEE 42nd Annual Computer Software and Applications Conference (COMPSAC), pp. 664–669 (2018). https://doi.org/10.1109/COMPSAC.2018.10315
9. Garant, D., Lu, W.: Mining botnet behaviors on the large-scale web application community. In: Proceedings of 27th IEEE International Conference on Advanced Information Networking and Applications, Barcelona, Spain, 25–28 March 2013
10. Lu, W., Miller, M., Xue, L.: Detecting command and control channel of botnets in cloud. In: Traore, I., Woungang, I., Awad, A. (eds.) ISDDC 2017. LNCS, vol. 10618, pp. 55–62. Springer, Cham (2017). https://doi.org/10.1007/978-3-319-69155-8_4
11. Lu, W.: An Unsupervised Anomaly Detection Framework for Multiple-Connection-Based Network Intrusions. Ottawa Library and Archives Canada (2007). ISBN: 9780494147795
12. Lu, W., Ghorbani, A.: Bots behaviors vs. human behaviors on large-scale communication networks. In: Lippmann, R., Kirda, E., Trachtenberg, A. (eds.) Proceedings of 11th International Symposium on Recent Advances in Intrusion Detection (RAID 2008), RAID 2008, LNCS 5230, pp. 415–416. MIT, Boston, USA (2008)
13. Feily, M., Shahrestani, A., Ramadass, S.: A survey of botnet and botnet detection. In: 2009 Third International Conference on Emerging Security Information, Systems and Technologies, pp. 268–273. Athens, Greece (2009). https://doi.org/10.1109/SECURWARE.2009.48
14. Leonard, J., Xu, S., Sandhu, R.: A framework for understanding botnets. In: 2009 International Conference on Availability, Reliability and Security, pp. 917–922. Fukuoka, Japan (2009). https://doi.org/10.1109/ARES.2009.65
15. Sinit: https://www.f-secure.com/v-descs/sinit.shtml. Retrieved on 27 Mar 2023
16. Phatbot: https://www.fortiguard.com/encyclopedia/ips/103350720. Retrieved on 27 Mar 2023
17. Nugache: https://www.usenix.org/system/files/login/articles/526-stover.pdf. Retrieved on 27 Mar 2023

18. Grizzard, J. B., Sharma, V., Nunnery, C., Kang, B.B., Dagon, D.: Peer-to-peer botnets: Overview and case study. In: Proceedings of the 1st USENIX Workshop on Hot Topics in Understanding Botnets, Cambridge, MA (2007)
19. Creality 3D printer: https://www.amazon.com/Comgrow-Creality-Ender-Aluminum-220x22 0x250mm/dp/B07BR3F9N6. Retrieved on 27 Mar 2023
20. Creality CR Touch Auto Bed Leveling Sensor Kit: https://www.amazon.com/dp/B09 P4YKRTD/ref=cm_sw_r_apan_i_6ZGJ0ATJ3JRJB55EKEPY?_encoding=UTF8&psc=1. Retrieved on 27 Mar 2023
21. HATCHBOX 1.75 mm Cool Gray PLA 3D Printer Filament: https://www.amazon.com/ dp/B015I1CYFE/ref=cm_sw_r_apan_i_N75E3SG3T7T1CPTMQ26C?_encoding=UTF8& psc=1. Retrieved on 27 Mar 2023
22. uxcell® M5 × 14 mm 316 Stainless Steel Metric Fully Thread Hex Socket Cap Screws: https://www.amazon.com/gp/product/B01LJROXK0/ref=ox_sc_saved_title_2? smid=A1THAZDOWP300U&psc=1. Retrieved on 27 Mar 2023
23. SanDisk SDSDQM-016G-B35A 16 GB Class 4 MicroSDHC Memory Card with SD Adapter: https://www.amazon.com/gp/product/B004G605OA/ref=ox_sc_saved_title_7? smid=ABYURLNKK9M7V&psc=1. Retrieved on 27 Mar 2023
24. Raspberry Pi Touch Screen Display: https://www.amazon.com/gp/product/B0153R2A9I/ref= ox_sc_saved_title_6?smid=A6EGA15UEFYEQ&psc=1. Retrieved on 27 Mar 2023
25. 5 mm to 5 mm Aluminum Flexible Shaft Coupling: https://www.amazon.com/gp/pro duct/B01EFFBM4I/ref=ox_sc_saved_title_1?smid=A26373IMBF4DLW&psc=1. Retrieved on 27 Mar 2023
26. 3D Printer Kits CNC Shield V3.0, Keyestudio R3 Board, Nema 17 Stepper Motor, 4PCS A4988 Driver & USB Cable, Heat Sink, Stepper Motor Controller Shield Kit: https://www. amazon.com/Tangxi-Printer-Stepper-Heatsink-Arduino/dp/B07SBDD4HL. Retrieved on 27 Mar 2023
27. Linear Ball Bearings, Linear Motion Ball Bearing Bushing for 3D Printer CNC Parts: https:// www.amazon.com/Linear-Motion-Bearing-Bushing-Printer/dp/B07K71FWMG/ref=dp_prs ubs_1?pd_rd_i=B07K71FWMG&psc=1. Retrieved on 27 Mar 2023
28. uxcell 2pcs 6 mm × 200 mm Metal Machine Turning Tool Rod Bar Lathe Round Stick: https:// www.amazon.com/uxcell-Metal-Machine-Turning-Tools/dp/B0BJ7D7V23. Retrieved on 27 Mar 2023
29. uxcell a16071500ux0127 M5 × 170 mm 304 Stainless Steel Fully Threaded Rod Bar Studs Fasteners. https://www.amazon.com/Uxcell-a16071500ux0127-Stainless-Threaded-Fasteners/dp/B01M4L8JDC. Retrieved on 27 Mar 2023
30. Raspberry Pi 3 Model B+ Board: https://www.amazon.com/ELEMENT-Element14-Raspbe rry-Pi-Motherboard/dp/B07P4LSDYV. Retrieved on 27 Mar 2023
31. uxcell M3 × 10 mm Thread 304 Stainless Steel Hex Socket Head Cap Screw Bolt: https://www.amazon.com/gp/product/B01MFA9YEP/ref=ox_sc_saved_title_1?smid= A1THAZDOWP300U&psc=1. Retrieved on 27 Mar 2023
32. Sina Booeshaghi, A., da Veiga Beltrame, E., Bannon, D., Gehring, J., Pachter, L.: Principles of open source bioinstrumentation applied to the poseidon syringe pump system. Sci. Rep. **9**, 12385 (2019). https://doi.org/10.1038/s41598-019-48815-9
33. poseidon: Open source bioinstrumentation. https://github.com/pachterlab/poseidon. Retrieved on 27 Mar 2023
34. Ultimaker Cura: https://ultimaker.com/software/ultimaker-cura. Retrieved on 27 Mar 2023
35. Argus ra 3.0.8: https://qosient.com/argus/man/man1/ra.1.pdf. Retrieved on 27 Mar 2023
36. Ghorbani, A., Lu, W., Tavallaee, M.: Detection Approaches, Network Intrusion Detection and Prevention: Concepts and Techniques, pp. 27–53. Springer Publisher (2009). ISBN-10: 0387887709

37. Lu, W., Traore, I.: A new evolutionary algorithm for determining the optimal number of clusters. In: Proceedings of IEEE International Conference on Computational Intelligence for Modeling, Control and Automation (CIMCA 2005), vol. 1, pp. 648–653 (2005)
38. Lu, W., Xue, L.: A heuristic-based co-clustering algorithm for the internet traffic classification. In: 2014 28th International Conference on Advanced Information Networking and Applications Workshops, pp. 49–54 (2014). https://doi.org/10.1109/WAINA.2014.16

A Strong Identity Authentication Scheme for Electric Power Internet of Things Based on SM9 Algorithm

Ji Deng[1]([✉]), Lili Zhang[1], Lili Jiao[1], Yongjin Ren[2], and Qiutong Lin[1]

[1] Police Officer College of the Chinese People, Chengdu 610213, China
490888186@qq.com
[2] Armed Police Meishan Detachment, Sichuan 620010, China

Abstract. With the development of 5G communication technology, the construction of the electric power Internet of Things relying on new technologies such as cloud computing, big data, and artificial intelligence has promoted the data integration of various electric power systems to achieve resource sharing services. With the continuous access of mass terminals to the Internet of Things, the attendant security authentication issues are becoming increasingly prominent. The traditional password based or digital certificate based authentication methods can no longer meet the access control of mass power terminals to power servers. In order to solve this problem, this paper proposes a strong identity authentication scheme based on SM9 cryptographic algorithm for the power Internet of Things without digital certificates and passwords, providing an effective solution for using intelligent terminals to achieve identity authentication for power Internet of Things systems in the Internet environment.

Keywords: SM9 algorithm · Internet of Things · Identity Authentication · Safety Certification

1 Introduction

With the development of 5G communication technology, facing the access of a large number of mobile terminals in the electric power Internet of Things, various types of electric power operation data are collected and analyzed, and a large amount of fragmented data is collected and provided to the data center for management through different devices. With the increasing frequency of information interaction between power systems, the security of access agents affects the reliability of data and instructions, and the identity authentication technology of devices will also play an important role in increasingly complex network environments.

L. Barolli (Ed.): CISIS 2023, LNDECT 176, pp. 200–209, 2023.
https://doi.org/10.1007/978-3-031-35734-3_20

2 Identity Authentication Technology

Traditional identity authentication methods mainly include passwords, tokens, and clio-metrics. In recent years, in order to improve the security of identity authentication, researchers have made various improvements to identity authentication schemes, rang-ing from multifactor and public key cryptography and more biometric schemes [1, 2]. In order to resist various attacks and solve problems such as key escrow, these schemes typically employ a third-party (CA) certification authority [3]. However, these schemes require creating a digital certificate for each terminal and establishing a key manage-ment center to provide key generation and storage for the authentication process [4]. At the same time, these schemes require the exchange of a large number of digital cer-tificates, placing a heavy burden on the authentication management system. Therefore, lightweight and convenient identity authentication mechanisms have become one of the important research directions.

With the vigorous promotion of domestic cryptographic algorithms, there has been more research on identity authentication schemes based on the national secret algorithm. Scheme [5] proposes a Kerberos identity authentication protocol based on the national secret algorithm, which requires a third-party (CA) authentication authority. Due to the fact that the State Security SM9 algorithm can implement identity authentication schemes without digital certificates, and has the advantages of flexibility, scalability, and simplicity compared to traditional PKI, SM9-based identity authentication schemes have attracted the attention of researchers. Scheme [6] proposes a distributed identity authentication scheme for distribution networks based on the SM9 algorithm, which can effectively achieve distributed control of power grid system identity authentication and has strong security. Scheme [7] proposes a mobile internet identity authentication scheme based on the SM9 algorithm, which can effectively resist replay attacks, phishing attacks, intelligent device loss, and other attacks, providing a direction for implementing strong identity authentication schemes in the mobile internet environment. Based on the research of Scheme 4, combined with the advantages of existing 5G technology and SM9 algorithm, this scheme proposes a strong authentication scheme suitable for network access the Electric Power Internet of Things, providing an effective solution for Internet of Things.

3 SM9 Algorithm

As a type of identity password, the public and private keys of National Secret SM9 are calculated by the Key Generation Centre (KGC) using the device identity, master key, master private key, and public algorithm parameters. It mainly uses various unique identi-fiers as public keys for data encryption and identity authentication, which is very suitable for e-mail protection, secure circulation of official documents, service and client security communication, identity authentication Applications such as Internet of Things security communication and cloud data protection have natural password delegation functions, which are very suitable for supervised electric-power application environments. They also have great advantages in monitoring the interconnection equipment of various duty stations [8].

3.1 Advantages of SM9 Algorithm

The advantages of SM9 algorithm are based on a comparative analysis of PKI and IBC technologies. Compared to PKI systems, the security application advantages of SM9 algorithm based on IBC technology systems in the field of the power networks are mainly manifested in the following aspects [9].

(1) The encryption strength of the State Secret SM9 algorithm is equivalent to the RSA encryption algorithm with 3072-bit keys, which cannot be cracked with existing computing power and has extremely high security.
(2) The national secret SM9 algorithm has a short key and better performance in terms of encryption, decryption, and verification signature efficiency.
(3) The SM9 algorithm does not require the construction of expensive infrastructure, nor does it require the issuance of certificates, storage of certificates, and exchange of certificates. Instead, it only requires the use of a public unique identity as a public key. Terminal computing costs are small, transmission consumption is low, key management is relatively simple, and deployment is relatively convenient. It is particularly suitable for the secure access of power network terminal devices.
(4) Due to the certificate less nature of SM9, the difficulty of key management will not significantly increase with the increase of users, and the size of system storage space and transmission speed will also increase less. Therefore, it can support large-scale horizontal expansion of the power network.
(5) Rich access control mechanisms that combine identity authentication and access control can effectively improve the security authentication and access control capabilities of the power network.

3.2 SM9 Signature

This section reviews the SM9 signature algorithm based on scheme in [10]. The algorithm description uses the symbols of scheme in [10].

(1) System Initialization

Given the security parameters λ, , respectively select the N (N is a large prime number and $N > 2^\lambda$) order additive cyclic group G_1 and G_2, and the multiplicative cyclic group G_T, which are the generators P_1, P_2 of the additive cyclic group G_1, G_2, with $P_1 = \psi(P_2)$, the key generation center KGC selects bilinear pairs e : $G_1 \times G_2 \rightarrow G_T$, and two cryptographic hash functions: $H_1 : 0, 1 * \times Z_N^* \rightarrow Z_N^*$, $H_2 : 0, 1 * \times Z_N^* \rightarrow Z_N^*$, and randomly selecting $s \in [1, N-1]$ as the system master key. Calculating the elements $P_{pub} = s \cdot P_2$ in the calculation group G_2, and $g = e(P_1, P_{pub})$ in the calculation group G_T, then selecting the function hid to generate the private key identifier by a byte. Last, Output the system public key mpk and secretly save the private key s.

$$mpk = (G_1, G_2, G_T, e, N, P_1, P_2, P_{puk}, g, H_1, H_2, hid)$$

(2) Key Generation

KGC calls the hid function to generate a pair of public and private key pairs for user A. The process is to calculate $t_1 = H_1(ID_A \| hid, N) + s$ on an elliptic curve finite field F_N. If $t_1 = 0$, the master key needs to be regenerated. Otherwise, calculate $t_2 = s \cdot t_1^{(-1)}$ and $d_A = t_2 \cdot P_1$, d_A as the user's signature private key, $Q_A = H_1(ID_A \| hid, N) \cdot P_2 + P_{puk}$ as the public key to verify the signature.

(3) Signature Generation

  For a message M ∈ $\{0, 1\}^*$ to be signed, signer A can digitally sign message M by following the steps below.

 ① Randomly generating a random number r ∈ $[1, N - 1]$.
 ② Calculating the elements $\omega = g^r$ in a group G_T.
 ③ Calculate an integer number h = $H_2(M \| \omega, N)$.
 ④ Calculate an integer number $\xi = (r - h) \bmod N$. if $\xi = 0$ then return to step ①.
 ⑤ Calculating the elements S = $\xi \cdot d_A$ in a group G_1.
 ⑥ Output the digital signature σ = (h, S) of message M.

(4) Signature Verification

  Given the signature message M' ∈ $\{0, 1\}^*$ and σ' = (h', S'), recipient B performs the following operation to verify the validity of the signature σ' = (h', S') of message M'.

 ① Calculating the element t = $g^{h'}$ in a group G_T.
 ② Calculate an integer number $h_1 = H_1(ID_A \| hid, N)$.
 ③ Calculate the element P = $h_1 \cdot P_2 + P_{puk}$ in the group G_2.
 ④ Calculate the elements π = e(S', P) and φ' = π · t in the group G_T.
 ⑤ Calculate the integer number $h_2 = H_2(M' \| \varphi', N)$ and then check $h_2 = h'$ whether it is true. If it is true, the verification passes, otherwise the verification fails.

4 Strong Identity Authentication Scheme for the Power Network Based on SM9 Encryption Algorithm

In the context of the power network, SM9 algorithm can easily achieve user authentication. However, due to the management of private keys on the user side, there is a significant potential for theft. Once the private key is leaked, it will pose a significant threat to the access verification of the power network. In order to prevent the risk of key theft, at terminals connected to the power network, additional biometric identification devices that support fingerprint, voice, pupil, face, and other biometric identification are added to obtain user ID information, achieving strong "password free" identity authentication for users, and improving the security of accessing the power network system.

4.1 Scheme Design

The scheme is divided into a power terminal and a power server. The power terminal is responsible for collecting legitimate user biological information, such as fingerprints, faces, voice prints, etc., generating user unique identifiers, and storing user private keys. The server is mainly composed of KGC, Identity Authentication System (AS), Registration Center(RS) and cloud DB to achieve key generation.

The security authentication processes mainly involves the following six steps, and the authentication model is shown in Fig. 1.

Step 1: Collect the biological information of legitimate users from the linkage terminal, and generate the user's unique and legal identity ID based on the user's biological information to apply for registration with KGC.

Step 2: KGC generates the public and private keys corresponding for the user, and stores the private keys in the private key storage module of the linked terminal. At the same time, it clarifies the operation permissions of the user, and enters relevant biological information.

Step 3: When the linkage terminal user wants to login into the linkage network server to access resources, they need to send an access request to the server.

Step 4: After receiving the access request, the resource server sends an authentication request and a random challenge value to the linkage terminal.

Step 5: The linkage terminal identifies the user's biometric information. If the identification passes, the challenge value is signed with a private key and sent to the security certification center.

Step 6: After receiving the signature, the security certification center verifies the signature and compares whether the random challenge value is consistent; If consistent, the authentication passes and the corresponding permissions of the user are opened; If inconsistent, access is denied.

4.2 Scheme Implementation

The implementation of this authentication scheme is mainly divided into four modules: system initialization, user registration, login and identity authentication, and key negotiation for resource transmission.

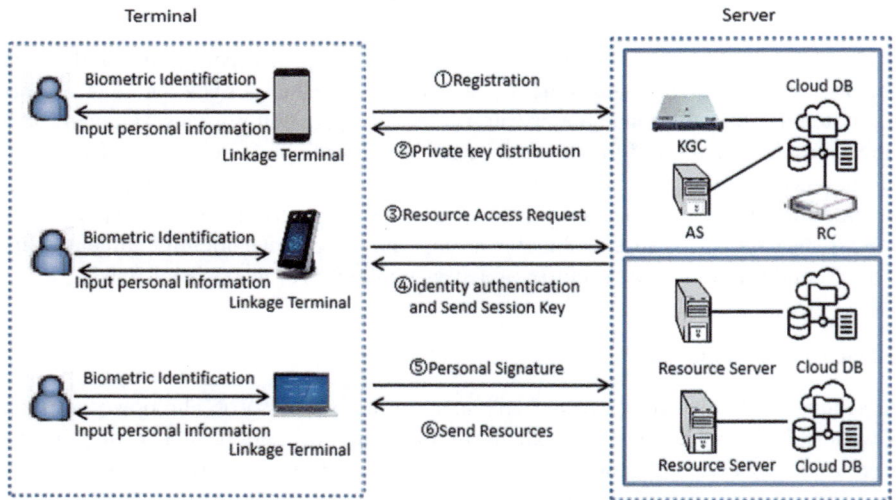

Fig. 1. Power identity authentication. Scheme.

4.2.1 System Initialization

The power server uses the key generation center (KGC) of the system to generate a random number $s \in [1, N - 1]$ as the system master key, and then calculates the element $P_{pub} = s \cdot P_2$ as the master public key in the group G_2. The KGC Saves private Key s and P_{pub} is publicly available.

4.2.2 User Registration

If a user requests to register as a legitimate user of the system through the linkage terminal, the following steps must be performed.

① User A obtains personal biological information through the terminal, and then invokes the terminal identifier management module to generate the user's unique and legal identity ID_A, and calculates $Q_A = H_1(ID_A \| hid, N) \cdot P_2 + P_{puk}$ to obtain the public key Q_A of user A.

② Call a random number generator to generate a random number R_{A1} for the first registration, which is not allowed to be generated frequently within a limited time to prevent the authentication service center from being attacked by malicious users who re register.

③ Obtain the current registration time t_{reg}^{A1} of user A.

④ Calculation $M_{log\_1} = \{ID_A \| Q_A \| R_{A1}\}$.

⑤ User A encrypts message M_{log_1} with the system's public key P_{pub} for obtaining the ciphertext $C = E_S(M_{log\_1})$, and sends the message $\{C \| t_{reg}^{A1}\}$ to RC through the mobile internet.

After receiving the registration request information $\{C \| t_{reg}^{A1}\}$ from terminal user A, RC of the system server needs to perform the following operations.

① Verify time t_{reg}^{A1} has not been exceeded the maximum interval time Δt which is agreed by the system. If the current system time is exceeded time $t_{reg}^{A1} + \Delta t$, the user registration request will be rejected.

② Check whether the user's ID_A has been registered. If registered, prompt the message "Registered" to the user and restrict the user to repeat registration within the specified time frame.

③ Use the private key s to decrypt C to obtain $M_{log\_1} = D_S(E_S(M_{log\_1}))$. So that, the registry obtains the public key Q_A and random number R_{A1} of user A, and stores the user information in RC as shown in Table 1.

④ The KGC calculates the private key d_A of user A based on ID_A and Q_A of user A, calculates to obtain $M_1 = H_1(ID_A \| R_{A1}) \oplus (R_{A1} \cdot d_A)$, confirms the registration time $t_{reg}^{A1'}$, and sends the message $\{M_1 \| t_{reg}^{(A1')}\}$ to user A.

After receiving the message $\{M_1 \| t_{reg}^{(A1')}\}$ from the linkage terminal user A, the terminal performs the following operations:

① Verify the registration time $t_{reg}^{A1'}$ is or not exceeded to the maximum interval time Δt. If it is exceeded, the received information will be discarded and re registered.

② The terminal calculates $d_A = M_1 \oplus H_1(ID_A \| R_{A1}) \oplus R_{A1}$ to obtain the private key d_A.

Table 1. Information stored by users in RC (considering users A, B, and C as examples).

User ID	User Public Key	Random number of user registration or authentication	Time of User registration or authentication
ID_A	Q_A	R_{A1}	t_{reg}^{A1}
ID_B	Q_B	R_{B1}	t_{reg}^{B1}
ID_C	Q_C	R_{C1}	t_{reg}^{C1}
…	…	…	…

③ Calculate $d_A' = d_A \oplus H_1(ID_A||R_{A1})$ and $R_{A1}' = R_{A1} \oplus ID_A$), and store the hash value $H_1(ID_A||R_{A1})$, d_A', t_{reg}^{A1} to the terminal. Registration is complete.

4.2.3 Login and Identity Authentication

If a user requests to login into the system to gain access to resources, he needs to be authenticated by AS, and the resources can only be accessed after passing identity authentication from AS.

To initiate a request to access system services, the user needs to perform the following steps.

① The terminal obtains the biological information of user A and invokes the terminal identifier management module to generate the user's unique and legal identity ID_A.

② Calculate $R_{A1} = R_{A1}' \oplus ID_A$ to obtain R_{A1}.

③ Calculate $H_1(ID_A||R_{A1})$, and then verify $H_1(ID_A||R_{A1})$ is the same as the hash value that already is existed on the terminal. If it is different, the user's access is denied. If the number of rejections has exceeded the set number, the application interface will be locked.

④ Invoke d_A' and calculate $d_A = d_A' \oplus H_1(ID_A||R_{A1})$ to obtain the private key of the terminal user A.

⑤ User A calls the random number generator to generate a new random number R_{Ai} and the current request for authentication time t_{reg}^{Ai}, calls the registration time t_{reg}^{A1}, and calculates $M_2 = \{ID_A||R_{A1}||R_{Ai}||t_{reg}^{A1}\}$.

⑥ User A signs M_2 to obtain the signature information $D_A(M_2)$ with his own private key pair, then calculates $M_3 = \{Q_A||D_A(M_2)\}$, and encrypt M_3 to obtain $M_4 = E_{P_{pub}}(Q_A||D_A(M_2))$ by using the public key of the system server. Send the message $\{M_4||t_{reg}^{Ai}\}$ to the server for requesting authentication.

After the server receives the information $\{M_4||t_{reg}^{Ai}\}$, it needs to complete the following steps.

① Verify the time t_{reg}^{Ai} for certification has exceeded the maximum interval time Δt. If it has exceeded, the user's request is rejected.

② Decrypt the information $D_s(E_{P_{pub}}(Q_A||D_A(M_2)))$ with the private key s to obtain $\{Q_A||D_A(M_2)\}$, and then verify the signature $D_A(M_2)$ with the public key of user A. If decryption and verification have passed, thereby obtaining ID_A, Q_A, R_{A1} and R_{Ai} of user A.

③ Verify those ID_A, Q_A *and* R_{A1} of user A in the terminal registry are equal to these ID_A, Q_A and R_{A1} of user A. If not, the authentication of user A is rejected.

④ The server provides the authenticated user A with access rights or corresponding resources based on the preset user permissions, and replaces R_{Ai} with R_{A1} in the registry.

After the user passes the authentication, perform the calculation $d'_A = d_A \oplus H_1(ID_A || R_{Ai})$ and replace d'_A with d_A as important information for the next identity authentication request.

4.2.4 Key Negotiation

After the identity authentication between the user and the server, the public and private key pairs generated by the SM9 algorithm can be used to achieve information ciphertext and transmission, or the random number used for authentication can be used as the symmetric encryption key for temporary sessions. The one-time encryption method can also ensure the security of the transmitted information.

5 Safety Analysis and Certification

5.1 Correctness Analysis

This paper verifies the correctness of this scheme based on the properties of bilinear pairs.

The properties of bilinear pairs are:

① $\forall P_1, P_2 \in G_1, Q \in G_2$, there is a conclusion that $e(P_1 + P_2, Q) = e(P_1, Q) \cdot e(P_2, Q)$.

② $\forall Q_1, Q_2 \in G_2, P \in G_1$, there is a conclusion that $e(P, Q_1 + Q_2) = e(P, Q_1) \cdot e(P, Q_2)$.

③ For any number $\alpha, \beta \in [1, N-1]$, $\forall P \in G_1, Q \in G_2$, there is a conclusion that $e(\alpha \cdot P, \beta \cdot Q) = e(P, Q)^{\alpha \cdot \beta} = e(\beta \cdot P, Q)^{\alpha} = e(P, \alpha \cdot Q)^{\beta} = e(\beta \cdot P, \alpha \cdot Q)$.

The correctness of signature verification for this scheme.

$h_2 = H_2(M' || \varphi', N) = H_2(M' || e((r-h) \cdot t_2 \cdot P_1, H_1(ID_A || hid, N) \cdot P_2 + s \cdot P_2) \cdot t, N)$

$= H_2(M' || e((r-h) \cdot t_2 \cdot P_1, (H_1(ID_A || hid, N) + s) \cdot P_2) \cdot t, N)$

$= H_2(M' || e((r-h) \cdot s \cdot t_1^{-1} \cdot P_1, (H_1(ID_A || hid, N) + s) \cdot P_2) \cdot t, N)$

$= H_2(M' || e((r-h) \cdot s \cdot t_1^{-1} \cdot P_1, t_1 \cdot P_2) \cdot t, N) = H_2(M' || e(P_1, P_2)^{(r-h) \cdot s \cdot t_1^{-1} \cdot t_1} \cdot t, N)$

$= H_2(M' || e(P_1, P_2)^{(r-h) \cdot s} \cdot t, N) = H_2(M' || e(P_1, s \cdot P_2)^{(r-h)} \cdot t, N) = H_2(M' || g^{(r-h)} \cdot g^h, N)$

$= H_2(M' || g^r, N) = H_2(M' || \omega, N) = h'$

5.2 Safety Analysis

(1) Terminal loss security analysis

The random number $R'_{A1} = R_{A1} \oplus ID_A$ and $d'_A = d_A \oplus H_1(ID_A || R_{A1})$ are stored on the linked terminal device are all encrypted data. Even if the terminal device is lost, an attacker may be unable to provide correct biometric information, thereby failing to obtain the user's identity information to decrypt the secret information. Therefore, the system can resist the security threat of information theft attacks and ensure the security of user information when terminal devices are lost.

(2) Identity phishing attacks

The core information $D_A(M_2)$ for identity authentication in this scheme is that the sender user A signs with a private key pair, if the counterfeiter C completes the fake signature $M_4' = E_{P_{pub}}(Q_C||D_A(M_2))$ with Q_C and P_{pub} to forge information, and send the message $E_{P_{pub}}(Q_C||D_A(M_2))$ to the server. The server uses the public key of impersonator C to verify the signature of user A, the validation definitely failed, access is denied.

(3) Replay attack

During registration and authentication, this scheme appends the time t_{reg}^{A1} for registration or time t_{reg}^{Ai} for authentication which have the effect of a timestamp. Even if the attacker tampers with the time, the server can identify and determine the replay information based on the random number and other information in the received information after decrypting with the private key, thereby effectively avoiding replay attacks.

(4) Key security

The system user's private key is calculated and generated by KGC based on the user's identity and the server does not store the user's private key, but the private key stored by the terminal device is an encrypted key $d_A' = d_A \oplus H_1(ID_A||R_{A1})$. Even if the private key d_A' is compromised, the real key d_A cannot be recovered due to lack of $H_1(ID_A||R_{A1})$, so that this scheme can devoid both the risk of disclosure and theft.

5.3 Strength Analysis

(1) Certification efficiency

This scheme has tested the computing time cost of each operation on the Inter Core i7-9700 platform. The XOR operation is very small and can be ignored. The most time-consuming operation is SM9 signature and verification. High performance computers can be used to improve the computing power of the server, greatly reducing the time cost of the server and further improving the authentication efficiency.

(2) High face recognition rate

Currently, face recognition algorithms based on convolution neural networks can achieve a recognition rate of over 99%, with a recognition time of less than 0.05 s for a single face, and are robust against interference such as lighting differences, facial expression changes, and presence or absence of obstructions, improving registration and login efficiency.

6   Conclusions

This scheme utilizes biometric technology and the State Secret SM9 algorithm to implement a strong identity authentication scheme for the power private network. This scheme adopts a password free security authentication method, avoiding the dilemma of identity authentication failure caused by password forgetting. Although the SM9 algorithm can cause significant server time overhead, it does not affect the user experience in practical applications. Through correctness and security analysis, this scheme can withstand multiple attack modes such as replay attacks, password avoidance attacks, replay attacks, phishing attacks, and intelligent device loss attacks, achieving security authentication of one-time encryption, ciphertext transmission, and two-way authentication, ensuring the first line of defense for the security of the power network.

References

1. Zhang, F., Wang, Q., Song, L.: Research on unified identity authentication system based on biometrics. Netinfo Secur. **19**(9), 86–90 (2019)
2. Zhang, X., Liu, J.: Multi-factor authentication protocol based on hardware finger print and biometrics. Netinfo Secur. **20**(8), 9–15 (2020)
3. Digi CertInc. Patent Application, Dynamic certificate generation on a certificate authority cloud, Published on line (USPTO20190356651). Medical Patent Business Week (2019)
4. Yin, R., Gao, Y., Li, K.: Application of identity authentication technology in mobile applications in power industry. China New Commun. **22**(10), 116 (2020)
5. Huang, D., Zhang, Z., Liu, J.: Improvement and analysis of kerberos identity authentication protocol based on SM algorithms. J. Jinling Inst. Technol. **38**(2), 1–8 (2022)
6. Liao, H., Yu, G., Ban, G.: Research on identity authentication technology in power internet of things based on SM9 algorithm. Shang Dong Electr. Power **47**(275), 1–4 (2020)
7. Zhang, Y., Sun, G., Li, Y.: Research on mobile internet authentication scheme based on SM9 algorithm. NeTinfo Secur. **4**(1), 1–8 (2021)
8. Wang, M.-D., He, W.-G., Li, J., Mei, R.: Optimal design of R-ate pair in SM9 algorithm. Commun. Technol. **9**(53), 2241–2244 (2020)
9. Dong, Y.-X., Quan, J.-B., Wang, M.-R., Luo, M., Yang, Y.: Research on the application of SM9 algorithm in the security field of Internet of things. China Acad. J. Electron. Publishing House **9**(35), 22–27 (2022)
10. Lai, J., Huang, X., He, D.: Security analysis of SM9 digital signature and key encapsulation. Sci. Sin. Inform. **51**, 1900–1913 (2021). (in Chinese)
11. Zhang, C., Peng, C.: Searchable encryption scheme based on china state cryptography standard SM9. Comput. Eng. **7**(48), 159–166 (2022)
12. Tang, F., Gan, N.: Anti malicious KGC certificate less signature scheme based on block-chain and domestic cryptographic SM9. Chin. J. Netw. Inform. Secur. **8**(6), 9–18 (2022)
13. An, T., Ma, W., Liu, X.: Aggregate signature scheme based on SM9 cryptographic algorithm in VANET. Comput. Appl. Softw. **37**(12), 280–284 (2020)

CPU Usage Prediction Model: A Simplified VM Clustering Approach

Rebeca Estrada$^{(\boxtimes)}$ (ID), Irving Valeriano (ID), and Xavier Aizaga (ID)

Information Technology Center and Electrical Engineering and Computer Science Department, Escuela Superior Politecnica del Litoral, ESPOL´, Campus Gustavo Galindo Km 30.5 V´ıa Perimetral, Guayaquil, Ecuador
`{restrada,ivaleria,xafraiza}@espol.edu.ec`

Abstract. Machine learning algorithms play an important role in resource management, allowing the improvement of the efficiency of resource usage in data centers (DCs) by predicting workload trends. In this paper, we propose a simplified system to predict the CPU usage of virtual machines (VMs) in a DC using Linear Regression Models while performing VM clustering based on common statistical characteristics of VM time series, which facilitates grouping VMs with similar behaviors and establishing clusters based on these characteristics. For each cluster, three representative VMs are established based on the time series of the closest VM to the cluster centroid, averaged time series for the cluster, and concatenated time series. Then, training of representative VMs is performed to finally choose the one with the lowest mean error per cluster. Simulation results show that, by performing clustering and training the model with representative time series, it is indeed possible to obtain a low mean error while reducing the local training time per VM.

Keywords: Clustering · Data Center · Machine Learning · Time series · Prediction

1 Introduction

Datacenter (DC) offers data storage and processing power based on users' demands and availability. Cloud computing should cope with two basic requirements: scalability and elasticity in order to handle the increasing user demand. Scalability is the system's ability to handle increased workloads and elasticity refers to the ability of the system to adjust its resources in response to the dynamic changes in workload.

Resource monitoring is an important process for cloud computing because it guarantees the quality of service (QoS) and the performance parameters of the services provided to users (e.g. scalability, utilization, time, workload management and cost) [1]. This process requires collecting data of CPU and memory usage of a DC. With this data, monitoring tools keep a centralized assessment of the status, availability and DC performance parameters. Currently, there is an interest to develop novel monitoring frameworks. Moreover, virtualization allows delivering computing resources to workstations in the same network. These resources are packed into a virtual machine (VMs).

L. Barolli (Ed.): CISIS 2023, LNDECT 176, pp. 210–221, 2023.
https://doi.org/10.1007/978-3-031-35734-3_21

However, there are risks associated with trying to manage a large number of VMs. Thus, one critical task is to detect, identify, and generate alerts regarding anomalous events of VMs in DC to avoid any malfunctioning of other users' VMs [9].

Virtual Machine (VM) dynamic deployment is one key technique for virtual resource management enabling the prediction of the resources used by the virtual machine and managing them accordingly. Virtual Machine resource dynamic deployment focuses on the fine-grained resource adjustment strategy. Therefore, there is a need of predicting the usage of resources for the virtual machine to avoid delays or over/under-provisioning. Time series prediction models can be used for these predictions because the VM resources are dependent on time. Various types of resources such as CPU usage, memory usage, disk read times, disk write times, and network usage are predicted using the above models.

In the literature, some approaches have been proposed to predict one metric or several metrics of interest based on the dataset. Moreover, it is possible to take preventive actions with these predictions. Based on these predictions, the warning rules can be improved in case of new unexpected events and the thresholds of the previously established metrics can be also adjusted [17]. This motivates our work to analyze historical data logs to predict CPU consumption.

In this paper, we propose to analyze several traditional machine learning models along with clustering techniques that can improve the prediction of a preferred metric (e.g. CPU usage). Using statistical characteristics of VM's time series and applying K-means clustering, we determine the clusters of VM with similar performance. Then, we select the best ML prediction models for the time series of different proposed types of representative VM belonging to a cluster.

Owing to the fact that several VMs may have very similar behavior due to the services they run, we propose to evaluate clustering methods for VM before selecting the prediction model for any virtual machine. Thus, it is possible to use a single prediction model per cluster and to void the computational effort of performing the learning process for every single VM within clusters. Moreover, we propose to analyze several linear regression models along with k-means clustering techniques that can improve the prediction of CPU usage. In the event that a new virtual machine is created in the data center, it should be added to the cluster with similar characteristics in terms of the mean and standard deviation of CPU usage and the mean and standard deviation of memory usage. In addition, the training process needs to be performed again if the VM modifies the time series for the representative VMs and the new best representative VM should be determined for the cluster. The main contributions of this paper are:

- A clustering algorithm for virtual machines in a DC together with a mechanism to determine three representative VMs per cluster for monitoring and predicting the CPU usage.
- An evaluation of various linear regression prediction models and the selection of the most appropriate for each cluster based on the root mean square error.

The rest of the paper is organized as follows. Section 2 presents the related work taking into account the prediction of computational workloads in DCs. Section 3 describes the proposed model to predict CPU usage based on the clustering of VMs. In Sect. 4, numerical results are presented using the dataset. Finally, Sect. 5 concludes our work and proposes future work.

2 Related Work

One of the most important challenges in DC is the efficient usage of computing resources in order to increase available resources and improve resource utilization. For this reason, this has been investigated using machine learning methods for DC resource consumption predictions[4]. In this section, we briefly present some time series based predictive models that can be found in the literature.

There have been many efforts to predict DC resource utilization based on time series solutions with several purposes: i) to assign computing resources to each job based on the resource availability [5], to share the resources between multiple different cluster computing frameworks [11], to improve the resource utilization [12, 23] and to retain the quality of SLAs [14]. Other approaches focused only on the CPU consumption prediction using time series such as in DC [16]. For example, AWS forecast offers 6 types of forecast models which include both the classic time series models based on moving average and machine learning models. The prominent forecasting models used are: Prophet[6], ARIMA [13, 24], SARIMA [18], moving average [15], exponential smoothing, and recurrent neural networks[2].

Linear regression models were used to predict CPU consumption at the host level in [8] to solve task over/under loading. In addition, a VM migration mechanism is proposed to reduce the average SLA violations compared to threshold-based migration plans. [25] proposes the use of auto-correlations of time series to characterize VM resource usage as input for a neural network to predict of peak consumption providing an online updating module to monitor the predictions errors of the workload changes.

Classical time series models and neural networks have been explored for short and long-term horizons. For instance, the performance of LSTM and SARIMA models were compared using Azure cloud dataset [18]. Artificial Neural Network was also used with historical data and tested on real-world datasets [3], and a performance comparison was presented with the existing classical time forecasting methods. In [10], the authors used Multivariate LSTM models to predict the future workload. In [19], deep belief networks are used to predict the resources extracting high-level features from the workload time series.

The performance of several prediction models of CPU usage using time series was compared in [20]. These models are Holt-Winters, ARIMA, and LSTMs and it was shown that Holt-Winters presents less precision due to the lack of stationary and tendency in the data of CPU usage while ARIMA does not make a good prediction in real time if the dataset is long and LSTM model presented the best performance and lowest RMSE.

To the best of our knowledge, the idea of combining historical data from several VMs with similar behavior has not been yet investigated. This can reduce the required training time for each individual VMs and enable the resource management system to cope with real-time decisions to improve the resource utilization of DCs.

3 System Model

Here, we present the proposed methodology to predict the CPU usage of overloaded VMs that can assist data center personnel in their operation and maintenance using a dataset previously used in [21]. Figure 1 presents a diagram summarizing the prediction module inside each cluster. In the following subsections, we first describe the clustering method used to discover VMs that share similar statistical characteristics. Three different representative VMs are defined for each cluster named closest, average, and concatenated VMs, for which the best ML prediction model is identified as the one providing the lowest RMSE error of the testing data. Finally, the best representative VM is selected as the one that provides the lowest RMSE when evaluating the prediction model with the time series with all VMs belonging to the cluster.

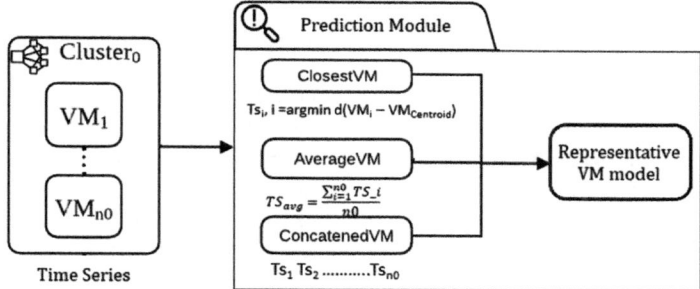

Fig. 1. Cluster-based Prediction Module

3.1 Dataset Information

The dataset used is GWA-T-12 Bitbrains, which contains performance metrics for 1,250 virtual machines from a distributed data center of Bitbrains[22]. Bitbrains is a service provider specializing in managed hosting and enterprise computing for businesses. This dataset consists of one file with the performance metrics for each VM. These performance metrics in the dataset that we used in our work are shown in Table 1.

For simplicity, we only use two metrics (i.e. CPU usage and memory usage) to represent the $n - th$ VM tuple with mean and standard deviation of these metrics, i.e. $Y_n = (\mu_{CPU}{}^n, \sigma_{CPU}{}^n, \mu_{RAM}{}^n, \sigma_{RAM}{}^n, N)$, where N represents the length of the time series. Thus, it is possible to group the VMs with similar statistical behavior.

3.2 Clustering Algorithm

It is important to remark that although VMs may handle heterogeneous workloads according to the tasks they are running, some virtual machines can exhibit similar behavior. This could be due to some VMs sharing the load of a task, or tasks may be different in nature but have similar computational load patterns. Our goal is to take advantage of

Table 1. Used information of Dataset

Name	Description and units
Timestamp	Number of milliseconds
CPU cores	Number of virtual CPU cores provisioned
CPU capacity provisioned	Capacity of the CPUs in terms of MHZ (number of cores x speed per core)
CPU usage	MHZ
CPU usage	Percentage
Memory usage	Active memory used in terms of KB

the similarities of the different virtual machines to select groups that can be represented by a single representative time series since creating a prediction model for each VM could be too expensive.

Let's consider the scenario of a data center composed of N virtual machines which are monitored by any cloud infrastructure management tool. Our model uses CPU and memory usage metrics, which are available in the Bitbrains dataset. Thus, each virtual machine can be represented by a tuple consisting of the mean CPU usage (μ_{CPU}^{n}), the standard deviation of CPU usage (σ_{CPU}^{n}), the mean RAM usage (μ_{RAM}^{n}), and the standard deviation of RAM usage (σ_{RAM}^{n}). Then, the optimal cluster configuration is determined using the K-means algorithm and the elbow method based on the distance between the tuples of the virtual machines.

The clustering algorithm provides the centroid and the VMs belonging to each cluster, ($\{VM^{c}\}, c \in C$), which is used to obtain the representative time series for each cluster, c. Table 2 shows the result after running the clustering algorithm.

Table 2. Clusters

Cluster No	CPU Usage			RAM Usage			Size
	Min	Mean	Max	Min	Mean	Max	
1	1.388	11.47	54.40	0	10.2974	72.36	109
2	0	1.57	12.38	0.03	4.53	25.64	639
3	33.53	46.18	72.22	6.47	10.92	17.56	56
4	9.57	25.67	46.34	1.79	8.16	36.55	37
5	0.08	0.55	6.01	14.08	30.29	55.32	174
6	0	0.89	3.47	1.37	22.90	119.81	26

As we can see, the cluster size varies from 26 to 639 VMs. We propose to classify the clusters as small size(¡100 VM), medium size (100–300 VM), and large size (¿300

VM) to identify if the cluster size has an impact on the selection of the LR prediction model or the selection of the representative VM.

Figure 2 presents the CPU and ram usage for a given VM within clusters 1,2 and 6. These clusters were selected because they are the ones with different sizes, i.e. medium, large and small respectively. The red line represents the CPU usage (%) while the blue line the RAM usage (%). From Fig. 2a, the values of both metrics are lower than 10% with a flat behavior, which means that the majority of the VMs in the real dataset are not overloaded. It should be noticed that this cluster is the largest cluster. In Fig. 2b, one can observe that the VM from Cluster 1 (medium size) presents a CPU usage between 20% and 50% while RAM Usage is 15% and 20%. It can be also appreciated that there is a periodicity involved in the behavior in both metrics. Finally, in the VM from cluster 6 (small size), the memory usage is between 16% and 20% while the CPU usage is lower than 2%.

Our proposed method creates three representative virtual machines with a representative time series within a cluster. In words, we define three representative VMs named closest VM, averaged VM, and concatenated VM because their representative time series is defined by the time series of the closest VM to the cluster centroid, or the averaged time series of all time series for the VMs belonging to the cluster, or a concatenated time series of all time series for the VMs belonging to the cluster respectively. Based on the RMSE error of the percentage of the testing data, LR prediction models are evaluated to select the best LR prediction model using the corresponding time series for each representative VM (i.e. Closest, Averaged, and Concatenated) with the lowest mean square error. Finally, the representative VM is selected as the one that provides the lowest RMSE prediction error taking into account all VMs belonging to a cluster. This is the prediction engine for the CPU usage prediction module as shown in Algorithm 1.

We propose to use the selected prediction models to estimate the CPU usage for the individual VMs within a cluster. The use of a single model representing virtual machines allows the model to predict an overloaded virtual machine. The proposed methodology is not limited for the prediction of the CPU usage only for the virtual machines present during training but also for new virtual machines with similar characteristics.

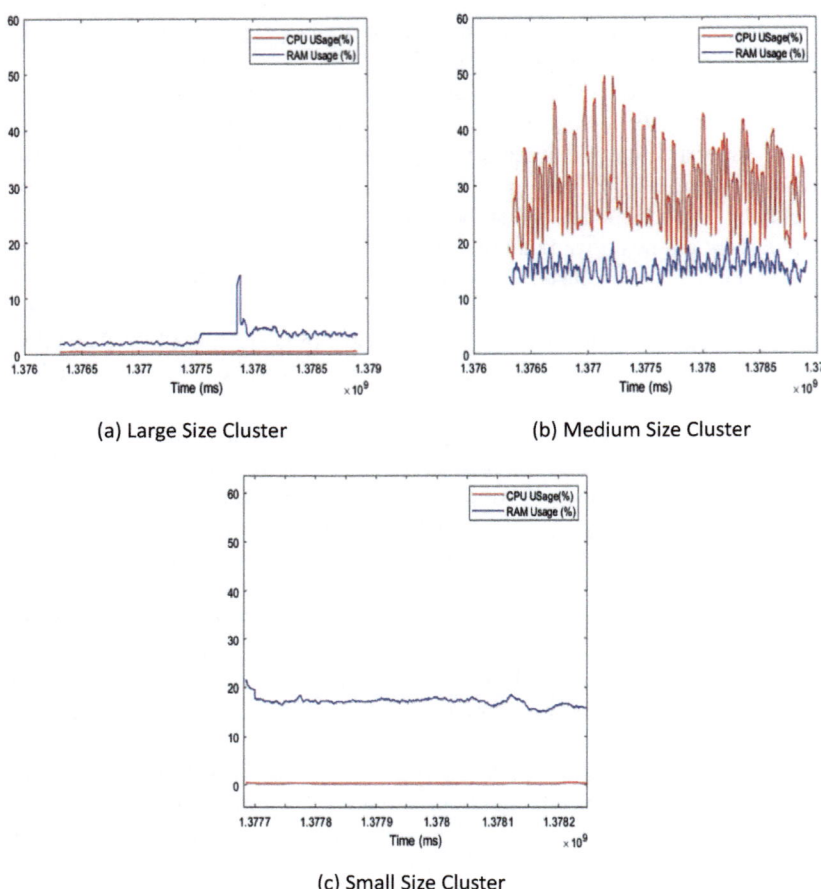

(a) Large Size Cluster   (b) Medium Size Cluster

(c) Small Size Cluster

Fig. 2. Example of the behavior of one particular VM inside a large and small cluster

3.3 Linear Regression Prediction Models

CPU usage value prediction is estimated using the Matlab *Regression Learner Toolbox* with the dataset presented in Sect. 3.1. This toolbox allows us to perform predictions with the following algorithms: *Linear Regression (LR)* - Linear, Interactions Linear, Robust Linear, and Stepwise Linear. These algorithms are used to model the relationships between CPU usage and the proposed features by means of an artificial intelligence algorithm. This approach models a linear function between a scalar response and various features [7]. The dataset is distributed as follows: 70% of the available data is used to train the algorithms, 15% is used to validate the prediction model and 15% is used to test the degree of accuracy of the prediction algorithms. Table 3 presents the numerical results of the RMSE and R^2 values obtained using seventeen different ML algorithms during the validation phase.

Algorithm 1: Clustering and Prediction Model Selection for representative VMs

---

Data: Time Series of VMS,
    Models for predictions per Cluster *Models*
Result: Selected Models per Cluster begin
    Models = [Set ML Prediction Models] Clusters =
    kmeansopt(VMs); for *each cluster c* ∈*Clusters* do
    $ts_{closest}$= select time series closestVM(cluster); $ts_{averaged}$=
    select time series averagedVM(cluster);

        $ts_{concatenated}$ = select time series concatenatedVM(cluster) ; for
        each model ∈*M L Models* do
            closest model$_i$ = train model($ts_{closest}$); average
            model$_i$ = train model($ts_{averaged}$); concatenated
            model$_i$ = train model($ts_{concatenated}$);
        end
        best closest = select best model(closest model); best average
        = select best model(average model); best concatenated =
        select best model(concatenated model);
    end
    for *each cluster c* ∈*Clusters* do
        Estimate the rmse error for all VMs in c using three trained models;
        best representative model = evaluate models(best closest, best concatenated, best average);
    end
    return best representative models;
end

Table 3. RMSE, R^2 and Training Times for ML Algorithms for Cluster 1

ML Model	Closest VM			Averaged VM			Concatenated VM		
	RMSE	$R2$	Training Time (sec)	RMSE	$R2$	Training Time (sec)	RMSE	$R2$	Training Time (sec)
LR-Linear	1.30E−08	1	8.56	0.0356	0.74	3.3414	7.02	0.64	3.64
LR-Interactions	2.20E−03	1	3.3345	0.035	0.74	0.91	3.26	0.92	2.10
LR-Robust	1.30E−08	1	8.28	0.037	0.66	1.92	7.95	0.54	4.38
LR-Stepwise	1.31E−08	1	7.81	0.0358	0.78	11.03	3.16	0.93	598.85

From this table, we conclude that the best ML model that fits the averaged and concatenated VMs is the LR-Stepwise model while for the closest VM is the LRRobust model in cluster 1, a cluster of medium size. The training time for these models is also estimated by the toolbox and they are also shown in Table 4 in the next section. For the concatenated time series, it can be observed that LR-Stepwise model presents the lowest RMSE, 3.16, but it requires higher training time (598 s) in comparison to LR-Interaction than provides RMSE error of 3.26 with R^2 of 0.92 but the training time is only 2.10 s. This should be taken into account when selecting the best model.

4   Results

In this section, we present the evaluation of the proposed methodology for predicting overloads in DCs at the VM level. First, we describe the scenario and evaluation metrics for the learning models. For the given dataset, we describe the behavior observed at different groups of VMs that form a cluster. Then, we evaluate the RMSE of the three representative models (i.e. closest, averaged, concatenated) using the same data from one random VM within each cluster. Finally, we compare the mean error for the three representative ML-trained models to choose the best model within a cluster.

4.1   Experiments' Settings

In our experiments, we focus on predicting CPU usage, knowing that CPU utilization is a critical resource that is usually a reference for generating DC alerts. We run the algorithm K-means together elbow method to form the best clusters of VMs with similar statistical characteristics (i.e. CPU usage and memory usage means and standard deviation). As it was described in Sect. 3.2, each cluster needs to select the best model to make predictions of CPU usage of any VM within the cluster and if the CPU usage predicted value is higher than the threshold (e.g. 50%), then, other actions need to be performed such as the alert generation.

To evaluate the performance of the ML algorithms, we used R squared (R^2) and mean square error (MSE). The former is generally used for explanatory purposes and provides an indication of the goodness or fit of a set of predicted values to the actual output values. RMSE determines the average of the squared differences (i.e. residual) between the actual and predicted values. RMSE is simply the square root of the MSE given by:

$$RMSE = \sqrt{\frac{\sum_{i=0}^{n}\left(Y_i - \hat{Y}_i\right)^2}{n}} \tag{1}$$

where Y_i and $\hat{Y}_i$ correspond the actual and predicted values respectively. For convenience, we decide to use the RMSE metric in order to select the best model in the cluster among the Closest, Average or Concatenated VMs.

4.2   Evaluation of Representative Models for Different VMs

Figure 3 shows the real value of CPU usage (%) compared with the predicted values using the trained models generated with the representative virtual machines Closest, Average and Concatenated for a medium size cluster (i.e. Cluster 1).

We can observe that the Concatenated model provides similar results to the original VM regardless of its distance to the centroid in the cluster with medium size, as shown in Fig. 3. From Fig. 3a, it can be remarked that the models trained with closest and concatenated representative VMs provide predicted values similar to the closest VM to the centroid of a medium size cluster. In the case of a virtual machine far from the

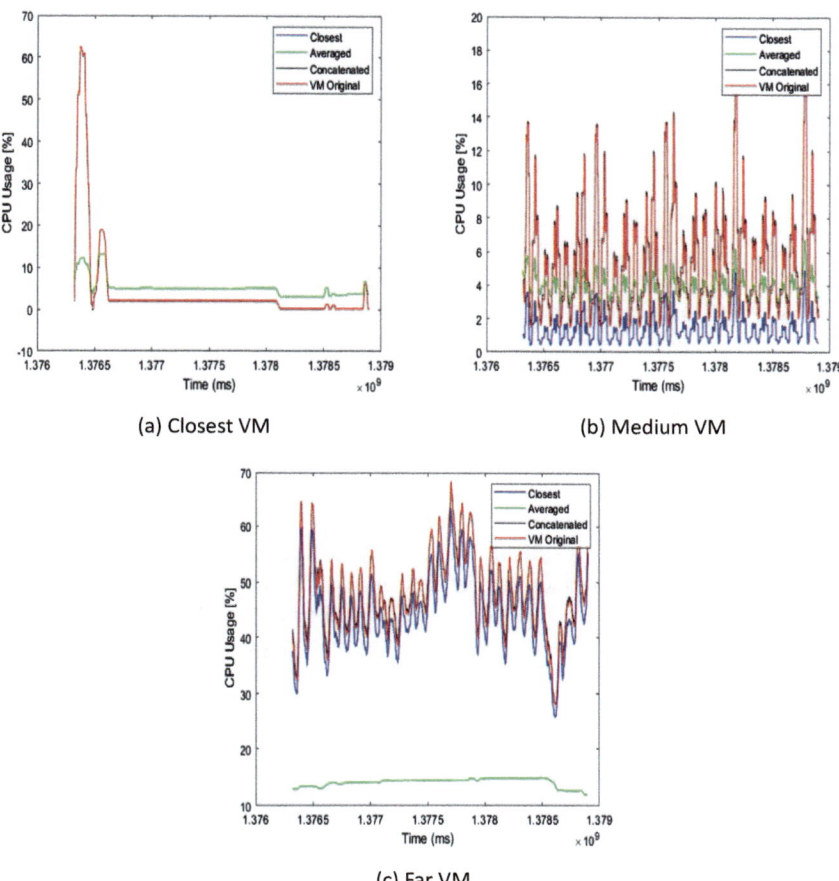

Fig. 3. Predicted CPU Usage using the three representative models for a Medium Size Cluster

centroid, one can notice that the trained models using the Closest VM or Concatenated VM provide predicted values close to the real CPU usage values, as shown in 3c.

Table 4 presents the RMSE values for three representative models for three cluster sizes (i.e. small, medium and large size). These values were calculated using the trained models for the representative VMs with other time series from three different VMs members of the cluster, which are at a near, medium and far distance from the cluster centroid.

Table 4. RMSE for Predicted CPU usage per cluster size

VM	Large Size			Medium Size			Small Size		
	Closest	Averaged	Concatenated	Closest	Averaged	Concatenated	Closest	Averaged	Concatenated
Near	0.00	0.03	0.00	0.00	4.94	0.06	0.00	0.0221	0.0036
Medium	0.16	2.21	0.08	3.98	3.66	0.34	0.06	0.01076	0.0063
Far	0.007	0.40	0.007	3.68	33.86	0.59	0.016	0.01587	0.2263

5 Conclusions

In this paper, we propose a system that consists of 3 stages: clustering of the virtual machines based on its own information (i.e. time series), evaluating several Linear Regression algorithms using the common features of VM time series such as the mean CPU and memory usage, standard deviation of CPU and memory usage and its length. This makes it easier to group VMs with similar behavior and establish clusters based on these features. Then, the training of the 3 representative models is performed to finally choose the one with the best results in each cluster. We conclude that the representative models allow the system to reduce the training of each individual VM while minimizing the prediction error for any VM in a given cluster. In future work, we want to evaluate more ML algorithms to see if it is possible to further reduce training times and cluster stability under dynamic workload changes in the DC.

References

1. Aceto, G., Botta, A., de Donato, W., Pescapè, A.: Cloud monitoring: a survey. Comput. Netw. **57**(9), 2093–2115 (2013). https://doi.org/10.1016/j.comnet.2013.04.001
2. Amazon: Time series forecasting principles with amazon forecast (2021)
3. Borkowski, M., Schulte, S., Hochreiner, C.: Predicting cloud resource utilization. In: IEEE/ACM 9th International Conference on Utility and Cloud Computing (UCC), pp. 37–42 (2016)
4. Box, G.E., Jenkins, G.M., Reinsel, G.C., Ljung, G.M.: Time Series Analysis: Forecasting and Control. John Wiley & Sons (2015)
5. Dai, L., Li, J.H.: An optimal resource allocation algorithm in cloud computing environment. Appl. Mech. Mater. **733**, 779–783 (2015)
6. Daraghmeh, M., Agarwal, A., Manzano, R., Zaman, M.: Time series forecasting using facebook prophet for cloud resource management. In: IEEE International Conference on Communications Workshops (ICC Workshops), pp. 1–6 (2021)
7. Estrada, R., Asanza, V., Torres, D., Bazurto, A., Valeriano, I.: Learning-based energy consumption prediction. Procedia Comput. Sci. **203**, 272–279 (2022)
8. Farahnakian, F., Liljeberg, P., Plosila, J.: LiRCUP: Linear regression based CPU usage prediction algorithm for live migration of virtual machines in data centers. In: Proceedings - 39th Euromicro Conference Series on Software Engineering and Advanced Applications, SEAA 2013, pp. 357–364 (2013)
9. Gill, P., Jain, N., Nagappan, N.: Understanding network failures in data centers: measurement, analysis, and implications. SIGCOMM Comput. Commun. Rev. **41**(4), 350–361 (2011)

10. Gupta, S., Dinesh, D.A.: Resource usage prediction of cloud workloads using deep bidirectional long short term memory networks. In: IEEE International Conference on Advanced Networks and Telecommunications Systems (ANTS), pp. 1–6 (2017)
11. Hindman, B., et al.: Mesos: a platform for fine-grained resource sharing in the data center. In: NSDI'11, pp. 295–308. USENIX Association, USA (2011)
12. Iqbal, W., Erradi, A., Mahmood, A.: Dynamic workload patterns prediction for proactive autoscaling of web applications. J. Netw. Comput. Appl. **124**, 94–107 (2018)
13. Janardhanan, D., Barrett, E.: Cpu workload forecasting of machines in data centers using lstm recurrent neural networks and arima models. In: 12th International Conference for Internet Technology and Secured Transactions (ICITST), pp. 55–60 (2017)
14. Kumar, J., Goomer, R., Singh, A.K.: Long short term memory recurrent neural network (LSTM-RNN) based workload forecasting model for cloud datacenters. Procedia Comput. Sci. **125**, 676–682 (2018)
15. Kumar, J., Singh, A.K.: Cloud datacenter workload estimation using error preventive time series forecasting models. Clust. Comput. **23**(2), 1363–1379 (2019). https://doi.org/10.1007/s10586-019-03003-2
16. Lindemann, B., Muller, T., Vietz, H., Jazdi, N., Weyrich, M.: A survey on long short-term memory networks for time series prediction. Procedia CIRP **99**, 650–655 (2021)
17. Mormul, M., Hirmer, P., Stach, C., Mitschang, B.: Dear: distributed evaluation of alerting rules. In: IEEE 13th International Conference on Cloud Computing (CLOUD), pp. 158–165 (2020)
18. Nashold, L., Krishnan, R.: Using lstm and sarima models to forecast cluster cpu usage. ArXiv abs/2007.08092 (2020)
19. Qiu, F., Zhang, B., Guo, J.: A deep learning approach for vm workload prediction in the cloud. In: 17th IEEE/ACIS International Conference on Software Engineering, Artificial Intelligence, Networking and Parallel/Distributed Computing (SNPD), pp. 319–324 (2016)
20. Rao, S.N., Shobha, G., Prabhu, S., Deepamala, N.: Time Series Forecasting methods suitable for prediction of CPU usage. In: 4th International Conference on Computational Systems and Information Technology for Sustainable Solution (CSITSS), vol. 4, pp. 1–5 (2019)
21. Sarikaa, S., Niranjana, S., Sri, K.V.D.: Time series forecasting of cloud resource usage. In: 6th International Conference on Computing, Communication and Automation (ICCCA), pp. 372–382 (2021)
22. TUDelft, D.U.o.T.: Dataset gwa-t-12-bitbrains. http://gwa.ewi.tudelft.nl/datasets/gwa-t-12-bitbrainss (2023)
23. U-chupala, P., Watashiba, Y., Ichikawa, K., Date, S., Iida, H.: Container rebalancing: Towards proactive linux containers placement optimization in a data center. In: IEEE 41st Annual Computer Software and Applications Conference (COMPSAC) 01, pp. 788–795 (2017)
24. Wang, J., Yan, Y., Guo, J.: Research on the prediction model of cpu utilization based on arimabp neural network (2016)
25. Xue, J., Yan, F., Birke, R., Chen, L.Y., Scherer, T., Smirni, E.: PRACTISE: robust prediction of data center time series. In: 11th International Conference on Network and Service Management (CNSM), pp. 126–134. IEEE (2015)

An Adaptive Virtual Node Management Method for Overlay Networks Based on Multiple Time Intervals

Tatsuya Kubo and Tomoya Kawakami[✉]

Graduate School of Engineering, University of Fukui, Fukui, Japan
`tomoya-k@u-fukui.ac.jp`

Abstract. IoT, where the low cost of sensor devices is a background for widespread use, handles much information closely related to location and time. To handle large amounts of sensor data efficiently, the authors have proposed a method to treat queries efficiently, mainly focusing on time ties. In the previous study, we proposed a method to improve efficiency by virtualization and changing the routing algorithm, but the effect was limited. Moreover, it did not consider differences in device performance, which is a characteristic of existing P2P networks. In this paper, we propose a method for adaptively allocating virtual nodes (computers). In the proposed, each physical node independently determines the number of virtual nodes based on its performance and the target size for the entire network. Experiments demonstrate that it can allocate nodes according to the processing capacity of nodes.

1 Introduction

From the prediction by Cisco, Machine to Machine (M2M) connections around the Internet will increase by 2 billion per year. The M2M is the basis of Internet of Things (IoT), and the traditional client-server system may not be able to deal with that much traffic because it will cause too much load, and the server will degrade or even stop working. Peer-to-peer (P2P) is a non-centralized and distributed system that each computer (node) runs an equivalent program. It will solve this overload problem because there is no centralized server.

The structured overlay network is a type of P2P systems which constructs a network with mathematical topology. Because the network always has a definite structure, it is easy to search for the data. Typical algorithms for the structured overlay networks like Chord [12] are called distributed hash table (DHT), and they use consistent hashing [5] to manage data for implement load balancing. On the other hand, data from sensors are frequently updated and deeply related to location or time. DHTs may not be suitable for these kinds of data, and they could cause redundant traffic or overload. Therefore, there are some overlay algorithms without hashing. For example, optimize for range queries [11]. In this work, however, we have focused on the time relation between sensor data and increased efficiency.

Sensor data, which plays a central role in IoT, is updated frequently and is deeply linked to position and time information. Therefore, it is necessary to have characteristics different from those of a common overlay network. There are various possible

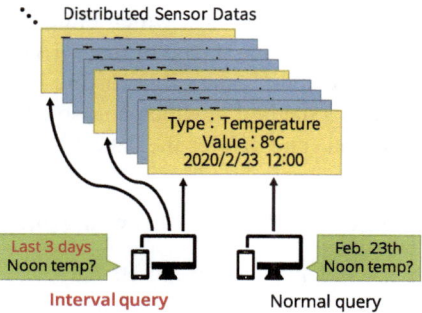

Fig. 1. Concept of the interval query

functions to be added, but here we describe in detail the research focusing on the connection between the data and time, which is the basis of this paper [8]. Network user may requests multiple data under the condition of a specific time interval due to the temporal connection of the sensor data. Examples are "July temperature every year" or "Every Sunday power consumption." Queries composed of such multiple queries are collectively called interval queries, and the outline is shown in Fig. 1. Interval queries are considered to be useful for big data and other uses such as obtaining statistical information from past data and making future predictions. In addition, we used node virtualization for load balancing [7].

The load-balancing method in Ref. [7] was all assumed to aim at equalizing the load and allocated the same number of virtual nodes to all nodes. However, P2P networks are not composed of a small number of servers under the control of an administrator, but a myriad of devices. Therefore, as a more realistic condition, we proposed a method for adaptively allocating virtual nodes. In the proposed method, each physical node independently determines the number of virtual nodes based on its performance and the target size for the entire network.

The related work such as a Chord network and overlay networks based on multiple different time intervals is described in Sect. 2. The proposed method is described in Sect. 3, and its evaluation is summarized in Sect. 4. The discussion is described in Sect. 5, and the conclusions are presented in Sect. 6.

2   Related Work

2.1   Chord

Chord overlay network has ring topology that makes algorithm easier [12]. Each node holds three types of route information: Successor, Predecessor, and Finger Table. Chord is known as the most representative algorithm of DHT and there are many related studies. For example, optimized for specific devices [13], extending the search to be performed on quadtrees and [14] generalized n-branch trees [1], allowing flexible selection of route tables [10], and hybrid networks combined with unstructured overlay networks [3].

Chord[#] [11] is one of the related studies, which eliminates the function of DHT. In Chord[#], keys are the raw value of the data, not the SHA-1 hash. If there is a lot of data for a specific key, the nodes that responsible for the keys will be overload. As a countermeasure, Ref. [11] uses a method in which a node with a low load dynamically enters next to an overloaded node and divides the data [6].

2.2 Overlay Networks Based on Multiple Different Time Intervals

To handle continuous data efficiently, we proposed the same ring network as Chord using the date as a key [8]. As with Chord[#], this method does not hash keys. The interval query is different from the simple range query because the request contains various units and the interval is longer than the unit time on the network. Therefore, we replaced Chord's Finger Table with one based on multiple time intervals assuming interval queries in this method.

In the methods not DHTs, the specific nodes must store a lot of data and will be overloaded if the data is biased to the specific time. Therefore, we used node virtualization for load distribution [7]. Virtualization means that a machine (referred to as a "physical node" when necessary) has not only one but multiple nodes (virtual nodes) on the network. This allows each physical node to have a larger area of responsibility in the network, and the impact of a single data on the network is relatively small. If the total number of virtual nodes is sufficient and if they are evenly and uniformly allocated to physical nodes, the load will be equalized stochastically. This is the opposite of DHT, which distributes the input data of the network by hash values, and aims to achieve the same effect by distributing the network side.

2.3 Adaptation to Virtualization

2.3.1 Disadvantages of Virtualization

The disadvantage of virtualization mentioned in the previous section is that the number of nodes on the network increases as a single physical node has multiple nodes. As the number of nodes increases, the size of the network grows, and the same query must be transmitted far beyond many nodes. Therefore, the overall number of hops for each query increases, and the total number of messages including relays also increases, which may increase the load on the entire network and the number of times each node forwards and replies to messages. In particular, since interval queries are always composed of multiple queries, the impact of an increase in the number of messages will be bigger, and countermeasures are necessary.

2.3.2 Improving Routing Efficiency

To mitigate this, we optimized the routing assuming virtual nodes [9]. Specifically, we proposed a routing method in which the transfer destination candidates selected during query transfer are expanded not only to virtual nodes but also to physical nodes. When a virtual node receives a query, it does not immediately determine the forwarding destination from its routing information but performs routing according to all information from its virtual nodes.

In this way, this method also adds the viewpoint from the higher-level physical node to the routing. In step 2, the virtual node only references its routing table and does not communicate with the outside. The processing to refer to the routing table increases, but it is not a big burden because it is the processing of $O(n)$ and the routing table has a fixed number of elements. If the physical node wants to send a query in step 3 from the different virtual nodes in step1, it can process by the program itself. It completely bypasses the network and shortens the route.

2.4 More Realistic Load Balancing

As in Sect. 2.2, where load balancing was performed by virtualizing nodes, load balancing methods in DHTs can be roughly divided into "optimization of node allocation," "node virtualization," and "popular data distribution" [4]. Optimization of node allocation is a static incorporated method into the network mechanism. It aims to equalize the load by eliminating bias in the keys, like using a hash function. For virtualization, in addition to the method which avoids overload by smoothing node placement in the key space, there is also a method that allows an arbitrary number of virtual nodes per physical node [2]. This makes physical nodes store data according to their resources and effectively use their performance.

3   Proposed Method

3.1   Overview

The load-balancing methods addressed in the previous studies [7] were all assumed to aim at equalizing the load. Therefore, the research in Sect. 2.2 allocated the same number of virtual nodes to all nodes.

However, P2P networks are not composed of a small number of servers under the control of an administrator, but a myriad of devices. Therefore, like servers, they are not systematically deployed to provide services, and the performance of each node differs in terms of computational processing power, network bandwidth, storage capacity, and so on. In other words, distributing the load evenly is not the same as distributing the load equally. Even if the load is evenly distributed, some physical nodes may become overloaded or, conversely, have too much performance. Therefore, as a more realistic condition, we assumed a case in which the performance of each physical node differs, and introduced the concept of varying the allocation for each physical node.

In this case, the first problem is to determine the performance evaluation index. This problem is complex. For example, mobile terminals have finite power resources, unlike stationary terminals. Here we assume that the performance metric is simply expressed as the number of virtual nodes that can be held. A physical node joins the network with an estimate of the number of virtual nodes it can have, based on the processing performance and bandwidth. To avoid exceeding this limit, more virtual nodes are placed on high-performance physical nodes and fewer on low-performance physical nodes. This method is a simple dichotomization model but can adequately reflect performance in many cases. Thus, the number of virtual nodes that each physical node can have is maintained, but it cannot have nodes up to these entire numbers due to the overhead of having too many virtual nodes. Instead, the network is adjusted so that a target number of virtual nodes is on the network.

3.2   Virtual Node Management

Here, we explain the specific allocation method. Suppose that a physical node V_k can have N_k virtual nodes join the network. V_k receives T, the number of target nodes in the network, and N_{Sum}, the total number of virtual nodes currently on the network. Equation (1) indicates the number of virtual nodes that V_k has.

$$\frac{n_k}{N_{Sum}} \cdot T \tag{1}$$

Since the actual number of virtual nodes is an integer, it is necessary to round the value of the (1), whose answer is a real number, to an integer. Three types of floor (2), ceiling (3), and round (4) are considered here.

$$V_k^F = \left\lfloor \frac{n_k}{N_{Sum}} \cdot T \right\rfloor \tag{2}$$

$$V_k^C = \left\lceil \frac{n_k}{N_{Sum}} \cdot T \right\rceil \tag{3}$$

$$V_k^R = \left\lfloor \frac{n_k}{N_{Sum}} \cdot T + 0.5 \right\rfloor \tag{4}$$

Among the variables used, N_{Sum}, in particular, is a network-wide indicator and cannot be obtained directly in a distributed environment. However, structured overlay networks require periodic communication with other nodes to maintain the routing table, which can be used to obtain this information. Specifically, the number of nodes is added when a new node is added, and information on the number of nodes is exchanged with other nodes when the routing table is periodically updated. Since each physical node always has one virtual node, the number of physical nodes is removed from T, i.e., T is the number of nodes to be increased from the number without virtualization.

3.3   Look-Ahead of Interval Query

The routing optimization shown in Sect. 2.3.2 was only aimed at preventing duplication of a single query in the forwarding path. However, an interval query consists of multiple consecutive queries. Although redundant communication within a single query has already been prevented, the same physical node may be responsible for the data again in subsequent queries, and redundant communication that has not been removed here may remain. We introduced a mechanism to look ahead across multiple queries to solve this.

A summary is shown in Fig. 2. Figure 2 shows an interval query consisting of five queries, where the number indicates the key, i.e., the time. Assuming that the key is in hours, the query requests data every two hours. Suppose that virtual node A_1 of physical

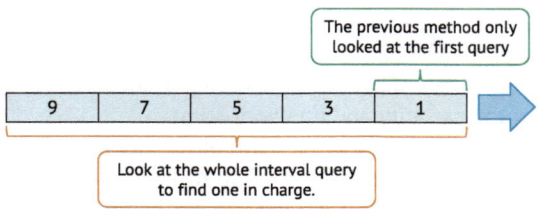

Fig. 2. The concept of look-ahead

Table 1. Total capacity in each distribution

	Exponential	Gaussian	Uniform
Sum	874	879	873
Average	4.37	4.39	4.37
Margin	9.25%	9.87%	9.13%

node A receives this interval query and that A has other virtual nodes $A_2, A_3, A_4, \cdots$. So far, the following procedure has been used to process the query:

1. A determines whether $A_1, A_2, A_3, \cdots$ is the range for which the first query 1 is responsible.
2. If A is in the range of responsibility, respond to the query and remove it from the interval query
3. Determine which of $A_1, A_2, A_3, \cdots$ is the most efficient place to send the first query, then send

However, for example, if A_3 is responsible for the data of 7 later, the interval query will be returned to A again without being noticed as a query that can be processed here.

To reduce this redundant communication, we changed procedure 1. as follows:

A determines if $A_1, A_2, A_3, \cdots$ are each in the range of responsibility of each element $(1, 3, 5, \cdots$ in this case) in the interval query

This operation prevents a query from going through the same physical node twice.

4 Evaluation

4.1 Simulation Environments

To verify whether it is possible to allocate virtual nodes according to the performance difference between physical nodes, we conducted a simulation that allocates 800 virtual nodes to 200 physical nodes. With the existing method, every physical node would be assigned four virtual nodes.

The performance of the 200 physical nodes follows three probability distributions: exponential, normal, and uniform. To adjust the total number of virtual nodes to near

800, we set the parameters so that the mean value is slightly larger than 4. Table 1 shows the total capacity for each distribution, and there is a margin of approximately 9% over the target of 800 virtual nodes. Therefore, with proper allocation, the load on all nodes should be kept below the capacity and no overloading.

We then evaluated the three rounding operations, rounding off, rounding down, and rounding up, corresponding to the formulas 2 to 4, which determine the allocation of virtual nodes for all distributions. It is assumed that it is difficult to determine the exact value of N_{Sum} in this equation, and each node obtains a value with an error of 5% from the true value.

We used the number of virtual nodes allocated and the load as evaluation indices. The load means the actual number of allocations divided by the allowable number of physical nodes. If the value exceeds 1.0, the node is overloaded. Conversely, a low value means that the performance of the physical node is not being utilized and that the node has a wasting capacity. Therefore, the goal is to keep the value as close to 1.0 as possible while avoiding exceeding 1.0.

4.2 Simulaiton Results

4.2.1 Capacity and Actual Allocations

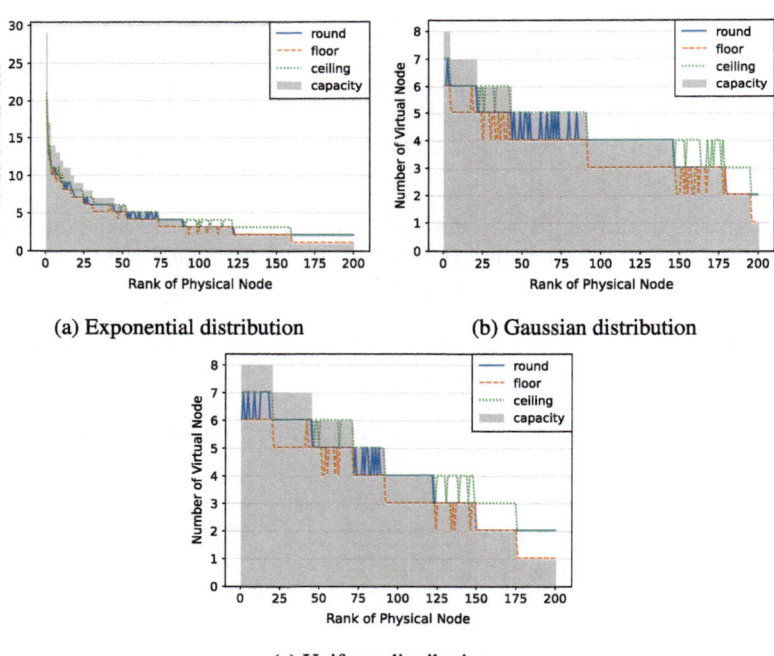

(a) Exponential distribution

(b) Gaussian distribution

(c) Uniform distribution

Fig. 3. Capacity and the actual number of allocated

First, Fig. 3 shows the capacity of the physical nodes in each of the three distributions and the number of virtual nodes allocated to each distribution. The horizontal axis corresponds to each physical node, sorted from largest to smallest capacity. The gray bars in the background indicate the capacity of each physical node, and the three broken lines indicate the actual number of allocations by round, floor, and ceiling, respectively. In these graphs, the existing method is omitted for clarity. If the existing method has shown, all nodes are assigned four virtual nodes, so a horizontal line appears at 4.

Figure 3a shows the case where the capacities of the physical nodes follow an exponential distribution. This distribution is a situation in which some nodes have high performance while the most of nodes have low performance, like Pareto's law. It can be seen that the allocation generally follows the distribution for both rounding methods. The allocation of virtual nodes varies mainly among the 50 to 125 physical nodes due to N_{Sum} errors, but this does not cause a significant increase in node load.

Among rounding methods, the ceiling, which allocates more nodes, always assigns one more node than the node capacity starting at about the 100th node. So the round and floor are superior in keeping the nodes from overloading. In particular, the round is superior in terms of effective utilization of performance, as it allocates more nodes even to the nodes with the highest capacity. On the other hand, two virtual nodes were continuously allocated to the node with the lowest capacity of 1, and only the floor could adapt in this section. In addition, there is almost no difference among the allocation methods for the 1–25 nodes with particularly high capacities. But there is a slight tendency that the allocation methods cannot keep up with a large amount of capacity allowed.

Figure 3b shows the case when the capacities of the physical nodes follow a Gaussian distribution, and Fig. 3c shows the uniform distribution. In both cases, the variance is smaller than the exponential distribution, so the performance difference is little, and more nodes have capacities near the mean value. Therefore, there is no clear indication that the allocation method can not keep up with the higher-performing nodes, and the utilization efficiency of the higher-performing nodes is increasing.

As in Fig. 3a, round and floor cause overloading of the nodes at the bottom end of the performance range. The ceiling made the best use of node performance for the high-performing nodes. However, in the Gaussian distribution, the floor allocated 3 virtual nodes for every node with a capacity of 4, which accounted for a quarter of the total. This reduced the utilization rate for many nodes.

4.2.2 Load on Physical Nodes

Next, Fig. 4 shows the load distribution on the physical nodes. The horizontal axis is the actual number of allocations divided by the allowable number of physical nodes, and the vertical axis is the empirical cumulative distribution. A load exceeding 1 indicates an overload, which should be suppressed. On the other hand, close to 0 means that the node is not effectively utilized. Therefore, it is ideal to keep the load as close to 1 as possible while avoiding exceeding 1. The figure also shows the existing method, in which all nodes are allocated four nodes. First, it can be seen that in all cases, the worst-case value of the proposed method is 2.0 times the load, which is lower than that of the existing one (4.0 times).

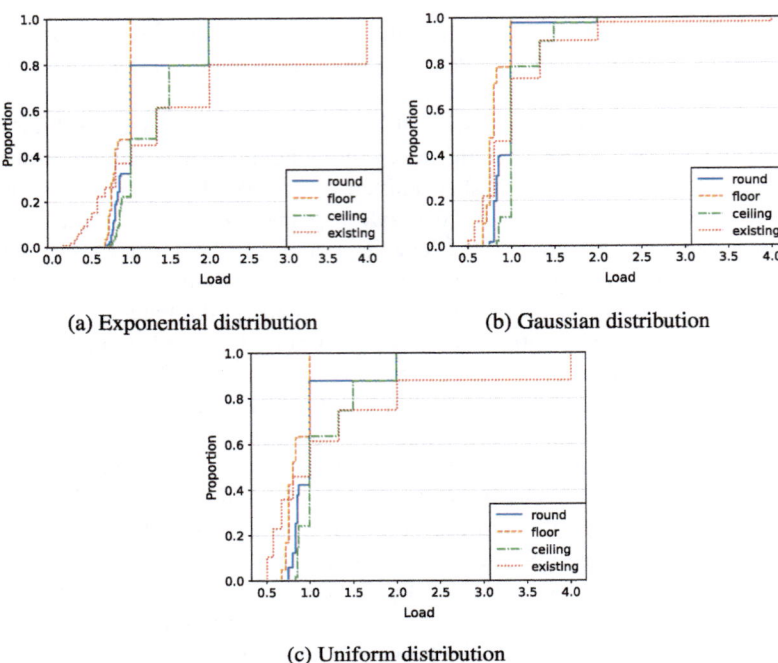

(a) Exponential distribution   (b) Gaussian distribution

(c) Uniform distribution

Fig. 4. Cumulative distribution of load

The exponential distribution of Fig. 4a is the most effective of the proposed method due to its large bias. First, while more than 50% of the nodes are overloaded by the existing method, the round could reduce to 20%, and the floor made no overload. Next, focusing on high-performance nodes with low loads, more than 10% of the nodes in the existing method have a load of 0.5 or less. In contrast, the proposed method made all nodes more than 70% loaded, indicating that the proposed method pulls the node performance more. The degree to which overloaded nodes are generated depends on the rounding method, and in the case of the ceiling, there is not much difference from the existing method. On the other hand, there is almost no difference among the rounding methods for the nodes with high performance, and the proposed method can increase the utilization ratio compared to the existing methods.

In the Gaussian distribution (Fig. 4b) and uniform distribution (Fig. 4c), where the difference in performance is small, the floor method, which allocates a smaller number of nodes, has a lower utilization rate than the existing methods. In particular, more than half of the nodes which are not overloaded in the Gaussian distribution are not utilized than the existing method. Only about 20% of the nodes have a load of 1, which is less than the other methods. This can be attributed to the fact that fewer virtual nodes are allocated to nodes that account for a large percentage of the total, like in the previous section. Therefore, the floor reduces load effectively but is less efficient in terms of utilization.

5   Discussion

The proposed method allocates a small number of nodes to low-performing nodes to avoid overloading them and a large number of nodes to high-performing nodes to increase their utilization, thus achieving the desired adaptive allocation of the number of virtual nodes. We examined several methods for rounding the number of nodes and found that flooring does not cause overloading but results in low node utilization, while ceiling tends to cause overloading but maintains high node utilization. Thus, there is a trade-off between node load and efficiency.

The choice of method to use depends on the importance of the node's performance indicators. In other words, if the performance indicator is only a reference that it is desirable to keep the load below this level in consideration of power efficiency and other factors, then the utilization efficiency should be increased. On the other hand, if the performance indicator is an absolute one that operation will be affected if the load exceeds this level, then priority should be given to reducing that. However, since the truncation method had some points where the utilization rate was lower than the existing method, it is not practical as it is and requires improvement by changing the criteria, etc.

We also found a difference of about 10% between the target number of virtual nodes and the number of virtual nodes allocated by the rounding method. However, we think it is unnecessary to strictly match this target number of nodes to the target value. This is because Chord, the foundation of this method, is capable of routing $O(\log N)$ messages for N nodes, and a change in the number of nodes of about 10% is not expected to have a significant impact on performance. Therefore, we believe that the node performance indicators are more important.

As an additional experiment, we measured the number of messages under the same conditions as in the previous study [9] with and without adaptive virtual node allocation, respectively. And we found a slight reduction in the number of messages, although the routing was unchanged. This is thought to be because of the routing optimization method of existing research described in the Sect. 2.3.2 became more effective due to the concentration of many virtual nodes on the high-performance node. This suggests that this method is also effective in reducing the load on the network, which is a significant advantage.

6   Conclusion

In this paper, we proposed an adaptive virtual node management method that considers differences in node performance to adapt to more realistic conditions and improve routing efficiency. Adaptive allocation was achieved by reducing the load on the nodes with low performance and increasing the utilization of the nodes with high performance.

As for future work, obtaining parameters such as N_{Sum} should be studied for virtual node allocation.

Acknowledgment. This work was partially supported by JSPS KAKENHI Grant Numbers JP22K12009 and community contribution projects for FY2023, University of Fukui.

References

1. Alima, L.O., El-Ansary, S., Brand, P., Haridi, S.: DKS(N, k, f): a family of low communication, scalable and fault-tolerant infrastructures for P2P applications. In: Proceedings of the 3rd IEEE/ACM International Symposium on Cluster Computing and the Grid (CCGRID 2003), pp. 344–350 (2003)
2. Dabek, F., Kaashoek, M.F., Karger, D., Morris, R., Stoica, I.: Wide-area cooperative storage with CFS. ACM SIGOPS Oper. Syst. Rev. **35**(5), 202–215 (2001)
3. Duan, Z., et al.: Two-layer hybrid peer-to-peer networks. Peer-to-Peer Netw. Appl. **10**, 1304–1322 (2017)
4. Felber, P., Kropf, P., Schiller, E., Serbu, S.: Survey on load balancing in peer-to-peer distributed hash tables. IEEE Commun. Surv. Tutor. **16**(1), 473–492 (2014)
5. Karger, D., Lehman, E., Leighton, T., Panigrahy, R., Levine, M., Lewin, D.: Consistent hashing and random trees: distributed caching protocols for relieving hot spots on the World Wide Web. In: Proceedings of the 29th Annual ACM Symposium on Theory of Computing (STOC 1997), pp. 654–663 (1997)
6. Karger, D.R., Ruhl, M.: Simple efficient load balancing algorithms for peer-to-peer systems. In: Proceedings of the 16th Annual ACM Symposium on Parallelism in Algorithms and Architectures (SPAA 2004), pp. 36–43 (2004)
7. Kawakami, T.: A node virtualization scheme for structured overlay networks based on multiple different time intervals. Appl. Sci. **10**(8596) (2020)
8. Kawakami, T.: A structured overlay network scheme based on multiple different time intervals. J. Inf. Process. Syst. **16**(6), 1447–1458 (2020)
9. Kubo, T., Kawakami, T.: An enhanced routing method for overlay networks based on multiple different time intervals. In: Proceedings of the 2nd IEEE International Workshop on Advanced IoT Computing (AIOT 2021) in Conjunction with the 45th Annual International Computer, Software and Applications Conference (COMPSAC 2021), pp. 1560–1565 (2021)
10. Nagao, H., Shudo, K.: Flexible routing tables: designing routing algorithms for overlays based on a total order on a routing table set. In: 2011 IEEE International Conference on Peer-to-Peer Computing (P2P 2011), pp. 72–81 (2011)
11. Schütt, T., Schintke, F., Reinefeld, A.: Range queries on structured overlay networks. Comput. Commun. **31**(2), 280–291 (2008)
12. Stoica, I., et al.: Chord: a scalable peer-to-peer lookup protocol for Internet applications. IEEE/ACM Trans. Netw. **11**(1), 17–32 (2003)
13. Tetarave, S.K., Tripathy, S., Ghosh, R.K.: V-Chord: an efficient file sharing on LTE/GSM network. In: Proceedings of ICDCN 2018 (2018)
14. Wu, J.G., Jiang, N., Zou, Z.Q., Hu, B., Huang, L., Feng, J.L.: HPSIN: a new hybrid P2P spatial indexing network. J. China Univ. Posts Telecom **17**(3), 66–72 (2010)

Ride-Sharing Allocation System and Optimal Path-Finding Algorithm for Marine Taxies in the Setouchi Inland Sea Area

Shiojiri Ryota[(⊠)], Takegami Risa, Murakami Yukikazu, Tokunaga Hidekazu, and Kimura Yuto

National Institute of Technology, Kagawa College, 355 Chokushi, Takamatsu, Kagawa 761-8058, Japan
{st22413,h-tokunaga,kimura-y}@kagawa.kosen-ac.jp,
murakami@t.kagawa-nct.ac.jp

Abstract. Although marine taxis offer advantages in terms of time flexibility, they face challenges owing to their chartered nature, resulting in high unit costs for reservations and limited passenger capacity. This study aims to facilitate the sharing of reservations for multiple marine taxi journeys while optimizing the routes to all destinations. The first approach proposes an algorithm that efficiently pools reservations based on three key factors: departure and arrival ports, and number of passengers. Thereafter, we developed an algorithm that calculates optimal routes for marine taxis by considering the impact of factors such as tidal currents and wave magnitudes on the vessel.

1 Introduction

The islands of the Seto Inland Sea are now attracting attention from Japan and abroad as world-famous art sites. At the Setouchi Triennale, various artworks are exhibited on each of the islands scattered throughout the Seto Inland Sea. [1] This has increased the demand for marine transportation to visit more islands in the Seto Inland Sea [2]. One means of marine transportation is the sea taxi, a small chartered service. Sea taxies are usually chartered, which means that there are many empty seats compared to the capacity of the ship, especially when a small number of people make reservations. To solve this problem, we propose an algorithm that allows multiple reservations to be combined into a "shared-ride" system for efficient transportation. Sea taxi services are chartered by small vessels, which are susceptible to tidal currents and waves. Therefore, we also propose an algorithm to find the optimal route from the departure port to the destination port, taking into account environmental factors.

2 Ride-Sharing Allocation System

2.1 Problem Definition

Consider the scenario depicted in Fig. 1, where the captain of a marine taxi company aims to allocate multiple reservations to the vessels in their fleet. Temporal considerations are not deliberated, and it is assumed that all reservations are for simultaneous boarding.

© The Author(s), under exclusive license to Springer Nature Switzerland AG 2023
L. Barolli (Ed.): CISIS 2023, LNDECT 176, pp. 233–242, 2023.
https://doi.org/10.1007/978-3-031-35734-3_23

Each reservation is composed of three components, namely the number of passengers, departure port, and arrival port. In this problem, it is feasible to assign several reservations to the same vessel, which we refer to as "carpooling" in this study. However, not all reservations can be shared, and carpooling is only permissible under the following two conditions.

- The total number of passengers allocated to the vessel does not surpass the vessel's capacity.
- The assigned departure/arrival port combinations fulfill specific criteria. The objective of the agent is to allocate as many reservations as possible per vessel without violating these conditions.

The aforementioned issue is being addressed within the context of a reinforcement learning framework. The term "environment" is used to refer to the "ship allocation scenario". To be precise, it pertains to the circumstance wherein each booking is assigned to a particular vessel. The "agent" modifies the aforementioned "ship allocation status" in a multitude of ways and contemplates distinct ship assignments. The reinforcement learning algorithm employed for this purpose was PPO [3].

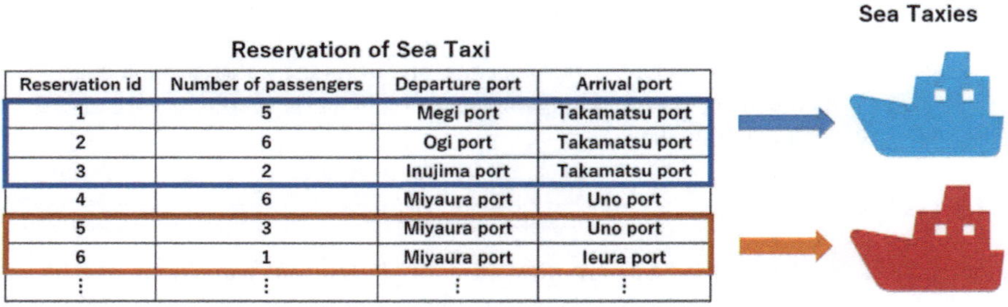

Sea Taxies

Reservation of Sea Taxi

Reservation id	Number of passengers	Departure port	Arrival port
1	5	Megi port	Takamatsu port
2	6	Ogi port	Takamatsu port
3	2	Inujima port	Takamatsu port
4	6	Miyaura port	Uno port
5	3	Miyaura port	Uno port
6	1	Miyaura port	Ieura port
⋮	⋮	⋮	⋮

Fig. 1. Problem Image Diagram

2.2 Program Implementation

Reservation Model

In this study, reservations made by passengers are considered as records comprising three distinct elements: the number of passengers, the departure port, and the arrival port. Each of these elements can be expressed in non-negative integer values. An example of a reservation record is presented in Table 1. The value assigned to the "number of passengers" is the total count of passengers. The values designated for "Departure Port" and "Arrival Port" relate to the ports and points that can be traversed by marine cabs, respectively. It is worth noting that the "Departure Port" and "Arrival Port" variables have different numerical values.

In this case, the serial number "reservation id" assigned to the reservation is defined as j,

$$j = 1, 2, \ldots, J. \tag{1}$$

Table 1. Example of Reservation record

Reservation id	Number of passengers	Departure port	Arrival port
1	1	1	2
2	11	1	2
3	1	2	3

Consider a natural number j, where the maximum value J represents the total count of reservations processed by the algorithm. Let r_j denote the number of passengers associated with the j-th reservation, t_j denote the arrival port, and f_j denote the departure port. The variables are defined as follows: p is the index assigned to each port; P denotes the total number of ports; r_j denotes the count of passengers; t_j denotes the arrival port index; and f_j denotes the departure port index.

$$r_1, r_2, \ldots, r_j, \ldots, r_J > 0. \tag{2}$$

$$t_j = 1, 2, \ldots, p, \ldots, P. \tag{3}$$

$$f_j = 1, 2, \ldots, p, \ldots, P. \tag{4}$$

Ship Model

Define the "ship id," k possessed by the captain as follows:

$$k = 1, 2, \ldots, K. \tag{5}$$

where K is the total number of vessels owned by the captain. Each ship has a capacity, which is the maximum number of people that can be on board. The capacity C_k of ship k is defined as follows.

$$C_1, C_2, \ldots, C_k, \ldots, C_K > 0. \tag{6}$$

State - Action - Reward

We formulate the state, action, and reward of reinforcement learning. The "state" is a matrix that indicates " the vessels which are allocated reservations".

$$\mathbf{S} \in \{0, 1\}^{J \times K}. \tag{7}$$

The binary matrix that represents the assignment of reservations to ships has a dimension of J rows by K columns, where each entry can take a value of either 0 or 1. Specifically, when the j-th reservation is assigned to the k-th ship, the corresponding entry S_{jk} in the matrix takes the value of 1. Conversely, when all entries in row j are 0, it implies that the j-th reservation has not been assigned to any ship. The "action" referred to in this context is the matrix that captures the process of "assigning a reservation to a ship".

$$\mathbf{A} \in \{0, 1\}^{J \times K}. \tag{8}$$

It is analogous to the term "state" and is denoted by a J-by-K binary matrix with binary values of 0 or 1. The value of $a_{jk} = 1$ indicates that the action of assigning k to reservation j has been carried out, whereas the value of $a_{jk} = 0$ implies that no operation has been performed on the corresponding reservation. Consequently, the action executed on S_{jk}, based on the value of a_{jk}, is defined as follows.

$$\begin{cases} S_{jk} \leftarrow 1, S_{jk'} \leftarrow 0 \ a_{jk} = 1, \\ S_{jk} \leftarrow S_{jk} \quad\quad\quad a_{jk} = 0. \end{cases} \tag{9}$$

However, it should be noted that, in this equation, k' is not equal to k. Finally, we contemplate the notion of "reward," which is quantified as a score that is determined by the number of reservations assigned to the vessel and the number of passengers on board. The reward can be defined as follows:

$$R = 100 \times \sum_{j,k} S_{jk} + 10 \times \sum_{j,k} S_{jk} r_j. \tag{10}$$

The greater the number of allocated reservations, the more substantial the reward; and if the number of reservations allocated is identical, the reward is proportional to the number of individuals on the vessel. Here, the objective is to maximize this score, as it is a measure of good ship allocation.

3 Optimal Path Finding Algorithm

3.1 Sea Area Setting

Typically, marine vessels are not engineered for extensive voyages, and it is deemed appropriate for a solitary journey to cover a distance of less than 62.1 [ml]. Furthermore, as the Seto Inland Sea is relatively tranquil with low wave amplitudes, our simulation did not incorporate a rigorous computation of the influence of wave amplitudes on the velocity of the ship. Instead, we regulated the speed of the ship in response to the wave amplitudes. The impact of waves on a ship during a prolonged voyage requires contemplation of factors such as the bow angle of the ship with respect to the direction of the waves, the mass of the ship, and the contact area between the ship and the seawater. Therefore, during the simulation of long-distance voyages, the discrepancy with actual navigation becomes considerably large. To account for these factors, a restricted region within the Seto Inland Sea, where marine cabs are commonly used, was designated as the area of study. Specifically, the geographical extent of the study area spanned from 34.1° to 34.6° north latitude and 132.6° to 134.4° east longitude.

3.2 Analysis Method

Consideration of Tidal Currents and Waves
To minimize the distance between the departure and destination ports, the direction of travel is determined by resolving the simultaneous equations of motion, which are Eqs. (11) to (16).

$$V_x = V\cos\theta - u_x. \tag{11}$$

$$V_y = V \sin \theta - u_y. \tag{12}$$

$$\theta_t = tan^{-1} \frac{V_y}{V_x} \tag{13}$$

Initially, the solution for the simultaneous Eqs. (11), (12), and (13) must be derived to determine the optimal heading for the vessel. Herein, V represents the velocity of the vessel sans the impact of waves and currents, and θ is the angle between the location and destination of the vessel. To attain V_x and V_y, which correspond to the speed of the vessel in the x and y directions after adjusting for the speed of the current u_x and u_y, respectively, the said currents must be subtracted from V. Subsequently, the angle θ_t that the ship should assume can be determined as follows,

$$V_t = \sqrt{V_x^2 + V_y^2} - v. \tag{14}$$

$$v = 0.04B + 0.07B^2. \tag{15}$$

$$B = \left(\frac{H}{0.0697}\right)^{\frac{1}{2.1}}. \tag{16}$$

Using the information derived from Eqs. (11), (12), and (13), the speed of the vessel can be obtained using Eqs. (14), (15), and (16). The deceleration of the vessel resulting from waves, v, is evaluated based on the wave height, H, and deducted as outlined in Eq. (12) to establish the speed of the vessel, V_t, after accounting for the impact of currents and waves [4]. The effects of currents and waves on the ship are calculated using these equations, as it heads in the desired direction, allowing it to proceed in the intended direction.

Obstacle Avoidance
In practical navigation, it is common to encounter obstacles such as islands and reefs that hinder direct passage between departure and destination ports. In such scenarios, it becomes necessary to chart a course that circumvents these impediments. The map is first treated as a two-dimensional coordinate system, with a straight line connecting the departure and destination ports represented as a linear function. If the path intersects an obstacle, a linear function for the normal perpendicular to the surface of the obstacle is considered instead. The maximum distance to the left and right of the path, respectively, is then identified in the direction of travel. Subsequently, the path is divided into two, one circumnavigating the obstacle on the left and the other on the right, and each path is explored separately. By repeatedly performing this procedure recursively whenever an obstruction is encountered along the path, multiple routes to the destination are generated. The route with the shortest distance is then assumed to be optimal before accounting for currents and waves. Finally, the direction of travel is determined by computing the effects of currents and waves, as described in the previous section.

Routing Recalculation at Regular Intervals

Upon determining the optimal course from the starting port to the destination port, it is not guaranteed that the ship would necessarily follow it. Furthermore, discrepancies exist between the external variables that affect the ship and the calculated variables. Hence, we incorporated a mechanism that minimizes these errors by computing the optimal route at regular intervals, allowing us to always be aware of the optimal course from the current location. In this investigation, the frequency of recalculating the optimal course was set at an interval of 30 s.

3.3 Sea Driving Test

Test Overview

We conducted a sea trial using a rescue boat from the National Institute of Technology Kagawa College between Takamatsu Port and Megijima Island and made a comparison between the measured values and the simulated ones. We compared the time taken by the driver to navigate to the destination port with the time taken when the boat advanced at the optimum angle calculated by simulation. The distance between Takamatsu Port and Megijima was 2434.8 m, and the engine speed was maintained at 2000 rpm. The tidal current forecast used in the simulation was sourced from MIRC Marine Information, and the wave height forecast was obtained from Yahoo!

Test Result

Table 2 shows the results of the test.

Table 2. Elapsed time for conventional and proposed methods (min: sec)

	Takamatsu Port → Megijima Island	Megijima Island → Takamatsu Port
conventional method (measured value)	10:17.6	10:17.9
Proposed Method (1st measured value)	09:41.0	10:12.2
Proposed Method (1st predicted value)	09:51.7	10:08.2
Proposed Method (2nd measured value)	09:40.1	10:40.8
Proposed Method (2nd Predicted value)	09:40.8	10:13.3

Initially, the conventional method that solely relies on human senses consumed around 10 min and 17 s for both the Takamatsu Port-Megijima and MegijimaTakamatsu Port routes. However, the proposed method, which adopted the same conditions to determine the optimal angle, proved to be more efficient than the conventional method for both routes, taking less time to complete. The measured time of the proposed method was shorter than that of the conventional method under identical circumstances. Notably, for each of the four measured times using the proposed method, the predicted time was longer than the measured time for the Megijima-Takamatsu Port route, while the predicted time for the Takamatsu PortMegijima route was longer than the time measured using the proposed method. One probable reason for the difference between the simulated and measured values using the proposed method is the inadequate calculation of wave effects. For future improvement, we suggest reducing the error by accounting for the angle of the head to the direction of the waves, the weight of the vessel, and the surface area where the vessel comes in contact with seawater while calculating the amount of deceleration, in addition to the constant speed deceleration according to wave height.

3.4 Future Issues

To operationalize the simulator, we propose the incorporation of GPS technology to obtain the current location of the vessel. We suggest the use of Google Maps JavaScript API Elevation service to discern between land and sea when navigating near islands. The Elevation service returns positive values for land coordinates and negative values for sea coordinates, which provides a basis for determining the terrain. Additionally, the simulation was conducted in Euclidean space, but the spherical shape of the Earth introduces errors when navigating. Hence, we recommend modifying the simulation program input in this study to operate based on latitude and longitude instead of Euclidean space coordinates.

4 Simulation Results

4.1 Ride-Sharing Allocation System

Twenty reservation records were generated, and a scenario was assumed where two vessels were allocated to carpool these reservations. The optimal carpooling patterns for each of the generated reservations were as follows. Although several combinations may lead to such patterns, the reward in both cases was 700, which represents the optimal solution. The test was executed with 100 training sessions, 10 operations per episode, and 1,000 trials. Table 3 displays the top three rewards that were frequently obtained after three runs of the program, each comprising 1,000 trials. The figures in parentheses specify the number of times the rewarding combination was chosen throughout the trials.

These outcomes suggest that the present program can allocate vessels to some degree, while it remains challenging to achieve an optimal resolution.

4.2 Optimal Path Finding Algorithm

An instance of a simulation that searches for an optimal route for a ship is presented, where the optimal route from the port of Naoshima to the port of Shodoshima is obtained.

Table 3. Results with 100 studies

	1st run	2nd run	3rd run
1	460(313)	450(271)	430(727)
2	430(267)	240(238)	450(196)
3	280(140)	230(165)	570(41)

During the analysis, a map image which depicts the departure and destination ports is loaded. Yahoo! Maps was used in the analysis. The pixel size in the Euclidean space was considered as 1 in the computation. Dummy data were employed for the tidal currents, where $(u_x, u_y) = (-1.8, 0.2)$ [m/s], and the east and north directions were consistently positive for all coordinates owing to the simulation area being small in the Seto Inland Sea. The wave height was uniformly set to $H = 1.5$ [m] for all coordinates as dummy data.

When searching for a potential route for the vessel, two paths are explored, one to the right and the other to the left, every time it maneuvers around an island. Hence, as the sailing distance increases and the obstacles between the departure and destination ports become more complex, an increasing number of potential paths will be searched. If some of the solutions diverge owing to obstacles or the vessel fails to reach the destination port altogether, or when a detour route is being investigated, the search may become protracted. Thus, an upper limit is established for the number of turns and solutions that do not converge or detours from the candidate optimal routes are excluded. In this simulation, when the number of turns reaches 30, the path is removed from the list of candidate shortest paths as an unsuitable path, and the next candidate path is sought. Figure 2 exhibits the simulation outcomes.

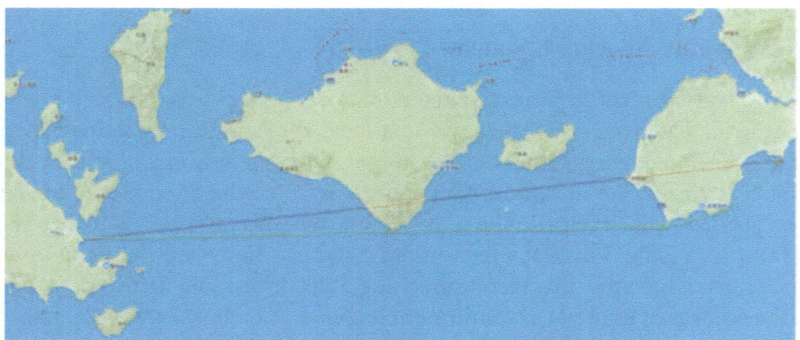

Fig. 2. Optimal route from Naoshima to Shodoshima

The upper arrow in Fig. 2 represents the straight line between the departure port and the destination port, whereas the lower arrow represents the optimal route between these two ports. The simulation outcomes are deemed accurate as they are equivalent to the optimal path that can be anticipated based on their appearance. Table 4 illustrates the distance to be covered, elapsed time, and direction to be taken by the vessel at regular intervals once the optimal path has been established. While the optimal path is determined every 30 s in actual simulations, this paper showcases the simulation results of the optimal path at 5-min intervals. The last line of the Table 4 shows the time, distance covered, and direction of travel upon the vessel arriving at the destination port on Shodoshima Island.

Table 4. Simulation result

Elapsed time(min)	Distance traveled (km)	Direction of travel(°)
0	0	1.9678
5	2.9564	1.9678
10	5.9128	1.9678
15	8.8607	1.1696
20	11.7992	1.1696
25	14.7377	15.3546
27.9998	17.3100	41.8631

5 Related Work

This chapter presents related research. Firstly, we describe related work on The Sharing Allocation System. Kurozumi et al. [5] investigated a Min–Max type cab dispatching problem, considering shared-riding. In their study, they introduced a standard deviation into the objective function to minimize the longest route taken by shared cabs. To solve this problem, we proposed a new neighborhood operation, combined with the tabu search method. Our approach differs from the tabu search method in that we use the optimal combination for carpooling, rather than individual cars. While Kurozumi et al. evaluated combinations based on the shortest path distance, we evaluated a combination that satisfies constraint conditions as a superior combination, with a larger number of passengers on board.

Next, we discuss related research on the Optimal Path Finding Algorithm. Hukuchi et al. [6] used a time-optimal control method to find the voyage time of a ship, which minimizes the evaluation function. They obtained the optimal path by considering the effects of waves and ocean currents. In comparison, Hagiwara et al. [7] developed a simulation program using forward dynamic programming to calculate the optimal route of an aircraft. Specifically, they studied the flight performance of a Boeing 747,400 flying between Narita and San Francisco, and simulated optimal routes by clarifying the relationship between flight altitude, airspeed, and fuel consumption relative to aircraft weight. To find the optimal path, they set transit points at regular intervals between destinations and recalculated the optimal path while the aircraft flew towards the next transit point, rather than the original destination. Similarly, we applied this method to a small vessel, a sea taxi, in our research.

6 Conclusions

In this study, an algorithm to identify optimal routes from departure to destination ports was presented, aiming to enhance the efficiency of marine transportation. Initially, a route which circumvents obstacles between the departure and destination ports was planned, and subsequently, the direction of navigation was computed after compensating for environmental factors, such as currents and waves, along the route. In future studies, an additional function will be incorporated to enable the collection of real-time data such as tidal currents and waves, through an API, for practical applications.

References

1. Setouchi Triennale Committee: Setouchi Triennale 2022. https://setouchiartfest.jp/about/out line2022.html
2. "PortRait Takamatsu Port, a hub of logistics and liveliness connected to the world", https://www.umeshunkyo.or.jp/209/309/index.html, "Japan Landfill Dredging Association 2020"
3. Schulman, J., Wolski, F., Dhariwal, P., Radford, A., Klimov, O.: Proximal policy optimization algorithms. arXiv preprint arXiv:1707.06347
4. Akira, H., Akihiko, H.: Optimal Routing to the Destination. Japan Institute of Navigation **83** (1990)
5. Suzuho, K., Rei, E., Hiroyuki, E.: min-max type cab dispatching problem considering shared rides. Information Processing Society of Japan **84** (2022)
6. Op.cit: Optimal Routing to the Destination. Japan Institute of Navigation **83** (1990)
7. Hagiwara, H., Suzuki, R., Ikuta, Y.: A Study on the Optimization of Aircraft Route. Japan Institute of Navigation **91** (1994)

Issues and Challenges When Metaverse Replaces the Workplace

Ardian Adhiatma(✉), Nurhidayati, and Olivia Fachrunnisa

Department of Management, Faculty of Economics, UNISSULA, Semarang, Indonesia
{ardian,olivia.fachrunnisa}@unissula.ac.id

Abstract. This study aims to develop a conceptual framework that describes the essentials needed when an organization uses the metaverse as an alternative to virtual offices. Recent discussion in the existing literature widely concludes that hybrid working offers high productivity, wellbeing, and employee mental health. However, not all work types can be done through metaverse, while certain parts of work might be finished in the metaverse. Another discussion also offers that the metaverse as a place for leisure rather than work. Drawing from a review of the current literature and interviews with three senior leaders, we provide detailed insight ranging from essential needs to strategy for maximizing the metaverse as a virtual office, an action list to be taken by the top management team, and modifications to organizational policies and practices that can be considered for implementation. The specific outcome to be targeted in this research is to understand the opportunities and challenges of the future of work and workplace.

Keywords: Metaverse · Future work · Future workplace · Mental health · Wellbeing · Virtual environment

1 Introduction: Metaverse as Future Workplace

Mark Zukerberg has imagined a future working in the metaverse, with people teleporting as holograms present at the office, banishing the need for commuting. This can bring about a sense of awkwardness, as this vision promotes the 'feeling' of being in the room together, having a shared sense of space, and making eye contact. However, there are other uses of VR in the workplace, such as technical training, socializing, and onboarding. For Meta, the challenge will be understanding and proving where VR will make the difference. This continues to be an issue that they have not been clear about. Various research has been conducted on the advantages of the metaverse, with varying conclusions. There are also discussions on what must be prepared in order to implement hybrid offices. Additionally, issues on how much one can handle anxiety, migraine, and nausea as a consequence of virtual or hybrid offices linger to be solved. As such, there are still opportunities for a wider discussion on the chances of utilizing the metaverse as an option for virtual offices.

Numerous studies have proven that hybrid work positively impacts productivity. For instance, [1] mentions that working in the metaverse can help reduce issues with

L. Barolli (Ed.): CISIS 2023, LNDECT 176, pp. 243–249, 2023.
https://doi.org/10.1007/978-3-031-35734-3_24

overpopulation and help maintain environmental sustainability. At the same time, [2] also suggests that the metaverse will accelerate the success of digital transformation in the workplace. Results from a literature survey conducted by [3] on employee experience in the metaverse provide a summary that the requirements on its infrastructure, such as text mining and analytics and data visualization, will become part of the hard infrastructure, and employee adaptability, as part of the soft infrastructure, is a vital elements in need of awareness. Aside from that, metaverse will be of frequent use in training programs and development [4], recruitment [5] and also within the area of customer service [6].

When various research and opinions state that the metaverse is an alternative virtual office that results in many benefits [7, 8], what do organizations need to prepare to respond to this phenomenon? This article aims to provide insights into what is needed in designing a metaverse as a hybrid or virtual office so that factors such as productivity and the mental health of each organizational member are well maintained.

2   Finding Insights: Methodology

To gain insight into these changes and how to prepare for them, we conducted a survey of 17 literatures with the theme of the future of workplace, metaverse as a workplace, the digital office, the virtual office, and essentials things towards hybrid works. In addition, we also analyzed the transcript of an interview with three senior consultants of leading executive search firms, published in YouTube and as a podcast. We did so understand how they perceived the changing role of leaders, given that they assist organizations in identifying the metaverse as an office alternative.

3   Essentials Need for Metaverse as Future Workplace

3.1   Digital Global Leadership

Global and digital leadership is an activity within cognitive competence that results in effective behavior that is effective in understanding digitalization prospects. Whereby they can carry out all their assignments with little to no obstacles due to the ease of digitalization. Each competent global leader will have a stronger instinct for managing relevant information, observing contextual signals, behaviors, and cultures, along with the capability to connect this information, resulting in a meaningful pattern. Global leaders can also develop a complex perception regarding their very own context of productivity. Global leaders differentiate between their work context and their work process, especially if a leader possesses the initiative to conduct change on a global scale.

3.2   Digital Ethics

The digital environment carries specific challenges for building communication and socialization ethics, which are then referred to as 'digital work ethics'. As expected, we cannot easily convert the similarity between work ethics in the offline environment and work ethics in the virtual environment. Digital ethics have been discussed many times in previous literature and are defined as a formulae system or rules in communication behavior for maintaining a relationship with other stakeholders according to their respective roles and positions both within the formal and informal relationship through the digital media. [9] defines digital ethics as individual values and ethical morals in using technology responsibly within the digital era.

In this research, we argue that leaders need to build digital ethics for spiritual engagement of employees. Since the possibility of work relationships in the metaverse is the limited of face-to-face interaction, this does not mean an ignorance of the importance of shaping employee engagement. By initiating 'digital work ethics' in the metaverse, we hope that employees will still engage with the organization's goals and values.

4 Strategy to Maximize Metaverse as Future Workplace

Our analysis from the existing literature and by observing a few transcripts on how to maximize metaverse as future workplace led us to the following three strategies that must be implemented by leaders:

a. Encourage the productivity

When organizations decide to carry out the hybrid working policy through the medium of a virtual office, leaders must assure the measurement of productivity and outcome that must be met by each member of the organization. Employees will be given freedom to finalize their work, utilizing the resources that are made available within the organization, and also make sure that work can be delivered in a timely manner.

b. Encourage the flexibility

The use of virtual offices offers flexibility in works schedule and pattern of executing work. As such, policies on the use of metaverse as a virtual office must be balanced with work flexibility. Research shows that work flexibility will work out when stabilized with work goal certainty, adequate facilities, and work characteristics that are capable of being completed from anywhere or working from anywhere. This work flexibility will make allow each member of the organization that collaborates in different location and different time zone to arrange a mutual decision in when and how they will carry out work.

c. Encourage the connectivity

However, each individual that is part of a virtual work team is a human. They possess the basic needs to socialize and connect with each other. When the metaverse becomes an option, leaders must make sure that sometime in the future, they can have physical rooms to socialize with each other and build emotional bonding. Nowadays, there are many applications and tools that can be utilized to facilitate affective connection in the virtual room. However, we argue that a physical office today is not only for completing work, but more than that, it is a room to socialize and fulfill connective needs both mentally and spiritually. Offices or work cubicles are not only defined as a place to complete work but more so as a 'value space and learning space', a place to increase the value that we have as human beings and also a place to continue learning to become a better self. As a 'value space' offices provide a social room for their employees, facilitate relationships, and create innovative collaborations.

5   Action List for Top Management Team

The following actions must be taken by the top management team when creating the metaverse as a virtual workplace:

a. Build shared understanding

Shared understanding is an effort to create a similar understanding and perception among all individuals within an organization regarding the foundational philosophy of why the top management team creates the metaverse as an alternative. This is due to the possibility that not all departments within an organization are befitting of carrying out their work within a virtual space. This shared understanding can also prevent the presence of misleading information or jealousy.

b. Build shared identity

This social identity in a virtual environment is important. Identity is a tool to understand individual actions, thoughts, and even individual feelings when joining a community. This virtual identity engagement will be beneficial for an organization because, firstly, when members of the organizations interact, they will share positive behavioral elements such as affection, motivation, and attributes that can increase performance, such as collective efficacy and a higher degree of team potential. Second, each member of the organization will compare their input and output within the organization. This is what is called social comparison. Each individual will participate in a competition to adjust their engagement in comparison with other members of the team. As such, it is clear that engagement built upon a shared identity in the virtual community will increase organizational performance.

c. Build Value Co-creation

The third element that an organization requires to go fully virtual or hybrid is ensuring that each individual holds the same values. This value is not the sum of each individual's values, but rather a co-creation of values between leaders and members of the organization. In our previous research, we argued that value creation is the desired goal of the organization, to help it understand the needs of its members. If value can be created, then this effort will support rapid learning because experience among members is an efficient way to create value.

6   Modification of Organizational Practices and Policies

With regard to converting to hybrid work, we conclude that an organization needs to modify some organizational practices and policies. These practices and policies correspond to how an organization manages its human capital.

6.1   Gamified Based HRM Practices

Aligning with the increase in discussions on the metaverse as an alternative style of work office, the management of employees based on technology through the gamification approach will be appropriately fitting to implement in an organization. The gamification approach is an approach to human resources management based on games supported by the use of technology that exude relaxedness for employees and also increase their competence to innovate and capability to reach organizational goals.

We recommend that efforts to develop gamification concepts, including *gamified training, gamified compensation,* and *gamified performance appraisal,* will help increase employee *engagement.* Furthermore [10] also mentions that gamification positively impacts work engagement for employees who collaborate from separate workplaces.

6.2 Establishing Fundamental Religious Work Values

In relation to the new understanding of the office as a medium to socialize and gather, we then propose that spiritual work value will need to be built by the leaders. Spirituality is one of the most influential factors that is significantly related to employee attitudes, values, and behavior. It specifically describes the effect on the problems that exist in an organization, including the approach and decisions of managers and employees. Based on the above understanding, it can be concluded that spiritual work values are the application of divine values or concepts as an order in the organization so that they have good moral and ethical principles and can distinguish between good and right [11–13].

The outcome of spiritual work values has a positive effect on the meaning of work [12]. Employees who incorporate spiritual values have their own experiences at work. Spiritual work values are an essential factor in activities in the world of work to remind employees to always behave based on spiritual values. This is reinforced by previous research by [14]. In conclusion, the more employees who have high spiritual values at work, the more meaningful work will be, fostering a sense of meaning.

6.3 Prepare Individual Readiness to Change

If today we face a change in work characteristics towards virtual or even using the metaverse, then it is only a matter of time before there are many changes that will impact on business models, offices, and procedures for carrying out work in the future. The success of an organization in finishing the job more or less depends on individual readiness to change. Individual readiness for change is defined as the extent to which individuals are prepared to participate in different organizational activities [15].

As such, in order to prepare for the metaverse as an alternative to hybrid work, a collection of HRM practices and organizational situations must be arranged to drive individuals to possess a high level of readiness, in various situations and conditions that will trigger the change. A summary of our research findings can be described and illustrated in Fig. 1.

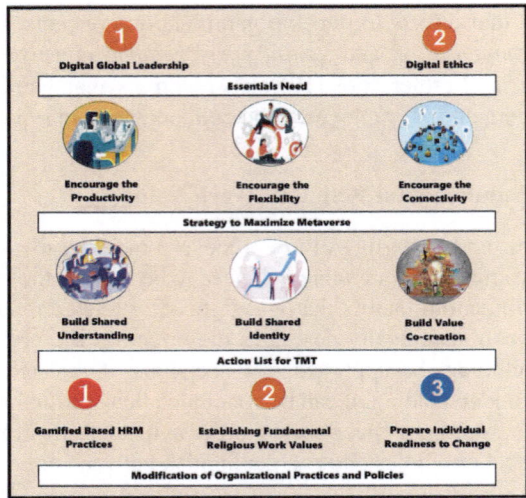

Fig. 1. Essential things for using the metaverse as a future workplace.

7 Conclusion

Our review offers insights into the essential needs for establishing the metaverse as a virtual workplace and provides some key strategies and conditions to be met. However, given the volatility of world work, we cannot guarantee that the pre-requisite conditions offered will remain the same for the next 5–10 years. The advance of virtual reality development may replace the leadership trait and style, the robot and machine may replace our whole work, ethical standards might be changed as well. However, we strongly recommend that spiritual values, especially those rooted in religious values, will take their place as basic fundamental values to catalyse any changes that occur within organization. Therefore, further research is needed to include how to select talent and prepare future leaders that are agile, adaptive, and empathetic to human needs in virtual environments.

References

1. Choi, H.Y.: Working in the Metaverse: Does Telework in a Metaverse Office Have the Potential to Reduce Population Pressure in Megacities? Evidence from Young Adults in Seoul, South Korea. Sustainability (Switzerland) **14**(6), (2022). Available at: https://doi.org/10.3390/su14063629
2. Hutson, J.: Social virtual reality: neurodivergence and inclusivity in the metaverse. Societies. mdpi.com **12**(102) (2022). Available at: https://www.mdpi.com/1714956
3. Carter, D.: Immersive Employee Experiences in the Metaverse: Virtual Work Environments, Augmented Analytics Tools, and Sensory and Tracking Technologies. Psychosociological Issues in Human Resource …. search.proquest.com (2022). Available at: https://search.proquest.com/openview/6fa26c8553f4bb6e9be9fef3b75ff6b2/1?pq-origsite=gscholar&cbl=2045093

4. Upadhyay, A.K., Khandelwal, K.: Metaverse: the future of immersive training. Strategic HR Review. emerald.com (2022). Available at: https://doi.org/10.1108/SHR-02-2022-0009

5. Durana, P., Krulicky, T., Taylor, E.: Working in the Metaverse: Virtual Recruitment, Cognitive Analytics Management, and Immersive Visualization Systems. Psychosociology. Issues Hum search.proquest.com (2022). Available at: https://search.proquest.com/openview/945bf80bb 6ca98077b588d5e73729b86/1?pq-origsite=gscholar&cbl=2045093

6. Batat, W., Hammedi, W.: The extended reality technology (ERT) framework for designing customer and service experiences in phygital settings: a service research agenda. Journal of Service Management. emerald.com (2022). Available at: https://doi.org/10.1108/JOSM-08-2022-0289

7. Browne, J., Green, L.: The Future of Work is No Work: A Call to Action for Designers in the Abolition of Work. In: CHI Conference on Human Factors in Computing Systems, April 2022. dl.acm.org, pp. 1–8 (2022). Available at: https://doi.org/10.1145/3491101.3516385

8. Hirsch, P.B.: Adventures in the metaverse. Journal of Business Strategy. Emerald Publishing Limited **43**(5), 332–336 (January 2022). Available at: https://doi.org/10.1108/JBS-06-2022-0101

9. Lemke, C., Monett, D., Mikoleit, M.: Digital ethics in data-driven organizations and AI ethics as application example. In: Barton, T., Müller, C. (eds.) Apply Data Science, pp. 31–48. Springer Fachmedien Wiesbaden, Wiesbaden (2023). Available at: https://doi.org/10.1007/978-3-658-38798-3_3

10. Wicaksono, K., Fachrunnisa, O.: Human resource based gamification and organizational support to increase employee engagement: a conceptual model. In: The 3rd National and International Conference 2022 'Proactive Management of Cooperative and Social Enterprises Reacting to the Challenges of New Social Dynamic', July 21–22, 2022, pp. 106–585 (2022)

11. Adhiatma, A., Fachrunnisa, O.: The Relationship among Zakat Maal, Altruism and Work Life Quality. International Journal of Zakat **6**(1), 71–94 (2021). Available at: https://doi.org/10.37706/ijaz.v6i1.255

12. Adhiatma, A., Nafsian, R.A.: A Model to Improve Islamic Community Social Identity and SMEs Performance. In: Proceedinh of 7th ASEAN Universities International Conference on Islamic Finance (AICIF) 2019 'Revival Of Islamic Social Finance To Strengthen Economic Development Towards A Global Industrial Revolution'. UNIDA Gontor Press, Gontor (2019). Available at: https://www.ptonline.com/articles/how-to-get-better-mfi-results

13. Sudarti, K., Fachrunnisa, O., Hendar, Adhiatma, A.: Religious value co-creation: a strategy to strengthen customer engagement. In: Barolli, L., Yim, K., Enokido, T. (eds.) CISIS 2021. LNNS, vol. 278, pp. 417–425. Springer, Cham (2021). https://doi.org/10.1007/978-3-030-79725-6_41

14. Milliman, J., Czaplewski, A.J., Ferguson, J.: Workplace spirituality and employee work attitudes: An exploratory empirical assessment. Journal of Organizational Change Management **16**(4), 426–447 (2003). Available at: https://doi.org/10.1108/09534810310484172

15. Huy, Q.: Intelligence, and Radical Change. The Academy of Management Review **24**(January 1999), 325–345 (1999). Available at: https://doi.org/10.2307/259085

Customer Engagement in Online Food Delivery

Alifah Ratnawati$^{(\boxtimes)}$ and Sri Wahyuni Ratnasari

Faculty of Economics, Universitas Islam Sultan Agung, Semarang, Indonesia
alifah@unissula.ac.id

Abstract. More and more customers prefer to purchase food via Online Food Delivery (OFD) service as they consider it more convenient, simple, and easy. This study aims to investigate and examine the role of Customer Engagement (CE) in mediating the effects of Consumer Satisfaction (CS) and Consumer Trust (CT) on Repurchase Intention (RI) of OFD services. A total of 237 consumers are selected as respondents. To examine the regression relationships of the four constructs, this study uses PLS-SEM. The findings show that CS is proven to have an effect on CE, trust and RI. Trust has no effect on CE but has an effect on RI, while CE has an effect on RI. Furthermore, CE is proven to be able to mediate the effect of CS on RI, but CE is unable to mediate the effect of Trust on RI. Trust is unable to mediate the effect of CS on CE, but trust can mediate the effect of CS on RI.

Keywords: Customer engagement · Repurchase Intention · Satisfaction · Trust

1 Introduction

Rapid development of communication and transportation technologies enables people to fulfil many aspects of their life more easily, including one related to services. One of the fastest growing services in recent times is online food delivery (OFD) service. The development of this service has been triggered by the previous COVID-19 pandemic forcing Indonesian consumers to buy products, including food, by online. Since then, more and more consumers have got used to using the OFD applications such as GoFood and GrabFood. Such applications provide consumers with conveniences in choosing, ordering and purchasing food online to be delivered to their home or offices and paying by Cash on Delivery (COD) system or using several e-wallets.

Shopping conveniences offered by OFD applications has attracted more consumers to use the applications. Moreover, the competition in OFD industry is becoming more intense especially when Shopee food and several other applications entering the market. In order OFD service providers to survive in the OFD industry, they have to maintain positive buying experiences by satisfying their consumers and winning their trust. Customers who have positive experiences with the company's products and services will feel satisfied and intent to repurchase [1]. Similarly, they will trust the company and intent to repurchase.

Several studies have suggested that consumer satisfaction (CS) will increase repurchase intention (RI) [1–3]. However, a study conducted by [4] found the opposite result arguing that CS does not influence RI. Moreover, studies conducted by [3, 5] reported that consumer trust (CT) will increase RI. A study by [2], however, has yielded a contradictive result suggesting that CT does not affect RI. Therefore, research on the impact of CS on RI as well the impact of CT on RI has shown inconsistencies results. To fill the research gaps, this study proposes customer engagement (CE) to bridge the relationships between CS and RI as well as between CT and RI. This variable is proposed as a solution considering customer management strategies have evolved from transaction marketing to relationship marketing and are now evolving to CE [6].

Although the OFD service industry seems very promising, the nature of this market is still poorly understood. In any service industry, providing superior service is very important to increase CS and CT which lead to RI. This may be done through CE. Based on the above explanation, OFD service providers must fully understand the impacts of CS and CT on RI and how to increase the occurrence of RI through the role of CE in online services. Such understanding requires in-depth analysis of the nature of regression relationships between the studied constructs. For that reason, this study was conducted, to offer solutions on how to increase RI based on CS and CT and mediated by CE in the context of OFD service providers.

2 Theoretical Studies

2.1 Customer Engagement

Engagement is defined as a state of being involved, directed, fully centered, or captivated of something (meaning constant attention), resulting in the consequences of a certain attraction or rejection. The more centralized an individual in approaching or moving away from a target, the more value is added or subtracted from the target [7]. [1] argue that CE is a mechanism for adding customer value to the company, either through direct or indirect contributions. Direct contributions may include customer purchases. Meanwhile, indirect contributions may consist of incentive references, conversations about brands on social media and feedback or suggestions provided by customers to the company.

In the field of Business, engagement is discussed in the context of contractual relationships, while in Management, as an organization's activity with its internal stakeholders. In the Marketing domain, engagement is associated with the level of active relationship between customers and the company and is referred to as customer engagement (CE) [6]. Understanding CE theory can improve our understanding on how CE differs from other customer relationship constructs. Engagement theory is developed from relationship marketing theory based on commitment and trust. CE is an interactive concept that arises during customer interaction with a company. CE reflects the customer's descriptions and can be used to continuously develop company in more variative ways [7]. It is characterized by [8]: *Enthusiasm, Attention, Absorption, Interaction* and *Identification*.

2.2 Satisfaction

Customer satisfaction is defined as "the level of person's feelings resulting from comparing perceived product performance". Currently, more and more studies have focused on investigating CS which is defined as "the quality level of service performance that fulfill customer expectations" [9]. In the context of E-commerce, E-commerce satisfaction refers to the degree to which individuals feel satisfied with all aspects of an e-commerce system. Meanwhile, purchasing behavior refers to the frequency and willingness to buy at a particular e-commerce site. Previous studies have examined various factors leading to e-commerce satisfaction. To measure CS toward OFD services, customer reviews on the services provided by vendors and drivers play a significant role. This satisfaction is an important e-commerce outcome which can lead to consumers' ongoing intention to visit e-commerce sites.

2.3 Satisfaction and Customer Engagement

[1] state that customers are engaged with the company when they are satisfied with their relationship and have an emotional bonding with the company. The CE theory states that if the customers are satisfied with the company and perceive emotional bonding, then they will engage with the company by purchasing (direct contributions), as well as providing referrals, influences and feedbacks (indirect contributions). Satisfaction produces direct contributions, and emotional bonding produces indirect contributions [1]. However, the relationships between satisfaction and direct contribution as well as between emotions with indirect contributions are moderated by various factors, such as industry types (service vs. product), company types (B2B (business to business) versus B2C (*business to customer*)), engagement rate, brand value, and comfort level.

Based on the above theory, it can be interpreted that when consumers are satisfied with OFD services, they will engage with the company (CE) by making purchases, providing referrals, providing feedback etc. From this perspective, the first hypothesis was proposed as follows:

H1. Increased satisfaction will result in increased customer engagement.

2.4 Trust

Consumers will develop trust in the brand as they expect the brand will act to meet their needs and desires. When consumers have trusted a brand, they will feel safe to wear the brand or subscribe to it. Sellers must maintain customer trust to prevent consumers from leaving the brand. This is in accordance with the opinion of [10] who state that brand trust is characterized by 1) the presence of trust in the brand which is sourced from the consumer's belief that the brand is able to meet the promised value, 2) consumers can rely on the brand, 3) the brand is able to provide honest information / as it is to consumers, 4) the brand is able to provide safety when consumers using the brand.

The use of online food delivery services requires consumers to fully trust the vendors (sellers) and the drivers in delivering their orders accordingly and on time. High level of trust in social commerce indicates that consumers perceive e-commerce and social media as integrated. It also indicates that social commerce features work reliably during shopping process and meet the needs of customers. Consumers will also assume that other consumers are reliable and care about their needs, thus promote positive perception of the Website. Consumers, therefore, consider e-commerce sites as a comfortable environment for shopping and feel satisfied with the sites. These full satisfaction and trust of consumers will give positive impacts on repurchase intention [11].

2.5 Satisfaction and Trust

Satisfaction, trust and commitment are the total quality of a relationship. The quality of relationship is defined as the overall customer evaluation on the strength of his/her relationship with the company. As mentioned earlier, satisfaction refers to a customer's overall evaluation of his/her experiences with the company's offerings. Trust is a necessary condition to build customer's commitment to a company. Along with commitment, trust is placed at the heart of the relationship marketing framework. Trust in human relationships comes from mutual relationships. Satisfaction toward marketing relationships is considered the main prerequisite for long-term customer-company relationships, as well as a prerequisite for trust and commitment. Similarly, in CE models, CS is considered as a direct predictor of CT [12]. Buyer satisfaction occurs when products and services meet buyer expectations. This satisfaction will have an impact on consumer trust.

In relation to OFD services, consumer satisfaction with OFD services in facilitating food purchases will have an impact on increased trust. Based on that statement, the second hypothesis is proposed as follows:

H2. Increased satisfaction will result in an increase in trust.

2.6 Trust and Customer Engagement

CE may be a relatively new concept compared to customer satisfaction or loyalty. More than twenty studies, however, have offered relevant perspectives regarding CE on social media [8, 13]. Consumers are using social commerce more frequently as a source of product-related information. They are participating in commercial activities supported by social media such as customer reviews, sharing, recommendations, and discussions. The definition of trust is selected, consistent with the perpective that trust is resulted by predictability. Following this definition, trust needs to be measured from a multi-dimensional perspective, including different guardians in social commerce. Trust is an important concept in e-commerce and can support significant positive outcomes, such as purchase intention, word of mouth, loyalty, and revisit intention [11].

Customer trust plays an important role in creating a strong bonding between service providers and customers. Trust leads to customers maintaining an ongoing committed relationship with the brand. Customers with a high level of trust tend to build solid relationships with the company. The theory of social exchange states that customers tend to interact with service providers they trust. CE can increase positive and reciprocal exchanges between customers and service providers, which strengthen their trust relationship. CE can increase CT. Highly engaged customers tend to keep their trust in the seller [14]. When consumers feel full trust in a product or service, the emotional bonding with the company becomes stronger [13].

Trust is a key success factor in interpersonal and business interactions. This statement is supported by a large number of research efforts in other disciplines such as Sociology, Social Psychology, Economics, and Marketing. Trust is critical in interactive exchanges among stakeholders as customers are anticipated to pay for services they have not received or experienced. CE can also be seen as a customer-to-enterprise relationship centered on the behavioral aspects of the relationship. Researchers also argue that CE includes customer co-creation in which customers may decide to use voice (communication behavior aimed at expressing their experience) or exit (behavior planned to limit or grow their relationship with the brand) [15]. CT and pleasure are closely related. By trusting the provider of services, satisfied consumers will intent to repurchase. Trust is also found to influence customer satisfaction, and that satisfaction positively influences the intention to repurchase [16]. Increased consumer trust will result in increased customer engagement [17]. In the context of OFD services, consumers who trust in OFD services will increase their engagement in the form of making purchases, providing referrals, providing feedback etc. Based on the above discussion, the third hypothesis is proposed as follows:

H3: Increased trust will result in increased customer engagement

2.7 Repurchase Intention

Re-purchase intention is the consumer's desire to repurchase a certain product because it has benefits and quality based on previous purchase*s* [18]. Repurchase intention is a consumer's perception in related to satisfaction with the brand that may result in continuous information processing [19].

RI is manifested in the form of consumer's willingness to continue buying or using the same products, to buy or use the products in the future, and consumer's intention to buy or use the product again [20]. In line with this, [21] also suggests that RI is indicated by the consumer's willingness to return, willingness to recommend, intention to return and the likelihood of repurchasing.

2.8 Satisfaction and Repurchase Intention

Customers who have positive experiences with the company's products and services will feel satisfied and are more willing to repurchase [1]. Satisfaction is an important and crucial construct in marketing. It is an essential factor to increase repurchase intention and build a long-term relationship between the company and the customer [22]. The study conducted by [20, 23] concluded that customer satisfaction will have an impact on repurchase intention. Regarding OFD services, customers who are satisfied with OFD will increase their repurchase intention. On the basis of this understanding, the fourth hypothesis is formulated as follows:

H4. Increased satisfaction will result in an increase in repurchase intention

2.9   Trust and Repurchase Intention

Trust has an impact on consumer's RI when measuring trust at the post-purchase stage. Post-purchase trust differs from initial trust. In the post-purchase phase, consumers have substantial and hands-on prior experiences to help making decision on whether to enter into future transactions with the same seller. In this repurchase situation, consumers tend to evaluate products or services based on their actual performances perceived after consumption. Using their hands-on experiences, customers tend to reevaluate their trust perception.

When relationship-based trust has been built, consumers acknowledge past purchasing experiences and may hesitate to switch to a new online store because switching requires learning costs. When online sellers act in a way that build consumer trust, their consumers may perceive lesser risks when using their websites. Such circumstance may encourage their customers to make repeat purchases [24].

In e-commerce literature, trust is a key factor influencing customer's purchase intention. Trust is a belief that the transaction partners will behave in a good will and in a favorable manner. Previous studies on trust in online shopping show that trust is positively related to repurchase intention [25]. In line with that study, it is understood that in OFD services, an increase in trust will result in an increase in repurchase intention. For this reason, the fifth hypothesis is proposed as follows:

H5. Increased Trust will result in increased repurchase intention.

2.10   Customer Engagement and Repurchase Intention

Customer engagement can be described as the process of developing cognitive, affective, and behavioral commitments to establish an active relationship with the company. Thus, CE can be manifested into four prominent dimensions: interaction, activity, behavior, and communication in a given context. The relationship also occurs in the management of e-commerce where engagement helps customers to make better online repurchase decisions [26].

CE behavior has been found to be a key driver of success in the online shopping environment and online brand community. A customer's willingness to participate in the brand's community can be seen in their willingness to repurchase merchandises or services from the brand. In the process of interaction, customers will establish close relationships with brand-related communities, and these relationships will encourage customers to purchase brand-related merchandises or services on a recurring basis [27]. Studies conducted by [8, 27, 28] have proven that increased customer engagement can result in increased repurchase intention. In accordance with those studies, hypothesis 6 is proposed as follows:

H6. Increased customer engagement will result in an increase in repurchase intention.

3   Research Method and Findings

Respondents in this study were online food delivery services' consumers in the city of Semarang, Indonesia. Data were obtained by distributing questionnaires to 237 respondents directly. There are four constructs in this study, namely: satisfaction, trust, customer engagement and repurchase intention. PLS-SEM was used to test regression relationships of the four constructs.

The results of causality relationships testing can be shown in Fig. 1 and Table 1 below:

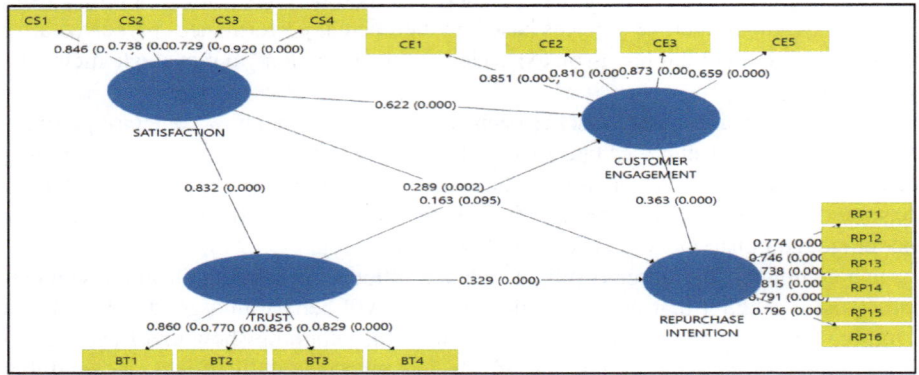

Fig. 1. Full model

Figure 1 shows the full model of the results of testing the causality relationship. Each variable has an indicator with outer loadings above 0.65 (p-value < 0.01) so that it meets convergent validity.

Table 1. The results of hypothesis testing

Hypothesis	Influences	*Path* Coefficients	P *Value*	Descriptions
H1	*Satisfaction* → customer engagement	0, 622	0,000	Accept H1
H2	*Satisfaction* → *Trust*	0,832	0,000	Accept H2
H3	*Trust* → customer engagement	0,163	0,092	Refuse H3
H4	*Satisfaction* → repurchase intention	0,289	0,001	Accept H4
H5	*Trust* → repurchase intention	0,329	0,000	Accept H5
H6	customer engagement → repurchase intention	0,363	0,000	Accept H6

Figure 1 and Table 1 above show significant direct effects of Satisfaction on CE (p-value < 0.01), Satisfaction on Trust (p-value < 0.01), Satisfaction on RI (p-value < 0.01), trust on repurchase intention (p-value < 0.01), and CE on RI (p-value < 0.01). Meanwhile, the insignificant direct effect is shown by the regression between trust and customer engagement (p-value > 0.01).

The role of mediation in the relationship between the four constructs were tested. In Table 2, it can be seen that CE is able to mediate the relationship between satisfaction and

RI. Trust is also able to mediate the relationship between satisfaction and RI. However, CE cannot mediate the relationship between trust and RI. Similarly, trust cannot mediate the relationship between CS and CE.

Table 2 Intervening tests

Intervening Models	P *Value*	Descriptions
Satisfaction → *Trust* → customer engagement	0,097	No mediation effect
Satisfaction → customer engagement → repurchase intention	0,007	Mediation Effect
Trust → customer engagement → repurchase intention	0,121	No mediation effect
Satisfaction → *Trust* → repurchase intention	0,000	Mediation Effect

4   Discussion and Conclusion

The findings of this study have confirmed the roles of satisfaction, trust and customer engagement in increasing repurchase intention. Customer satisfaction with OFD service will increase customer engagement. Customer engagement is proven to be a mediating variable between satisfaction and repurchase intention. This is consistent with the theory of customer engagement, that customers who are satisfied with the company's services and have an emotional bonding will be more engaged with the company. Customer engagement for OFD services can be seen from enthusiasm, interest and pleasure in using OFD service. In addition, customer engagement can also be seen from how often customers give "like" ratings and provide "comments" when the OFD marketplace posts content on social media.

Trust cannot mediate the relationship between satisfaction and customer engagement, but satisfaction has an effect on customer engagement. Trust cannot increase OFD customer engagement however customer satisfaction with OFD services can increase customer engagement.

Increasing OFD repurchase intention can be done directly through trust, satisfaction and customer engagement. Increasing repurchase intention can also be done indirectly, through satisfaction mediated by customer engagement, and through satisfaction mediated by trust. An increase in repurchase intention indicates that consumers will continue to use OFD, they prefer OFD as their first choice, and are willing to recommend OFD to others.

References

1. Pansari, A., Kumar, V.: Customer engagement: the construct, antecedents, and consequences. J. Acad. Mark. Sci. **45**(3), 294–311 (2017)
2. Ginting, Y.M., Chandra, T., Miran, I., Yusriadi, Y.: Repurchase intention of e-commerce customers in Indonesia: an overview of the effect of e-service quality, e-word of mouth, customer trust, and customer satisfaction mediation. Int. J. Data Netw. Sci. **7**(1), 329–340 (2023)

3. Purnamasari, I., Suryandari, R.T.: Effect of E-Service Quality on E-Repurchase Intention in Indonesia Online Shopping: E-Satisfaction and E-Trust as Mediation Variables. Eur. J. Bus. Manag. Res. **8**(1), 155–161 (2023)
4. Hellier, P.K., Geursen, G.M., Carr, R.A., Rickard, J.A.: Customer repurchase intention: A general structural equation model. Eur. J. Mark. **37**(11/12), 1762–1800 (2003). Dec.
5. Miao, M.: The influence of e-customer satisfaction, e-trust and perceived value on consumer's repurchase intention in B2C e-commerce segment. Asia Pac. J. Mark. Logist. **34**(10), 2184–2206 (2022)
6. Pansari, A., Kumar, V.: Customer engagement marketing. In: Palmatier, R.W., Kumar, V., Harmeling, C.M. (eds.) Customer Engagement Marketing, pp. 1–332. Palgrave Macmillan (2018)
7. Behnam, M., Hollebeek, L.D., Clark, M.K., Farabi, R.: Exploring customer engagement in the product vs. service context. J. Retail. Consum. Serv. **60** (2021)
8. So, K.K.F., King, C., Sparks, B.: Customer engagement with tourism brands: scale development and validation. J. Hosp. Tour. Res. **XX** (2012)
9. Afthanorhan, A.: Assessing the effects of service quality on customer satisfaction. Manag. Sci. Lett. **9**(1), 13–24 (2019)
10. Chaudhuri, A., Holbrook, M.B.: The chain of effects from brand trust and brand affect to brand performance: the role of brand loyalty. J. Mark. **65**(2), 81–93 (2001). Apr.
11. Lin, X.: Building E-Commerce Satisfaction and Boosting Sales: The Role of Social Commerce Trust and Its Antecedents. Int. J. Electron. Commer. **23**(3), 328–363 (2019)
12. Barari, M.: A meta-analysis of customer engagement behavior. Int. J. Consum. Stud. **45**(4), 457–477 (20219)
13. Santini, F. de O., Ladeira, W.J., Pinto, D.C., et al.: Customer engagement in social media: a framework and meta-analysis. J. Acad. (2020)
14. Li, M.W.: Unlocking the customer engagement-brand loyalty relationship in tourism social media: The roles of brand attachment and customer trust. J. Hosp. Tour. Manag. **44**, 184–192 (2020). no. Query date: 2023-02-23 09:20:11
15. Agyei, J.: Influence of trust on customer engagement: empirical evidence from the insurance industry in Ghana. SAGE Open **10**(1) (2020)
16. Ruswanti, E.: Word of mouth, trust, satisfaction and effect of repurchase intention to Batavia hospital in west Jakarta, Indonesia. Manag. Sci. Lett. **10**(2), 265–270 (2020)
17. Santini, F. de O.: Customer engagement in social media: a framework and meta-analysis. J. Acad. Mark. Sci. **48**(6), 1211–1228 (2020)
18. Ilyas, G.B.: Reflective model of brand awareness on repurchase intention and customer satisfaction. J. Asian Finance Econ. Bus. **7**(9), 427–438 (2020)
19. Fazal-e-Hasan, S.M.: The role of brand innovativeness and customer hope in developing online repurchase intentions. J. Brand Manag. **26**(2), 85–98 (2019)
20. Novitasari, D., Jeppri Napitupulu, B.B., Abadiyah, S., Silitonga, N., Asbari, M.: Linking between Brand Leadership, Customer Satisfaction, and Repurchase Intention in the E-commerce Industry. Int. J. Soc. Manag. Stud. IJOSMAS **03**
21. Huang, H.-C., Chang, Y.-T., Yeh, C.-Y., Liao, C.-W.: Promote the price promotion: The effects of price promotions on customer evaluations in coffee chain stores. Int. J. Contemp. Hosp. Manag. **26**(7), 1065–1082 (2014). Oct.
22. Ashfaq, M.: Customers' Expectation, Satisfaction, and Repurchase Intention of Used Products Online: Empirical Evidence From China. SAGE Open **9**(2) (2019)
23. Du, H.S., Xu, J., Tang, H., Jiang, R.: Repurchase Intention in Online Knowledge Service: The Brand Awareness Perspective. J. Comput. Inf. Syst. **62**(1), 174–185 (2022). Jan.
24. Sullivan, Y.W.: Assessing the effects of consumers' product evaluations and trust on repurchase intention in e-commerce environments. Int. J. Inf. Manag. **39**, 199–219 (2018). no. Query date: 2023-02-23 09:37:53

25. Liu, Y.: The effects of online trust-building mechanisms on trust and repurchase intentions: An empirical study on eBay. Inf. Technol. People **31**(3), 666–687 (2018)
26. Lim, X.J.: What s-commerce implies? Repurchase intention and its antecedents. Mark. Intell. Plan. **38**(6), 760–776 (2020)
27. Zheng, R., Li, Z., Na, S.: How customer engagement in the live-streaming affects purchase intention and customer acquisition, E-tailer's perspective. J. Retail. Consum. Serv. **68** (April, 2022)
28. Habib, S., Hamadneh, N.N., Hassan, A.: The Relationship between Digital Marketing, Customer Engagement , and Purchase Intention via OTT Platforms vol. 2022 (2022)

Human-AI-powered Strategies for Better Business Applications

Josef Mayrhofer[✉]

Medford, NJ, USA
Josef@performetriks.com

Abstract. This contribution presents strategies for intelligent knowledge transformation using qualitative analysis and Human-AI-powered remediation plans. First, we discuss classic maturity models and why they often fail in our dynamic world. Then, we present a better way towards knowledge modeling and mentorship, which is implemented in our Gobenchmark product. Last, we demonstrate a business case that underpins the relevance and opportunities of knowledge modeling.

1 Introduction

Software dominates many industries. History shows that if the software fails, businesses and consumers suffer. For instance, a severe outage of a central bank in Singapore in 2021 resulted in a financial charge of almost 700 million dollars [7]. As a result, regulatory agencies worldwide take system reliability seriously, and businesses must avoid failures in their mission-critical applications at all costs.

The design, implementation, and maintenance of resilient IT services remain an afterthought. There are almost no university programs teaching methods to design and validate software for high availability and reliability. On the other hand, no measurements were available to understand gaps in implementing applications with performance and reliability in mind. Due to these shortcomings, some enterprises invested years in finding out which methodical approach worked for them, while others failed and lost millions due to severe software outages.

For reliability to work, it must put us on a journey where we learn concepts as we are doing things. Following this core idea, we have invented a Human-AI-powered approach to collect and store knowledge in a reusable format to guide everyone on their journey to build fast and reliable business applications [1]. The benefit of this Human-AI-powered approach is that it removes the guesswork, gives immediate feedback in terms of a rating, and provides instructions on how to improve their practices to build better business applications.

After an introduction to existing process improvement frameworks, such as the Capability Maturity Model (CMM), we explain the Human-AI-powered approach for better business applications. In addition, we demonstrate the practical implementation of Human-AI-powered knowledge modeling in the Gobenchmark platform.

L. Barolli (Ed.): CISIS 2023, LNDECT 176, pp. 260–268, 2023.
https://doi.org/10.1007/978-3-031-35734-3_26

2   Maturity Models and Their Tradeoffs

In the past, maturity models were the most important technique for understanding where organizations are and how they adapt to efficient and effective practices. For example, the very famous CMM (capability maturity model) was developed by the Software Engineering Institute of Carnegie Mellon University in 1987 to understand a company's software engineering maturity model.

The core idea behind this CMM is that a better maturity level should increase the chances of developing and releasing high-quality software. CMM uses five levels built on each other and comes with a requirements catalog. If an organization fulfills the requirements on a certain maturity level, it also complies with all levels below.

The problem with classic maturity models is always that they are [5] static, and we try to apply them to dynamic variables such as customers, technology, and markets. Our world is changing very fast. We have new technologies, such as AI, containerization, and Kubernetes, to name a few. So our world is no longer a snapshot of the past but an ever-changing environment.

Advancements in technology, such as AI, bring new capabilities, the market is changing rapidly, and customers have heightened user experience expectations. Consequently, we need an intelligent way of collecting and sharing knowledge instead of static maturity models.

3   Human-AI-powered Strategies

The research agency Gartner put it very well: "There is an urgent need to fulfill heightened user expectations about the performance and availability of business-critical applications. To succeed, software engineering leaders must develop their teams' competency to mitigate reliability risks [4]".

Looking at the past few years and seeing the impact of the covid-19 pandemic on the speed of digital transformation, it is understandable that our population relies more and more on IT services. Nevertheless, software failures are expensive, so we must find strategies for creating better business applications.

On the one hand, every organization could find its way to develop and operate reliable applications. But on the other hand, we could intelligently share good practices and increase the chances that everyone would implement better business applications.

There is no one fit's all approach when we look at maturity models. It would set businesses at risk if they followed a try-and-error approach to find better strategies for creating reliable business applications.

In Gobenchmark, we combine Human-AI-powered knowledge with qualitative analysis, making the unmeasurable measurable and bringing flexibility for changes in markets, customers, and technologies (see Fig. 1).

The core elements of Gobenchmark are:

- Collect advice from industry experts
- Qualitative Analysis
- Score Comparison

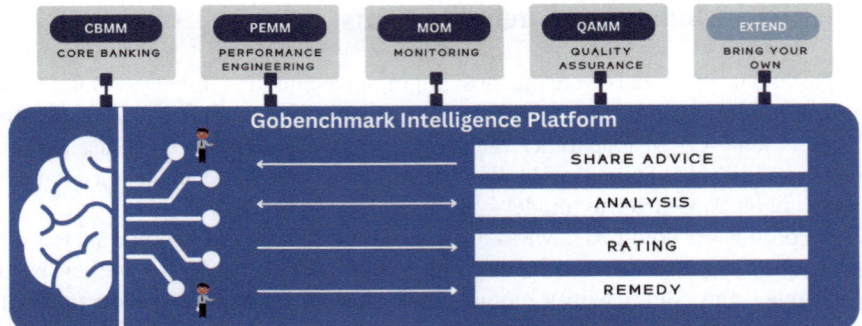

Fig. 1. The Human-AI-powered *Gobenchmark* platform. It's knowledge moduls for *CoreBanking, Performance Engineering, Monitoring, Quality Assurance*, and it's core features *Share Advice, Analysis, Rating and Remedy*.

- Remediation

The following chapters describe core aspects of Human-AI-powered strategies for better business applications implemented in Gobenchmark.

3.1 Intelligent Knowledge Sharing

Some successful businesses, such as Uber or Airbnb, are built on the sharing philosophy. A group or team sharing ideas is always better than individuals. For several reasons as mentioned above, there is no way to succeed with a static maturity-based approach. This is why we believe collecting knowledge from SMEs, storing it intelligently, and reusing it to guide needy customers is the only promising option.

In Gobenchmark, we created a share advice catalog [2] allowing every industry professional to share good practices and hints about their solutions and how they rate their practices and tools. Furthermore, we store such advice in a flexible and reusable format to ensure that our built-in AI can utilize this information when creating recommendations for a client's remediation plan.

3.2 Qualitative Analysis

In research, qualitative analysis is a method to analyze nontangible aspects of a subject to evaluate its characteristics, reputation, and performance. We can also describe it as a non-scientific approach to understanding things that are difficult to express in numbers.

The benefit of qualitative analysis is that it builds on knowledge collected over many years. For example, analysts used it initially to measure the risks involved in securities.

There are five methods for conducting a qualitative analysis such as

1. Content Analysis
2. Narrative Analysis
3. Grounded Analysis
4. Framework Analysis

5. Discourse Analysis

In Gobenchmark, we have decided to implement the framework-based analysis because it generates descriptive and explanatory conclusions. The interviewee walks through 27 questions structured in domains and practices. Each practice can be answered by choosing Always, Often, Rarely, or Never. All the answers are stored in the SaaS-based Gobenchmark platform at the end of the analysis.

3.3 Scoring and Comparison

In the early days of the stock market and securities, when no rating existed, people often lost their investments due to many unknowns in their purchased financial products. When rating agencies came in and rated, all financial products customers understood the risks [8] involved. As a result, they had much better chances to protect their investments by selecting products with a higher rating.

The same principles apply to business applications. These days we have no rating for the reliability of business applications. A high CMM level is no indicator of performant, secure, and well-designed IT services. Nevertheless, the outcome of our qualitative analysis can be transformed to a rating from C- to A and indicates how organizations or teams are adopting industry best practices.

By seeing the rating of practices and domains, we can identify blindspots, compare businesses to their peers, and set the foundation for generating a remediation plan.

In Gobenchmark, we show the rating immediately after the qualitative analysis on a scoring dashboard. There are several benefits of seeing the score of practices and domains:

- We understand gaps much faster.
- We can focus our efforts on critical blindspots.
- We have everything we need to show a comparison to industry standards, peers.
- We can build the remediation plan based on indentified gaps expressed by lower scores.

The benchmark in Gobenchmark is dynamic and will be re-calculated month by month. We break our benchmarks down to industries, domains and practices which is allowing us a fine grained comparison.

3.4 Remediation

As mentioned earlier, for reliability to work, it must put us on a journey where we learn concepts as we are doing things. Seeing gaps expressed by a rating does not help too much. If we leave organizations alone to solve these shortcomings, they might run into further issues, such as going in the wrong direction.

The AI-powered brain of Gobenchmark provides the expected guidance. It analyzes the qualitative analysis results, incorporates knowledge from shared advice by industry experts, and creates a remediation plan that shows how organizations can reach the next level by improving their practices, methods and using better toolings.

Our world is changing extremely fast, and we can't expect our current approach to work tomorrow or in several weeks ahead. This is where new knowledge from industry experts helps. But we also see challenges in getting relevant insights from SMEs. For this reason, we've integrated Open AI to acquire domain and score-specific advice, which we can incorporate into the AI-powered remediation plans that provide a much better mentorship for customers.

4   Business Case for Human-AI-powered Strategies

There are no indications that the speed of digital transformation might slow down over the next few years. No matter their size, businesses must focus on keeping system errors low to avoid expensive consequences. In this business case calculation we demonstrate how knowledge modeling techniques could result in outstanding financial benefits.

On the costs side of this business case, we have the following:

- Defects
- Consulting
- Tools
- Infrastructure
- Image of a company
- Customer retention and engagement
- Sales

On the savings side of this business case, we have the following:

- Reduction in defects
- Better image
- Reduce infrastructure
- Improve sales
- Improve user experience

We removed costs and savings related to user experience, infrastructure, and image to reduce the complexity in this business case (see Fig. 2).

Rating	# Issue Reduction	Costs $	Savings $	Saving Cumulative $
C	Y1 (40 %)	180,000.00	120,000.00	120,000.00
B	Y2 (50 %)	90,000.00	90,000.00	210,000.00
A	Y3 (50 %)	45,000.00	45,000.00	255,000.00
Savings after 3 years: 255,000.00				

Issues per year: **30**
Cost of a single issue per year: **$ 10,000**

Fig. 2. The Business Case calculation for Human-Ai-powered strategies. It shows how a better *Rating* reduces the number of issues and saved this organization *255,000 USD* over three years.

In this comparison, we have selected a business facing 30 defects on production per year due to technical depths and problems in their performance engineering practices. The average defect cost is [8] 30 times higer if they are detected in production and in this example we've used 10.000 USD as the cost of a single defect in production.

The organization we've used in the calculation of the business case decided to use the knowledge modeling techniques of Gobenchmark to improve their practices and cut down a high number of software performance issues in production.

Initially, their rating was C- because they had not adopted practices, tools, and methods to integrate performance engineering into their value stream. However, the remediation plan helped them to reduce problems leaking to production by 50%.

After another six months, they conducted their last qualitative analysis, which resulted in a B- rating. Thanks to the provided remediation plan, they cut their defect costs by another 40%.

Their target was a rating of A, so this business returned one year after they started their journey and conducted their final assessment. The implementation of their final remediation plan pushed them over their finish line. Gobenchmark saved 255.000 USD within 18 months by sharing knowledge how to build better, more robust business applications (see Fig. 3).

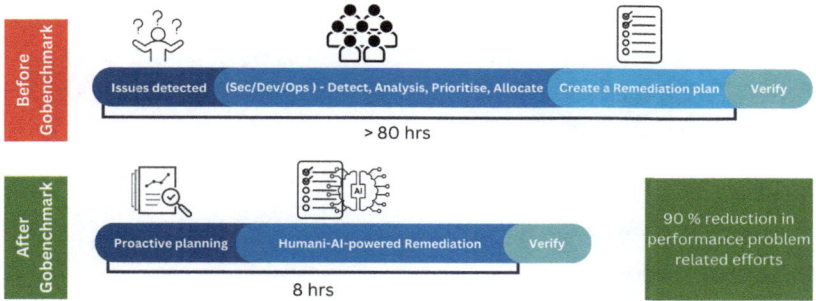

Fig. 3. A comparison indicating the how Human-AI-powered strategies reduced performance problem related efforts. Before this business was using Human-Ai-powered strategies they followed a re-active issue detection approach. After they used *Gobenchmark* to improve their ratings in *Performance Engineering* and implemented the remediation activities their problem related effort reduced by 90%.

5 Practical Implementation of Human-AI-powered Strategies in Gobenchmark

Collecting the latest information about methods and tools is crucial to avoid outdated knowledge. So we use a share advice feature to ask SMEs about their preferred practices and tools and their corresponding rating between 0 for not good and 10 for very good (see Fig. 4).

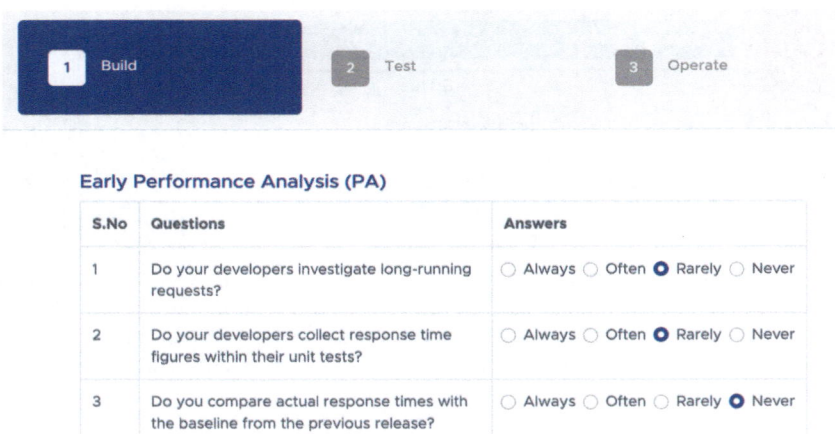

Practice

Questions

1. Performance Analysis by Developers

Please describe your approach

> Please describe your approach

Rating for Approach

Poor   3   Excellent

Please describe your tools

> Please describe your tools

Rating for Tools

Poor   3   Excellent

Fig. 4. Share Advice form in *Gobenchmark* to collect latest industry knowledge.

We use a Framework Analysis technique to understand the current challenges of businesses. A web-based form structured in three domains, nine practices, and 27 discovery questions discovers how our customers are accomplishing (see Fig. 5).

| 1 | Build | 2 | Test | 3 | Operate |

Early Performance Analysis (PA)

S.No	Questions	Answers
1	Do your developers investigate long-running requests?	○ Always ○ Often ● Rarely ○ Never
2	Do your developers collect response time figures within their unit tests?	○ Always ○ Often ● Rarely ○ Never
3	Do you compare actual response times with the baseline from the previous release?	○ Always ○ Often ○ Rarely ● Never

Fig. 5. Framework Analysis for domain *Build*, Practice *Early Performance Analysis* to calculate the rating in *domains* and *practices*

The outcome of the Framework Analysis is visualized in spider and bar diagrams. In addition, we compare candidates to peers, show their rating history, and highlight how they do in certain practices compared to the industry benchmark (see Fig. 6).

The outcome of the rating and the provided advice from industry experts is finally used to create a tailor-made remediation plan. In addition, it demonstrates the practices, methods, and tools to be adopted to reach the target rating (see Fig. 7).

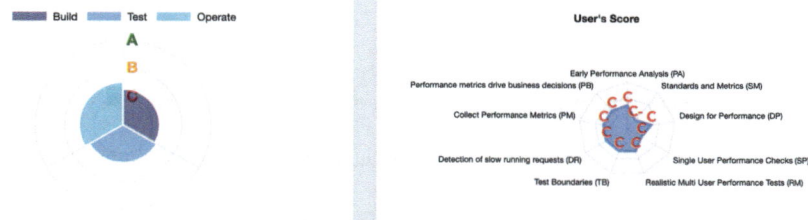

Fig. 6. Scoring dashboard to vislualize the rating of *domains* and *practices*.

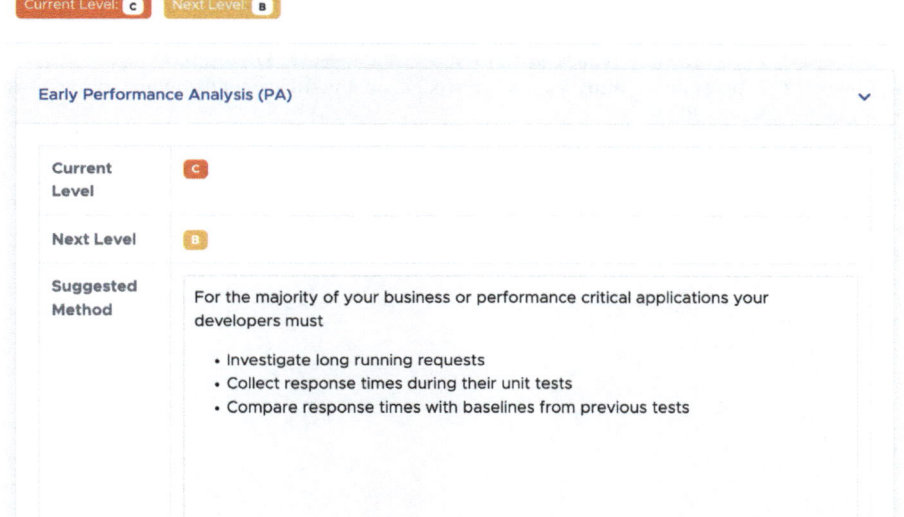

Fig. 7. A remediation plan for the practice *Early Performance Analysis (PA)* to guide customers how to close gaps.

6 Conclusion and Outlook

We propose to adopt more intelligent, Human-Ai-powered strategies for knowledge sharing in yet immature but business-critical disciplines such as performance engineering, monitoring, security, or software quality assurance. Our world is changing too fast, and classic static maturity models will become outdated quickly. The Human-AI-powered strategies outperform classic maturity models by their ability to include the latest industry knowledge and the mentorship they provide regarding tailor-made remediation plans.

We hope that the spirit of the sharing economy will gain more ground in IT over the next few years, and SMEs will understand that sharing good practices is essential to build a stable and reliable digitalized world.

At the same time, we motivate everyone working in system-critical industries such as banks, infrastructure provider, or government to learn more about the groundbreaking benefits of Human-Ai-powered strategies and explore the Gobenchmark [10] platform.

References

1. Mayrhofer, J.: United States Patent Office (2022)
2. Mayrhofer, J.: Performetriks.: https://www.performetriks.com/blog/categories/gobenchmark (2023)
3. Gartner: Report about Digital Transformation. https://www.gartner.com/en/information-tec hnology/insights/digitalization (2022)
4. Herschmann, J., Guttridge, K., Murphy, T.: Gartner Research. https://www.gartner.com/en/ documents/4007508 (2021)
5. Oreilly, B.: barryoreilly.com. https://barryoreilly.com/explore/blog/why-maturity-models-dont-work/ (2019)
6. Vaidya, D.: Wallstreetmojo. https://www.wallstreetmojo.com/qualitative-analysis (2023)
7. Reuters.: Marketscreener. https://www.marketscreener.com/news/latest/Singapore-s-cen bank-says-digital-services-outage-at-DBS-unacceptable--43363754/ (2023)
8. Finney, D.: Investopia. https://www.investopedia.com/articles/bonds/09/history-credit-rat ing-agencies.asp (2022)
9. Deepsource.: NIST. https://deepsource.com/blog/exponential-cost-of-fixing-bugs/ (2022)
10. Mayrhofer, J.: Gobenchmark powered by Performetriks. https://gobenchmark.io/# (2023)

The Role of Affective and Cognitive Engagement in Process of Knowledge Creation and Implementation

Hesti Widianti[1,2], Olivia Fachrunnisa[1(✉)], and Bedjo Santoso[1]

[1] Department of Management, Faculty of Economics, Universitas Islam Sultan Agung (UNISSULA), Jl. Kaligawe Raya Km 4, Semarang, Jawa Tengah 50111, Indonesia
`hestiwidianti@std.unissula.ac.id,`
`hesti.widianti@poltektegal.ac.id, {olivia.fachrunnisa,`
`bedjo.s}@unissula.ac.id`
[2] Politeknik Harapan Bersama, Tegal, Jawa Tengah, Indonesia

Abstract. This study aims to develop a new concept resulting from a synthesis of knowledge management practices with collective engagement process. Knowledge management is defined as managing knowledge effectively within an organization and treating knowledge as an organizational asset. The knowledge management field identifies two main types of knowledge, explicit and tacit knowledge, and includes four main sections: people, process, technology, and governance. Engagement is related to the understanding of an employee, about why and how providing optimal contribution continuously in knowledge production and its influence in implementing knowledge both sharing and utilizing. This study uses an integrated, comprehensive literature review. It concludes that Knowledge Quality Engagement (KQE) is defined as the quality of acquiring and implementing knowledge involving cognitive and affective aspects of individuals engagement. We also propose that KQE has four dimensions: (a) quality of knowledge acquisition, (b) quality of knowledge utilization, (c) cognitive involvement, and (d) affective engagement in seeking and sharing knowledge. In the future, we will empirically examine this new concept and propose that this new knowledge quality will improve human resource performance.

Keywords: Knowledge creation · knowledge implementation · affective engagement · cognitive engagement · knowledge quality

1 Introduction

Research on knowledge quality has been carried out by many previous researchers, however, some are still studying about practices and characteristics of knowledge creation [1–4]. In addition, previous research focuses on its body of knowledge, not yet on creating knowledge quality. It has yet to include an element of belief in assessing knowledge quality. In addition, knowledge is measured for its quality by identifying that tacit knowledge and variables related to knowledge quality must be evaluated from

© The Author(s), under exclusive license to Springer Nature Switzerland AG 2023
L. Barolli (Ed.): CISIS 2023, LNDECT 176, pp. 269–277, 2023.
https://doi.org/10.1007/978-3-031-35734-3_27

time to time, collected, processed, and used measurably [4]. Thus, defining, explaining, and assessing knowledge quality, as well as researching in-depth and comprehensive knowledge that is contextually relevant and can be implemented to encourage people to achieve better moral awareness for themselves and others, is interesting to study. Future quality of knowledge must realize the benefits of knowledge for knowledge users and knowledge creators so that the level of quality assured.

Moreover, current research also needs to pay more attention to the quality of obtaining and using knowledge. Great attention must be paid to how each individual or organization behaves while acquiring and using knowledge. One of the efforts that can be made to ensure that the process of obtaining and using knowledge will determine the quality of the knowledge itself is the individual's involvement in seeking, obtaining, and using knowledge. According to Ho et al. [5, 6], cognitive engagement is concerned with an individual's comprehension of the reasons why and methods for consistently producing their best work, as well as the impact this has on attaining organizational objectives. Absorption and attention are the two components of cognitive involvement, respectively. The cognitive absorption factor demonstrates the degree of employee involvement and ability to concentrate more on the duties of the position. Engagement level aims to raise the standard of work through innovation and creation (knowing how well their skills and function to support corporate objectives). Employees need a certain degree of cognitive resources to raise the quantity of work they do in order to be more productive. Affective engagement, on the other hand, is the sensation of being emotionally drawn to one's job duties, coworkers, managers, leaders, and the work environment [7]. Workers who believe their work is valued, for instance, feel a sense of belonging, perceive their work as significant, and experience feelings of safety and confidence. Given that the firm fundamentally seeks stability, affective engagement may be more appropriate [6].

In this research, we integrate the importance of engagement in knowledge acquisition process and knowledge utilization process to ensure the knowledge quality. People who engage seriously in acquisition and implementation process both in term of cognitive and affective will have higher responsibility to assure that the knowledge is accurate, useful and only for good things, such as innovation and creativity. In conclusion, there is still much further study left. Therefore, this research will internalize engagement process to create and use knowledge in content, process, and results.

2 Literature Review

2.1 Knowledge Quality

The most valuable asset in today's information age is knowledge. Knowledge is a powerful intellectual asset because it is unique and distinctive. Furthermore, today's competitive competition makes every company vie for a system to store and manage knowledge. Therefore, every organization's performance can be judged by its intellectual assets. According to Nonaka and Takeuchi [8], there are two types of knowledge (intellectual assets): explicit knowledge and tacit knowledge. Explicit knowledge is the knowledge that has been collected and translated into a form of documentation (written) so that it is easier for others to understand, for example, books, scientific articles, and so on. In contrast, tacit knowledge is knowledge in one's mind following other people's understanding and experience. Hence, it makes this type of knowledge unique and distinctive.

From an organizational perspective, knowledge management is the activity of acquiring knowledge from organizational experiences, organizational policies, and each other's experiences to achieve organizational goals; this activity is carried out by a combination of technology, organizational functions, and cognitive-based strategies (CBS/Cognitive based strategies) to gain knowledge and create new knowledge by improving the cognitive system in solving problems and making decisions. Furthermore, in developing employee competence and expertise, an organization must work on it holistically, meaning that its competence must be integrated and realized in teaching and learning activities. Therefore, competence is one of the fundamental factors in improving organizational performance.

One of the challenges for an organization is managing information and knowledge effectively and enabling the creation of knowledge-based value; this lies in how organizations get their employees to collaborate. The ultimate goal of implementing knowledge management is to increase entity intelligence or corporate IQ. A follow-up to creating new knowledge is the management of knowledge assets to be utilized. At this stage, there are usually two critical managerial activities. The first is managing the content (content management) of various intellectual capital or explicit knowledge recorded in documents. Secondly, managing tacit knowledge is still stored in the owner's mind. Both types of knowledge in the process require a different approach to management. Human resources (HR) will be willing to spend their time acquiring specific knowledge if this knowledge has value and benefits for them [9]. Knowledge quality is identified as a knowledge creation process that requires knowledge to be collected, processed, and used in a measurable way [10]. From the perspective of using knowledge, knowledge is acquired and integrates all different specialized knowledge sources [11]. In other research, knowledge quality is defined as relevant and valuable knowledge [12]. The quality of knowledge refers to how far knowledge is extended, adapted, and easily applied to tasks [12].

Many studies have examined knowledge quality, including research by [1], which explored the quality of knowledge as measured by frequency, usability, and innovation. Kulkarni et.al, [3] considers the quality of knowledge content as assessed by usability and various formats [4] conceptually develops dimensions of knowledge item quality such as accuracy, consistency, currency, data interpretation, level of context, level of detail, level of importance, share, usability, and volatility. While [13] measured knowledge quality through precision, fulfillment of needs, and accuracy. Knowledge quality is assessed from three dimensions: intrinsic knowledge quality, contextual knowledge quality, and actionable knowledge quality.

2.2 Creative Engagement Process

Engagement is a behavior that shows the degree to which the individual is moved to integrate into his work in an organization [14]. Individual engagement with tasks and group goals will positively affect the level of innovation and creation). Collective engagement is more than the aggregate sum of individual involvement [15]. The engagement has two basic components. First, employees must be aligned on common goals, and second, they must be committed to mutually supporting each other's efforts. When engaged employees are focused on achieving goals, the activities of sharing information, sharing values, and sharing vision values will occur to reinforce one another mutually. In the end, when employees are fully engaged and focused on organizational goals, a mutually supportive working relationship will generate group energy, enthusiasm, and focus on achieving common goals.

To achieve these results, leaders must understand how to translate individual engagement to collective engagement. Kahn [16] defines engagement as a more comprehensive description of the investment in affective attitudes, behavior and one's cognitive energy at work. Collective engagement is an organizational-level construct and indicator of a motivational environment within the organization (motivational aspect). Meanwhile, individual engagement is based on a person's engagement with the organization, so at this level, the evaluative part dominates. Moreover, the antecedents of collective organizational engagement are motivating work designs, HRM Practices, and CEO Transformational Leadership Behaviors [6]. The three organizational resources are to fulfill the adequacy of the needs of meaningfulness, psychological safety, and psychological availability. Meaningfulness is influenced by the characteristics of tasks and job roles; psychological safety is an individual's comfort in his role in the organization without fear of negative consequences on self-image, status, or career [16]. Operationally, motivating work design can be carried out from the first time an individual joins an organization by providing meaningful tasks and challenges.

In knowledge management practices, engagement can be defined as the degree of involvement in mental and psychological aspects while searching, creating and utilizing or implementing knowledge for managerial practices. This mental and psychological involvement will help increase personal responsibility to ensure the quality of the knowledge obtained and shared. Moreover, engagement in knowledge creation and utilization will ensure that the quality of expertise created meets the requirements for the goodness and benefits of that knowledge. Therefore, it is important to note that engagement process is main elements in knowledge management process.

3 Knowledge Quality Engagement: An Integrative Review

3.1 Concept Development

In order to encourage human resource performance and complement the limitations of previous research on the meaning and dimensions of knowledge quality, this research will be conducted by internalizing creative engagement in process to create and implement knowledge in terms of content, process, and results. This study uses the basic theory of knowledge quality. Through the knowledge quality process, personal knowledge can be transformed into organizational knowledge. The purpose of knowledge quality is basically to manage knowledge effectively in an organization where members of the organization are users of the system and are given specific roles and responsibilities. Knowledge quality is an integral part of knowledge quality, ensuring content is useful, accurate, and relevant. Knowledge must be contextual because this is what distinguishes knowledge from information. Organizations should consider knowledge an asset and capture it because of its importance and usefulness [11].

Existing research related to knowledge quality only focuses on the body of knowledge, not on the process of creating knowledge. It has yet to include elements and engagement in process to assess knowledge quality. This is where the need to integrate engagement quality both in affective and cognitive. Knowledge quality engagement is the quality of acquiring and implementing knowledge that involves individual cognitive and affective aspects of engagement. Based on the research review of knowledge quality, there is still an opportunity to improve it; further evaluation and research are needed. Figure 1 describes an integrative review of knowledge quality and collective engagement process in knowledge management practices.

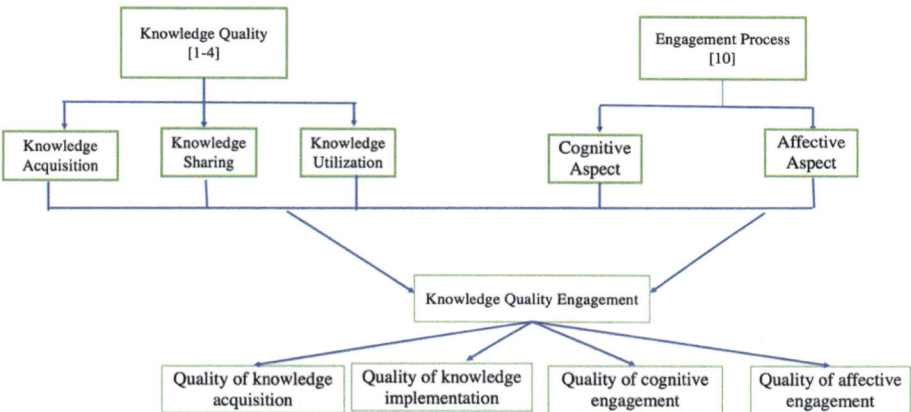

Fig. 1. An integrative review of knowledge management practices and collective engagement process.

3.2 Measurement Development

We develop and propose a measurement of each dimension of Knowledge Quality Engagement as follows:

1. **Quality of Knowledge Acquisition**

   We define the quality of knowledge acquisition as the acquisition of knowledge as 'internalization', which reflects that knowledge is based on understanding various sources of knowledge. This process is then stored in the human brain with various elements or elements such as expertise, knowledge transfer, reasoning, rules, and the ability to understand. The process of seeking knowledge must be comprehensive. The knowledge acquisition process requires extracting, structuring, and organizing knowledge from various sources of knowledge. Acquiring knowledge is natural in human civilization, which can be achieved through acquiring knowledge through programmed activities and scientific activities (experience). In addition, knowledge will acquire its quality if in the process of acquiring knowledge through three sources/tools; senses, mind, and heart. The process of acquiring coherent knowledge will produce knowledge that can positively change individual thinking processes so that mastery of a collection of knowledge and skills will lead to the formation of character as a result of the learning process and the commitment shown in acquiring this knowledge and skills will have an impact on mastering knowledge optimally which includes cognitive, practical, and psychomotor skills. The dimension of quality of knowledge acquisition is measured through the indicators of coming from multiple sources, accuracy of source, degree of context, preciseness, and meeting needs.

2. **Quality of Knowledge Implementation**

Knowledge implementation starts from creating, storing, sharing, and updating knowledge by involving people, processes, technology users, and knowledge sharing. The key to the success of knowledge implementation is to utilize one's knowledge in creating or using old knowledge. Therefore, one of the benefits of knowledge implementation is preventing loss of knowledge possessed by acquiring, integrating, storing, sharing, applying and assessing knowledge quality implementation that is measured through indicators of multiple formats, usefulness, innovativeness, accuracy, consistency, sharing, and preciseness.

3. **Quality of Cognitive Engagement**

The stage of processing information includes reaching the activities of cognition, intelligence, learning, problem-solving, and concept formation. This process involves the cognitive/soul of doing something repeatedly, continuously, and deeply so that the result is firmly embedded in the soul; this is the result of learning related to primary and teaching factors. The soul is the source of good deeds, and it also suggests that the soul is the center for acquiring and storing knowledge. Humans can think and engage in perception through soul power. The soul influences the body through action and perception. When a person acquires knowledge and continues to do it consistently, it will be ingrained in the soul and become a habit that will be quickly recovered. His intelligence is enhanced by greater exposure of knowledge and craft to the soul. We measured through the indicators of quality of communication, quality of knowledge content, creative thinking, critical thinking, and collaborative.

4. **Quality of Affective Engagement**

Quality of affective engagement is an affective learning process that includes the original basis for and is a form of attitude, emotional encouragement, individual interests and attitudes, affective aspects improve cognitive and psychomotor aspects in optimizing results. The quality of knowledge is not only cognitive and psychomotor but also faith as an integral part of a human being. Faith as the highest value must be a substance that is internalized through various fields of study so that the subject of actual knowledge is good, both in the sense of increasing knowledge and faith. This dimension will be measured through indicators of social attitude, positivity/harmony, connectivity with knowledge source, and obedience to community's law.

The results of a comprehensive and integrative review of the existing literature to build measurements of Knowledge Quality Engagement is presented in Table 1.

Table 1. Measurement development.

Previous Research Findings	Quality of Knowledge Acquisition	Quality of Knowledge Implementation	Quality of Cognitive Engagement	Quality of Affective Engagement
1. usefulness 2. multiple formats [3]	1. Multiple source 2. Accuracy of source 3. Degree of context 4. Preciseness 5. Meeting needs	1. Multiple formats 2. Usefulness 3. Innovativeness 4. Accuracy 5. Consistency 6. Sharing 7. Preciseness	1. Quality of communication 2. Quality of knowledge content 3. Creative thinking 4. Critical thinking 5. Collaborative	1. Social attitude 2. Positive/Harmony 3. Connectivity with knowledge source 4. Obey community's law
1. frequency, 2. usefulness, and 3. innovativeness [2]				
1. accuracy 2. consistency 3. currency 4. data interpretability 5. degree of context 6. degree of detail 7. degree of importance 8. sharing 9. usefulness 10. volatility [14]				
1. preciseness 2. meeting needs 3. accuracy [12]				

4 Conclusions

This research proposes a new concept related to knowledge quality as part of knowledge management practice, as we termed Knowledge Quality Engagement (KQE). Knowledge quality is the quality of the process of acquisition and implementation of knowledge, while engagement process involves the cognitive and affective aspects of the individual, the effort to create and utilize knowledge which is seen from the content, process, and results. Defining, explaining, and assessing knowledge quality and researching in-depth and comprehensive knowledge that is contextually relevant and implementable will encourage people to achieve better moral awareness for themselves and others.

The knowledge quality which come from good engagement process in creating and production will provide awareness for every individual seeker and user of knowledge to believe that their knowledge benefits themselves and others. In addition, individuals will have a sense of responsibility to share and use their knowledge only for good. Finally, we propose knowledge quality engagement as a quality of knowledge implementation as a result from acquisition which involve both affective and cognitive aspects of engagement. It has four dimensions: quality of knowledge acquisition, quality of knowledge implementation, quality of cognitive engagement, and quality of affective engagement. Our future research will empirically examine this concept using a quantitate method to validate the proposed measurement.

References

1. Soo, C.W., Christinesooutseduau, E.: The role of knowledge quality in firm performance. In: Third European Conference Organization Knowledge Learning Capability, p. 23 (2003)
2. Soo, C.W., Devinney, T.M., Midgley, D.F.: The role of knowledge quality in firm performance. In: Organizations as Knowledge Systems Knowledge, Learning Dynamical Capability, pp. 252–275 (2016). doi: https://doi.org/10.1057/9780230524545_12
3. Kulkarni, U.R., Ravindran, S., Freeze, R.: A knowledge management success model: theoretical development and empirical validation. J. Manag. Inf. Syst. 23(3), 309–347 (2006). https://doi.org/10.2753/MIS0742-1222230311
4. Yoo, D.K., Vonderembse, M.A., Ragu-Nathan, T.S.: Knowledge quality: antecedents and consequence in project teams. J. Knowl. Manag. 15(2), 329–343 (2011). https://doi.org/10.1108/13673271111119727
5. Ho, V.T., Wong, S.S., Lee, C.H.: A tale of passion: linking job passion and cognitive engagement to employee work performance. J. Manage. Stud. 48(1), 26–47 (2011). https://doi.org/10.1111/j.1467-6486.2009.00878.x
6. Fachrunnisa, O., Adhiatma, A., Tjahjono, H.K.: Cognitive collective engagement: relating knowledge-based practices and innovation performance. J. Knowl. Econ. 11(2), 743–765 (2018). https://doi.org/10.1007/s13132-018-0572-7
7. Armstrong, S.J., Fukami, C.V.: Self-assessment of knowledge: a cognitive learning or affective measure? Perspectives from the management learning and education community. Acad. Manag. Learn. Educ. 9(2), 335–341 (2010)
8. Nonaka, I., Takeuchi, H.: The Knowledge Creating Company: How Japanese Companies Create the Dynamics of Innovation. Oxford University Press, New York (1995)
9. Adhiatma, A., Fachrunnisa, O., Tjahjono, H.K.: A value creation process for sustainability of knowledge based-society. In: AISC, vol. 1194, Springer International Publishing (2021)
10. Oleksenko, V.: Features of knowledge quality of university students in Ukraine. Int. Lett. Soc. Humanist. Sci. 76, 36–42 (2017). https://doi.org/10.18052/www.scipress.com/ilshs.76.36
11. Kaun, C.G., Jhanjhi, N.Z., Wei, G.W., Sukumaran, S.: Quality model for knowledge intensive. J. Eng. Sci. Technol. 16(3), 2696–2718 (2021)
12. Yoo, D.K.: Substructures of perceived knowledge quality and interactions with knowledge sharing and innovativeness: a sensemaking perspective. J. Knowl. Manag. 18(3), 523–537 (2014). https://doi.org/10.1108/JKM-09-2013-0362
13. Durcikova, A., Gray, P.: How knowledge validation processes affect knowledge contribution. J. Manag. Inf. Syst. 25(4), 81–108 (2008). https://doi.org/10.2753/MIS0742-1222250403
14. Zhang, X., Bartol, K.M.: Linking empowering leadership and employee creativity: the influence of psychological empowerment, intrinsic motivation, and creative process engagement. Acad. Manage. J. 53(1), 23 (2010)

15. Bakker, A.B., Xanthopoulou, D.: The crossover of daily work engagement: test of an actor–partner interdependence model. J. Appl. Psychol. **94**(6), 1562–1571 (2009)
16. Kahn, W.A.: Psychological conditions of personal engagement and disengagement at work. Acad. Manage. J. **33**(4), 692–724 (1990)

Improving Business Success Through the Use of Business Capital Management and Accounting Information

Luluk Muhimatul Ifada[1](✉), Rita Rosalina[2], and Chrisna Suhendi[1]

[1] Department of Accounting, Faculty of Economics, Universitas Islam Sultan Agung, Semarang, Indonesia
{luluk.ifada,chrisnasuhendi}@unissula.ac.id
[2] Faculty of Economics, Universitas Islam Sultan Agung, Semarang, Indonesia
ritarosalina02@unissula.ac.id

Abstract. The micro, small and medium enterprise (MSME) sector is the largest economic support sector in Indonesia, but along with the development of the times, there are still many problems that occur and have not been resolved, one of which is the low business success of MSMEs. This study aims to offer a conceptual framework for the effect of using accounting information and business capital management on business success. This study uses a quantitative approach with research data in the form of primary data through a questionnaire. The population in this study are owners or managers of micro, small and medium enterprises (MSMEs) in Java Island, Indonesia. The sampling technique used is non-random sampling with the purposive sampling method. The data analysis technique in this study will use multiple linear regression analysis. Several factors that influence business success proposed in this conceptual article are the use of accounting information and business capital management.

Keywords: Use of Accounting Information · Business Capital Management · Business Success

1 Introduction

1.1 Background

Indonesia is a country whose economy is based on a populist economy marked by the presence of Micro, Small, and Medium Enterprises (MSMEs). Hasibuan (2020), stated that the management of MSMEs, which is easy and does not require a lot of money, makes the role of MSMEs quite large, both at the regional and national levels. In the national economy, MSMEs have proven their role, especially in the aspect of the income distribution, increasing job opportunities, increasing non-oil and gas exports, and economic development in rural areas. Based on data from the Ministry of Cooperatives and Small and Medium Enterprises (Ministry of Cooperative SMEs). The number of MSMEs in Indonesia until 2020 reaches 65.4 million MSME units, which are divided into Micro Enterprises (UMi) of 64.6 million, Small Businesses (UK) of 700 thousand, and Medium Enterprises (UM) of 65 thousand.

L. Barolli (Ed.): CISIS 2023, LNDECT 176, pp. 278–291, 2023.
https://doi.org/10.1007/978-3-031-35734-3_28

Business success is managed success in realizing goals that are very important in business survival. Business success is marked by an increase in the amount of production, an increase in profits or profits, an increase in the number of sales, and stable business growth (Arlianto 2014). The success of the business cannot be separated from the role of the owner in running his business. The key to business success is making accurate managerial decisions and policies (Merdekawati & Rosyanti 2020).

Business success is influenced by several factors, one of which is the use of accounting information as an important factor in influencing business success. According to Mastura et al. (2019), business success is influenced by the role of management in utilizing accounting information. Every business activity requires accounting records so that all transactions that have occurred can be known with certainty and clarity.

Accounting information is defined as accounting financial records that can be used by business owners in knowing the amount of operating income received, the number of operating costs incurred, and the amount of loss or profit earned. Decision-making and policies on business management such as market development, policy determination prices, and so on are based on accounting information in the form of financial statements (Hasibuan 2020). The success of an MSME is determined by making the right decisions, so accounting information has an important role because it is used in decision-making considerations (Wibowo & Kurniawati 2016).

According to Nurwani & Safitri (2019), the majority of micro, small and medium enterprises in Indonesia have not used and utilized accounting information in managing their business, due to the many problems that arise, namely the application of financial accounting in MSMEs in Indonesia is still weak and low due to low education, low understanding of Financial Accounting Standards (SAK), as well as legal regulations that do not yet exist regarding the obligations of MSMEs to prepare financial reports. Many MSME actors cannot continue their business because of the many problems they face (Diansari & Rahmantio 2020).

MSME actors still ignore the importance of using accounting information such as bookkeeping in business activities that are just developing, so this becomes an obstacle for MSMEs in financial planning, knowing financial conditions, and borrowing money which will slow down business success (Merdekawati & Rosyanti 2020). MSME activities which are still considered traditional have many weaknesses, namely, they still carry out traditional accounting techniques that cannot distinguish between controlling family (personal) finances and finances from business activities, so many MSME actors still combine personal finance with business finance.

Another weakness is that the prospect of business progress which is increasingly complex over time is still ignored by MSME actors (Lazuardi & Greetings 2016). According to Wibowo & Kurniawati (2016), some factors hinder the success of MSMEs in Indonesia, namely: (1) the lack of business capital issued, (2) the lack of knowledge about the market owned by MSME actors, (3) the low technology used in business activities, and (4) bargaining power is still weak.

According to Candra et al. (2020), there are several external challenges faced by MSMEs, namely: (1) market competition is increasing in line with the presence of globalization, (2) regulation and law enforcement are still weak, (3) the level of trust in product quality is still low by consumers in the country and (4) development assistance that has not been spread evenly to MSME production centers. Therefore, to be able to develop MSMEs to support the Indonesian economy to increase the use of accounting information for MSMEs. In solving these problems, it is necessary to use accounting information in the process of determining the right policies for MSME owners to increase business success.

In addition to the use of accounting information, business success is influenced by other factors, namely business capital management. When building and running a business, business capital becomes a very important part because it is used to support operational activities and ensure business turnover (Diansari & Rahmantio 2020). The important role of business capital management causes business operational activities to be carried out properly, to seize investment opportunities to increase the profitability of MSMEs.

In connection with the realization of business success, it will result in increased effectiveness of business capital management, therefore MSMEs will update the way of taking the total adequacy of the availability of business capital used in achieving the level of business success. Business capital in an MSME requires good management so when MSMEs have good business capital management, it can make it easier for MSMEs to achieve business success. However, if MSMEs have low business capital management, it will slow down business success (Firdarini & Prasetyo 2020).

The problems that exist in MSMEs in Indonesia can be seen from various aspects. First, in terms of the use of accounting information, where MSME actors are still very low in the use of accounting information, they still ignore the importance of the role of accounting information in business, and MSME actors use traditional recording systems in business activities by combining business finance with personal finance. This causes MSME actors to find it difficult to find out the advantages and disadvantages they will get to run their business in the future so that the use of accounting information that is still low will slow down business success for MSMEs.

Furthermore, from another perspective, it is seen from business capital management. Many MSMEs are still not good at managing their business capital, and MSME actors are still having difficulties in managing business capital. This is because MSMEs still have a low understanding of the importance of business capital management in increasing business success. Whereas the role of business capital is very important, where business capital can be used as a reference to get investment opportunities and get money loans from outside parties.

Research on the use of accounting information on business success has been carried out by many previous researchers, namely: Firdarini & Prasetyo (2020); Mastura et al. (2019); Nurwani & Safitri (2019); Yulianthi & Susyarini (2017), which which states that there is a positive influence between the use of accounting information and business success, as well as research Fauzi (2020), which states that the use of accounting information does not affect business success.

The previous research on business capital management affects business success, namely: Firdarini & Prasetyo (2020) concludes that there is a positive influence between business capital management and business success. And according to research Fauzi (2020) stated that business capital has a negative influence on business success, and also according to research in Netty & Yustien (2019), the authors state that there is no influence between business capital and business success.

This research refers to research Diansari & Rahmantio (2020), which states that the use of accounting information and business capital has a significant influence on business success. This shows that if MSMEs use accounting information and good business capital management in managing their business, they can achieve and increase business success so that MSMEs in lending business capital will not experience difficulties (Diansari & Rahmantio 2020).

This study combines research models that have been carried out by previous researchers such as (Diansari & Rahmantio 2020), (Hasibuan 2020), (Firdarini & Prasetyo 2020), (Fauzi 2020), (Netty & Yustien 2019), (Nurwani & Safitri 2019), (Eve E 2019), (Apriliani & Widiyanto 2018), and (Yulianthi & Susyarini 2017). The combined model is synthesized to produce a new model that is different from the previous research model.

Based on the phenomena and research gap above, this research is interesting to do to examine the effect of the independent variable, namely the use of accounting information and business capital management on the dependent variable, namely business success in micro, small and medium enterprises in Java Island continuation of business success. This research is also very important for the level of business success so it is hoped that MSMEs can bounce back and the economy in Indonesia will also recover. So that the formulation of the problem in this study is as follows:

1. How to increase business success through the use of accounting information?
2. How does business capital management affect business success?

2   Literature Review

2.1   Stakeholder Theory

In 1984 R. Edward Freeman first proposed stakeholder theory, he defined and described it as a "separation thesis" in the workplace in business discussions. This theory was then popularized by Clarkson in 1994 who saw stakeholders have an interest in an organization based on moral or legal reasons. An organization has an obligation if there is a party that has legal rights to the organization. This can result in good implementation for an organization in maintaining good relations with stakeholders. An organization should be responsive to stakeholders.

Theory stakeholdersdefines a company not only to carry out operational activities for itself but also to be useful for shareholders. In research conducted by Azmi (2019) and Gray, Kouhy, and Adams (1994) it is stated that the survival of a company is influenced by support from shareholders or other parties. A result of activities or actions decided by the organization legally or morally, personally or collectively, is a right and interest of the organization. Stakeholders include shareholders, suppliers, creditors, employees, consumers, communities, and others.

Stakeholders provide an important role in MSMEs. This relates to the source of wealth provided by stakeholders for the company's operational activities such as donations to the company, loans, and government regulations. So the company should improve its performance to gain the trust of shareholders that the company can develop. To achieve this is to use accounting information and good business capital management can be used by management as material for business planning and control in making decisions that are useful for achieving success for the sake of business continuity.

2.2 Motivation Theory

The word "movere" is a Latin word for motivation which means force, encouragement, or driving, which causes action or action. The word "movere" is defined as motivation in English which has the meaning of generating motives, giving motives, and conditions that cause encouragement. Motivation is explained as an impulse that invites people to act and behave by motivational techniques based on the cause and effect of an action, namely aspects that make a person able to do or not do something.

In 1993, Bedard and Chi put forward a theory of motivation which was strengthened by Spilker in 1995. They stated that to increase the understanding of owners or managers in using accounting information in business, it is necessary to have motivation for owners or managers to understand accounting knowledge.

Business success in a company is not only for profit but motivation in developing a business must also be improved. Business actors, especially MSMEs, must have high motivation in each of them to continue to improve performance in the current business competition. Therefore, it can be concluded that the theory of motivation is the ability of the owner or manager to be able to motivate employees to have high knowledge of accounting information in its application and to manage business capital well, in its business activities, to achieve business success (Yolanda et al. 2020).

2.3 Business Success

Business success is a condition that exceeds other parallel conditions (Sustainable 2011). Business success is also defined as the achievement of goals and objectives by the company, which is not interpreted directly (Radzi et al. 2017). Some of the factors supporting business success in Micro, Small, and Medium Enterprises are a description of work motivation, business abilities which are illustrated through attitudes, knowledge, and skills, education levels, and relevant experience. The success of a business is described through profits or additional wealth obtained in business operations. Business success is not only felt physically but business success can be accepted by management in the form of inner satisfaction and individual calling.

2.4 The Use of Accounting Information

Use means a process or method of using something. According to I Cenik and Endro (2016) in Nurwani & Safitri (2019), information is defined as the output of data management that is useful for information users. Isaac and Arief (2015) in Nurwani & Safitri

(2019), define accounting as a service activity that is useful in making quantitative information and data, especially in the financial department of a company so that it is useful in making policies and decisions to determine the choice that is considered the most appropriate compared to other options.

Furthermore, accounting information is defined as information that is needed in regulating the company to avoid being there are problems related to the company's activities (Yousef 2013). According to Nwaigburu & Mark (2014), accounting information is a contribution that has a significant nature to activities that are useful in making decisions in a company. From the definition described above, it can be concluded that the use of accounting information is defined as a process of applying accounting information that can generate benefits in the form of quantitative and qualitative data needed by a company in an accurate decision-making process.

Priliandani et al. (2020) mention that in carrying out its functions, management requires information, such as quantitative information or qualitative information. Quantitative information that is widely used is in the form of accounting information. Financial statements must be prepared properly to be useful for internal parties or external parties of the company. Qualitative information is in the form of information about company policies such as in strategic preparation. In the process of management supervision, operational supervision, and strategic planning, using financial accounting information (Candra et al. 2020).

2.5 Business Capital Management

Business capital is one of the factors that must exist before carrying out production activities. According to Prawirosoentono (2007) in Apriliani & Widiyanto (2018), Business capital is defined as an asset that must be owned by a company to earn future profits and is usually expressed in units of value. Apriliani & Widiyanto (2018) revealed that the development and achievement of income of a business are influenced by the size of its business capital.

According to Rumerung (2018), the presence of business capital is very important in building and running a business. However, the events that often become a problem are how to manage business capital appropriately and optimally so that the business that is run will generate profits and can achieve its goals. For MSMEs, large and small business capital will be a problem in itself, because if the amount of business capital is too much compared to the business needs, it will result in a lot of loading costs, but on the contrary, if the amount of business capital owned is too small, it will make the business run will feel difficult. Business capital used in running a business must be under business needs.

3 Hypothesis Development

3.1 The Effect of Using Accounting Information on Business Success

Accounting information is a very important factor in an organization or business. An organization uses accounting information for the planning, management, or evaluation of an organization. The use of accounting information in the form of notes to record transactions related to business receipts and expenses. All business activities can run well due to the available accounting information that can influence achieving business success.

An MSME in achieving business success cannot be separated from the influence of the use of accounting information which is used as a basis for decision-making in running a business, such as being used for market development, pricing, and so on. So that MSMEs need the use of good accounting information for the success of their business. Business success for MSMEs will be the key to the success and survival of MSMEs in the future.

Christian & Rita (2016) revealed that accounting information has a very important influence on achieving business success so that it can be used as a basis for making decisions. A lack of knowledge of accounting information can cause financial bookkeeping activities to be hampered. Accounting information is used by owners or managers such as financial records to find out operational costs that must be incurred, find out how much income is earned, and find out the amount of profit/loss earned (Mastura et al. 2019). According to Suryana (2013) in Nurwani & Safitri (2019), business success is marked by increasing capital, increasing income, increasing sales volume, increasing production output, and increasing the number of workers.

Studies by Diansari & Rahmantio (2020), Nurwani & Safitri (2019), Hasibuan (2020), and (Yulianthi & Susyarini 2017), mention the use of accounting information has a significant effect on business success. This shows that when business actors, especially MSMEs use accounting information well in making decisions, achieving business success will be easier to achieve and increase, compared to MSMEs that do not use accounting information in their decision-making process.

Based on this explanation, the following hypothesis can be formulated:
H1 = The use of accounting information has a positive effect on business success

3.2 The Effect of Business Capital Management on Business Success

One of the important factors in running a business is business capital because a business cannot operate if there is no business capital. The success of a business is influenced by the size of the amount of business capital which is sufficient to increase the smoothness and facilitate the development process of a business (Feriansyah & Manullang 2015). Business capital is defined as the amount of money used to operate a business so that it can grow and run.

Capital in business can be viewed from several sides, such as capital to set up a business, capital for business development, and capital to operate daily business activities (Agustina 2015). (Diansari & Rahmantio 2020). An organization in determining the business capital needed by an organization needs to be determined precisely because it will be used to ensure the smooth running of business activities.

Previous research by Apriliani & Widiyanto (2018), Diansari & Rahmantio (2020), and Firdarini & Prasetyo (2020) reveals the influence between business capital management and business success. This means that if there is an increase in business capital, it must be balanced with an increase in business success. It can be concluded that if the amount of business capital owned is insufficient, it will cause problems in the production process, and vice versa if the business capital owned is excessive, it will cause investment opportunities to be hampered, because the business capital used is only for operational activities. Therefore, managers must decide how much business capital is appropriate and appropriate so that the company's operational activities can run smoothly and can capture investment opportunities to increase the level of profitability so that business success can be achieved.

A business, especially MSME, really needs a management role, especially in terms of business capital. Business capital is the driving wheel of MSME activities so good management is needed. The goal is that having good business capital management in an MSME will make it easier to achieve business success. On the other hand, if an MSME has poor business capital management, then MSMEs will find it difficult to achieve business success.

Based on this explanation, the following hypothesis can be formulated:
H2 = Business capital management has a positive effect on business success

4 Research Methodology

4.1 Types of Research

The type of research used is explanatory research. Explanatory research is a study that describes the relationship between variable X and variable Y. The research method used is a quantitative method using a questionnaire. According to Sugiyono (2019) in Blue & Regards (2016), the quantitative research method is defined as a method based on the philosophy of positivism which is used to examine the population and sample. Later quantitative methods are used to test a variable by using statistical tools in the form of numbers or scores which are generally obtained using data collection tools with answers in the form of questions that are given a range of scores or weights so that they will generate hypotheses and can explain the relationship between the independent variable and the dependent variable has the ability to develop an understanding of various things.

4.2 Population and Research Sample

Population
This study uses a population of SMEs in Java Island Indonesia. According to data from the Office of Cooperatives and Micro, Small, and Medium Enterprises (Diskop UKM) states that the total number of MSMEs in Micro, Small and Medium Enterprises (MSMEs) thrive in a number of areas. This can be seen from the data reported by the Ministry of Cooperatives and Small and Medium Enterprises (Kemenkop UKM), the total number of MSMEs in Indonesia will exceed 8.71 million business units in 2022. Java Island

dominates this sector. It was recorded that West Java became the MSME champion with a total of 1.49 million business units. Thin in second place is Central Java, which reached 1.45 million units. Third, there is East Java with 1.15 million units. The author took the population of Java Island, Indonesia because for the results of the study to be more accurate by using a larger population and wider range. If you only use the population in one or two districts or cities, the number of population in the district that has the criteria according to the provisions that have been made by the author is not sufficient, so the results of the study will be less accurate because it only uses a population with a small scope, but by using a population MSMEs in Java Island, Indonesia the author can take samples of MSMEs in various districts or cities in Java Island.

Sample

The sample is part of the total population that has certain terms and conditions (Sugiyono 2019). Determination of the sample size based on the number population used by the author, namely MSMEs located in districts or cities of Java Island, West Java with a total of 1.49 million business units. Central Java, which reached 1.45 million units. East Java with 1.15 million units. The following is a calculation of the overall sample used in this study according to the 1960 Slovin formula:

$$n = \frac{N}{1 + N(e)^2}$$

$$n = \frac{4.090.000}{1 + 4.090.000\,(0.05)^2}$$

$$n = 399, 96$$

$$n = 400\ respondents$$

So, the minimum number of samples used is 400 respondents.

4.3 Sampling Technique

This study used a non-random sampling technique using a purposive sampling method. A Non-random sampling technique is a sampling technique for each member of the population who is not given the same opportunity to be used as a research sample. The purposive sampling method belongs to the category of non-random sampling technique. The purposive sampling method is a technique with a method considering certain characteristics that will be used in determining the research sample (Sugiyono 2019). In the purposive sampling method, the author can choose research subjects and research locations with the aim of studying or understanding the main problems to be studied. The characteristics of the samples used are as follows:

1. MSMEs are located in West Java, Central Java and East Java.
2. Have a minimum of 3 employees.
3. The business has been running for at least 2 years.
4. Minimum monthly income of IDR 2,000,000.

4.4 Data Sources and Types

This study uses primary data types. Primary data is a source of data obtained by the author directly (Sugiyono 2019). Primary data were obtained directly from respondents through questionnaires using all original data collection methods. The primary data used in this study were obtained through questionnaires distributed directly (offline) or online using google forms. The respondents in question are owners or managers of MSMEs in Java Island, especially West Java, Central Java and East Java, both those who have not or have been registered with the Office of Cooperatives and Small and Medium Enterprises of each Province.

4.5 Method of Collecting Data

The data collection method in this study used a questionnaire or questionnaire. The questionnaire or questionnaire method is a data collection technique carried out by the author by distributing written statements and questions directly or indirectly to be answered by respondents (Sugiyono 2019). If the number of respondents is very large and widely distributed, then the questionnaire is suitable to be applied. In the questionnaire, some questions or statements are open or closed. From the questionnaire or questionnaire method, the authors prepared two methods of distribution, namely online using google forms or offline which were given directly using paper to several potential respondents, namely owners or managers of SMEs in Java Island as many as 400 respondents.

4.6 Research Period

This research will be carried out from December 2022–May 2023 by going through the research stages. Includes observation, submitting research proposals, making and testing research instruments, data documentation, and research data analysis.

4.7 Operational Definitions and Variable Indicators

Business Success

This study uses the dependent variable, namely business success. Business success is defined as the perception of the owner or founder of the business about the performance of his business compared to the goals to be achieved. The increasing number of sales, increased production, ever-increasing profits, and businesses that are always developing are signs of business success (Arlianto 2014) (Merdekawati & Rosyanti 2020). According to Firdarini & Prasetyo (2020), a sign of business success is the addition of the number of employees and an ever-increasing sales turnover.

Use of Accounting Information

The first independent variable used in this study is the use of accounting information.

Usage in the Big Indonesian Dictionary is a method and a process of using something (Priliandani et al. 2020). The definition of accounting information is information that has a quantitative nature related to the company as a basis for making decisions to choose the right alternative (Firdarini & Prasetyo 2020). It can be concluded that the use of accounting information is the process, methods, and actions of using and using accounting information for economic decision-making in determining many choices among alternative actions so that the decisions taken will be more appropriate (Hasibuan 2020). According to Belkaoui (2010) in Firdarini & Prasetyo journal (2020) There are three classes of accounting information, namely financial information, management information, and operating information.

Business Capital Management
The second independent variable used is business capital management. Management is defined as a systematic process within a company for the control and supervision process in achieving business goals. Furthermore, the understanding of business capital is the amount of money used in operating a business so that it continues to run and develop. The amount of business capital owned by a business must be following the required needs because less business capital will cause difficulties in running a business, while excessive business capital will make it difficult for businesses to find investment opportunities from outside parties. So business capital in running a business must be appropriate and appropriate and requires good business capital management.

4.8   Analysis Techniques

The analysis technique is a data processing technique using a computer program with Statistical Product and Service Solutions (SPSS) version 25.0 which can process statistical data accurately and quickly. The analysis technique in this study is used to examine the effect of using accounting information and business capital management on business success. Data analysis is made to make decisions from the processed data. In this research, descriptive statistical analysis, data quality test (validity test and reliability test), classical assumption test (data normality test, multicollinearity test, and heteroscedasticity test), multiple linear regression analysis, and hypothesis testing (F statistic test, statistical test were carried out). t, and the coefficient of determination test (R^2) (Table 1).

Table 1. Indicator variables.

No.	Variables	Indicators	Sources
1.	Business Success	1. Number of employees who have increased 2. Increased consumer orders 3. Increased turnover or income 4. Product promotion has increased 5. The selling price of the product has increased 6. Business capital that has increased 7. Higher level of sales 8. Production continues to increase 9. Production equipment upgrade	Nurwani & Safitri (2019)
2.	The Use of Accounting Information	1. Use of accounting information for projecting future funding needs 2. Controlled expenses 3. Well-measured business productivity 4. Using operation information 5. Using management accounting information 6. Using financial accounting information 7. Using statutory accounting information 8. Using budgetary information 9. Using additional accounting information 10. There is an increase in business productivity 11. The production process is always supported	(Diansari & Rahmantio 2020)
3.	Business Capital Management	1. Initial capital invested 2. Business working capital 3. Business operating capital 4. Challenges in obtaining capital	Diansari & Rahmantio (2020)

5 Conclusions

Based on the conceptual explanation of the research above, the use of accounting information and business capital management affects business success. The use of accounting information is useful in making the right decisions for business actors in achieving business success. Business capital owned by MSMEs has a very important role in a business because business capital is the driving wheel of MSME operational activities so good management is needed to achieve business success.

References

Apriliani, M.F., Widiyanto, W.: Pengaruh Karakteristik Wirausaha, Modal Usaha Dan Tenaga Kerja Terhadap Keberhasilan UMKM Batik. Econ. Educ. Anal. J. **7**(2), 761–776 (2018)

Arlianto, T.: Pengaruh Penggunaan Informasi Akuntansi Terhadap Keberhasilan UMKM (Studi Kasus Pada Industri Konveksi Desa Padurenan Kecamatan Gebog Kabupaten Kudus), pp. 1–47. Universitas Kristen Satya Wacana Salatiga (2014)

Christian, A.B.G., Rita, M.R.: Peran Penggunaan Informasi Akuntansi Dalam Pengambilan Keputusan Untuk Menunjang Keberhasilan Usaha Role of the use of accounting information in decision making to support business success. J. EBBANK **7**(2), 77–92 (2016)

Diansari, R.E., Rahmantio, R.: Faktor keberhasilan usaha pada UMKM industri sandang dan kulit di Kecamatan Wirobrajan Kota Yogyakarta. J. Bus. Inf. Syst. **2**(1), 55–62 (2020). https://doi.org/10.36067/jbis.v2i1.60. (e-ISSN: 2685–2543)

Fauzi, N.A.: Pengaruh Karakteristik Wirausaha, Modal Usaha dan Penggunaan Informasi Akuntansi Terhadap Keberhasilan UMKM Industri Shuttlecock di Desa Lawatan Kecamatan Dukuhturi Kabupaten Tegal. Jurnal Repository FEB Universitas Pancasakti Tegal (2020)

Feriansyah, I., Manullang, R.R.: Analisi Pengaruh Faktor Modal Usaha, Tingkat Pendidikan, Lokasi Usaha, dan Lama Kecil Menengah (Studi Kasus Pada Binaan Dinas Perindustrian dan Perdagangan Kota Pangkalpinang). Jurnal Ilmiah Progresif Manajemen Bisnis (JIPMB) **4**(2), 27–38 (2015)

Firdarini, K.C., Prasetyo, A.S.: Pengaruh Penggunaan Informasi Akuntansi Dan Manajemen Modal Kerja Pelaku Umkm Terhadap Keberhasilan Usaha Dengan Umur Usaha sebagai Variabel Pemoderasi (Studi Kasus Pada Industri Kreatif Di Yogyakarta). Jurnal Stie Semarang **12**(1), 19–32 (2020). https://doi.org/10.33747/stiesmg.v12i1.394

Hasibuan, H.T.: Pengaruh Penggunaan Informasi Akuntansi terhadap Keberhasilan Usaha Mikro Kecil. E-Jurnal Akuntansi **30**(7), 1872 (2020). https://doi.org/10.24843/eja.2020.v30.i07.p19

Kemenkopukm. Perkembangan Data Usaha Mikro, Kecil, Menengah (UMKM) dan Usaha Besar (UB) Tahun 2018–2019. Kemenkopukm.Go.Id (2019). https://www.kemenkopukm.go.id/uploads/laporan/1617162002_SANDINGAN_DATA_UMKM_2018-2019.pdf

Lazuardi, Y., Salam, F.A.: Pengaruh Penggunaan Sistem Informasi Akuntansi Terhadap Keberhasilan Usaha Kecil Menengah. Jurnal Ilmiah Akuntansi Peradaban **19**(1), 1–10 (2016)

Lestari, F.: Pengaruh jiwa kewirausahaan dan kreativitas terhadap keberhasilan usaha pada sentra industri rajutan binong jati bandung. Jurnal Ekonomi Bisnis Dan Akuntansi **1**(1), 14–27 (2011)

Mastura, M., Sumarni, M., Eliza, Z.: Peranan Infomasi Akuntansi terhadap Keberhasilan UMKM di Kota Langsa. Jurnal Ekonomi Dan Bisnis Islam **4**(1), 20–33 (2019). https://doi.org/10.32505/v4i1.1248

Merdekawati, E., Rosyanti, N.: Faktor-Faktor Yang Mempengaruhi Keberhasilan Umkm (Studi Kasus Pada Umkm Di Kota Bogor). JIAFE (Jurnal Ilmiah Akuntansi Fakultas Ekonomi) **5**(2), 165–174 (2020). https://doi.org/10.34204/jiafe.v5i2.1640

Netty, H., Yustien, R.: Pengaruh Modal, Penggunaan Informasi Akuntansi dan Karakteristik Wirausaha Terhadap Keberhasilan Usaha Kecil (Survei Pada Usaha Rumahan Produksi Pempek di Kota Jambi). Jurnal Ilmiah Akuntansi Dan Finansial Indonesia 3(1), 63–76 (2019). https://doi.org/10.31629/jiafi.v3i1.1582

Nurwani, N., Safitri, A.: Pengaruh Penggunaan Informasi Akuntansi terhadap Keberhasilan Usaha Kecil Menengah (Studi pada Sentra Dodol di Kec. Tanjung Pura). Liabilities (Jurnal Pendidikan Akuntansi) 2(1), 37–52 (2019). https://doi.org/10.30596/liabilities.v2i1.3332

Nwaigburu, K.O., Mark, B.U.: The use of accounting information in decision making for sustainable development in Nigeria: a study of selected tertiary institutions in Imo state. Int. J. Sci. Res. Educ. 7(2), 167–175 (2014)

Priliandani, N.M.I., Pradnyanitasari, P.D., Saputra, K.A.K.: Pengaruh Persepsi dan Pengetahuan Akuntansi Pelaku Usaha Mikro Kecil dan Menengah terhadap Penggunaan Informasi Akuntansi. Jurnal Akuntansi, Ekonomi Dan Manajemen Bisnis 8(1), 67–73 (2020). https://doi.org/10.30871/jaemb.v8i1.1608

Radzi, K.M., Nazri, M., Nor, M.: The impact of internal factors on small business success: a case of small enterprises under the FELDA scheme. Asian Acad. Manag. J. 22(1), 27–55 (2017)

Rumerung, D.: Analisis Tingkat Keberhasilan Usaha Usaha Kecil Mikro dan Menengah di Kabupaten Maluku Tengah. Jurnal SOSOQ 6(1), 75–92 (2018)

Sugiyono: Metode Penelitian Kuantitatif Kualitatif dan R&D (Fauzi (ed.)) (2019)

Wibowo, A., Kurniawati, E.P.: Pengaruh Penggunaan Informasi Akuntansi Terhadap Keberhasilan Usaha Kecil Menengah (Studi Pada Sentra Konveksi di Kecamatan Tingkir Kota Salatiga). Jurnal Ekonomi Dan Bisnis 18(2), 107 (2016). https://doi.org/10.24914/jeb.v18i2.269

Yousef, B.A.S.: The use of accounting information by small and medium enterprises in south district of Jordan, (An empirical study). Res. J. Finan. Account. 4(6), 169–175 (2013)

Yulianthi, D.A., Susyarini, A.W.N.P.: Pengaruh Penggunaan Informasi Akuntansi Terhadap Keberhasilan Usaha Jasa Penginapan Bertaraf Kecil. Prosiding Sentrinov 3, 2477–2497 (2017)

Psychological Achievement Leadership: A New Leadership Style Based on Psychological Work Contract and Achievement Motivation

Ratih Candra Ayu[✉], Olivia Fachrunnisa, and Ardian Adhiatma

Department of Management, Faculty of Economics, Universitas Islam Sultan Agung (UNISSULA), Semarang, Indonesia
ratihca_pdim5@std.unissula.ac.id, {olivia.fachrunnisa, ardian}@unissula.ac.id

Abstract. This paper aims to develop a new concept of leadership style based on psychological work contracts and achievement motivation. We used extensive and comprehensive literature to create a theoretical synthesis. The result shows that Psychological Achievement Leadership can be defined as leaders who can inspire and encourage members to always excel based on the psychological relational work contract that has been agreed upon. The ultimate goal to have the best work performance and achievement is psychological satisfaction by both leader and member. We also propose four dimensions to indicate psychological achievement leadership: Achievement Motivation, Inspirational Motivation, Affective Work Contract, and Cognitive Work Contract. In our future research, we empirically test the effectiveness of this concept to improve organizational performance.

Keywords: Transformational leadership · psychological contract · inspirational Achievement · motivational inspiration

1 Introduction

Motivational inspiration which is the second component of transformational leadership is a leader who is able to inspire and motivate their employees to do business and achieve success outside [1]. Inspirational motivation arises when leaders inspire their followers to achieve by setting high but reasonable goals for their followers and their organization. Recent research suggests an alternative characterization of leadership to employ a different leadership style that goes beyond the traditional, autocratic approach that once dominated organizations [1, 2]. Surveys show that, a supportive and participative leadership style for leaders who are open-minded in using non-traditional systems will improve performance [2]. The reconstruction of the transformational leadership style in this study is structured mainly on two main dimensions, namely inspirational motivation and intellectual stimulation by incorporating the dimensions of the psychological work contract and the need for achievement from elements of intrinsic motivation.

The theoretical gap in this research stems from studies on transformational leadership theory, which is a leadership style that identifies the changes needed, develops a vision

L. Barolli (Ed.): CISIS 2023, LNDECT 176, pp. 292–298, 2023.
https://doi.org/10.1007/978-3-031-35734-3_29

that will pave the way for changes to be made and implements the necessary plans for these changes to occur and takes the organization towards a better direction [3]. However, this transformation towards change only identifies mundane changes, not yet individual changes which in their psychological contract are caliphs who must lead their followers towards changes for the better.

The leadership's quick response to changes in the external environment is very important to prevent crimes that can arise at any time [4], but the orders of some leaders to subordinates have a low level of obedience which results in increased poor performance [5]. Leaders' low awareness to always be proactive about environmental changes and proactively share knowledge with subordinates is the main factor for low performance [5]. In addition, the leadership's adaptation to the new workplace environment and with the new team was not well established. A Leader is an officer who is ready to be placed anywhere, so adaptation learning is the key to success in leading however, the gap between rank and achievement of leaders and subordinates results in ineffective orders to be implemented [6].

A new leadership concept is needed which is a synthesis of the concepts of transformational leadership and the psychological contract, namely psychological achievement leadership so that organizations can quickly adapt to changes that occur then carry out generative learning to increase creativity and encourage organizations to adopt new ways to see old methods [4]. The application of psychological achievement leadership in the environment is expected to be the right step to carry out institutional changes, especially to improve the performance of leaders which is based on psychological rewards not on an economic orientation with organizational learning that is properly adapted from external changes.

2 Literature Review

2.1 Transformational Leadership: Intellectual Stimulation and Inspirational Motivation

Intellectual stimulation dan inspirational motivation are the two dimensions of transformational leadership [6]. Transformational leadership is concerned with developing individual potential and fully motivating them towards the greater good versus their own self-interest, within a values-based framework [5, 6]. The transformational leadership model contains four components: idealistic influence, inspirational motivation, intellectual stimulation, and individual consideration.

Leaders who demonstrate inspirational motivation push their employees to achieve more than ever before by developing and articulating a shared vision and high expectations that motivate, inspire, and challenge [1]. Leaders who embody intellectual stimulation help employees to question their own commonly held assumptions, reframe problems, commit morals, and approach things in innovative ways.

Several studies on motivation conclude that inspirational motivation is an important factor in the role of a leader in driving career success [6]. However, in the process individuals tend to see success from their personal orientation and ignore the importance of the ethical dimensions of motivation [7]. By incorporating an ethical dimension to achievement motivation will direct individuals to better management practices related to performance achievement. Intrinsic motivation is the spirit to do the best and make continuous improvements coupled with always asking and helping team members to do their best. Ethical intrinsic motivation has several unique characteristics, including being divided into two dimensions of time, namely affective and cognitive aspects. Then, having the passion to be better does not stop when performance standards have been met but continues to strive for continuous improvement. To achieve the best performance, other people are not considered as competitors to be fought against but are considered as partners and objects to be used as benchmarks. Furthermore, someone who has motivation does not only think about himself but also invites and helps others in an effort to do good [8].

In connection with this inspirational motivation, leaders will provide motivating inspiration to followers, where one of them is the motivation to achieve great work performance. However, work performance that is often encouraged or inspired by leaders is work performance which may be related to economic motives in the form of monetary rewards or individual awards that refer to the economic dimension. This is in line with the work contracts established between leaders and followers. The psychological contract is a series of unwritten expectations between each member and the manager. The psychological contract is an employee's individual belief that the organization has promised future returns and has an obligation to reward their contributions [9, 10]. The concept of ethical intrinsic motivation is the result of a synthesis of manifest needs theory and ethical values. This motivation has 4 (four) dimensions: voluntary helping and inviting others to socialize; voluntary helping and inviting others to help colleagues; voluntary helping and inviting others to join in; voluntary helping and inviting others to excel.

Transformational leaders intellectually stimulate their followers to challenge existing assumptions and solicit followers' suggestions and ideas. Finally, individual consideration occurs when leaders pay special attention to employees' needs for achievement and development; they provide the empathy, compassion, support, and guidance needed for employee well-being [1]. Overall, this transformational leadership must be inspired by followers to achieve higher levels of performance [9].

Leaders who embody intellectual stimulation help employees to question their own commonly held assumptions, reframe problems, and approach things in innovative ways [9]. Transformational leaders intellectually stimulate their followers to challenge existing assumptions and solicit followers' suggestions and ideas. Finally, individual consideration occurs when leaders pay special attention to employees' needs for achievement and development; they provide the empathy, compassion, support, and guidance needed for employee well-being [11]. Overall, this transformational leadership must be inspired by followers to achieve higher levels of performance [3].

2.2 Psychological Contract

Psychological Contract is the process by which these perceptions come to an agreement between the two parties [10]. The literature on the type of contract shows that the importance of the type of contract will influence employee behavior. The importance employees attach to relational obligations has a positive effect on employee behavior. In human life there is an obligation to do good and avoid bad/bad actions which are universal and constitute a moral imperative, based on human nature. All forms of human action refer to his view of good and bad. Values of good and bad will always be a source of reference (frame of reference) in carrying out various actions in life. Etymologically these terms are synonyms that have almost the same meaning to one another, but terminologically they still have different meanings.

3 Achievement Psychological Leadership

3.1 Concept Development

Based on the integration of transformational leadership from inspirational motivation, achievement motivation and work contract, dimension of Achievement psychologolical leadership can be developed. Figure 1 describes the integration of achievement psychological contract, along with its dimensions.

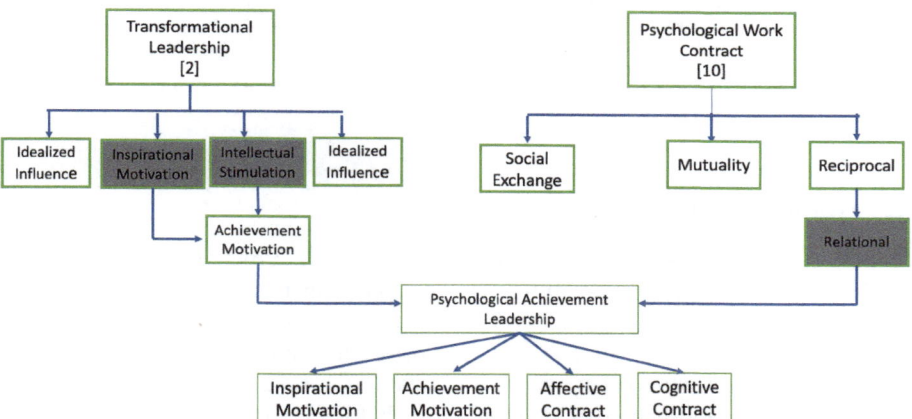

Fig. 1. Integration of Transformational Leadership Theory and Psychological Work Contract

Based on the integration of transformational leadership and psychological work contract, the dimensions and indicators can be compiled as described in Table 1.

It can be concluded that the integration of the dimensions of transformational leadership (intellectual stimulation and inspirational motivation) and psychological work contract theory are: inspirational motivation, achievement motivation, affective work contract and cognitive work contract.

Table 1. Dimensions of psychological achievement leadership.

No	Inspirational Motivation + Achievement Motivation	Psychological Work Contract	Psychological Achievement Leadership
1	Voluntary helping others to achieve the best performance	Relational	Stimulate obedience personnel in achieving the best achievement (inspirational motivation)
2	Voluntary to invite others to achieve the best	Reciprocal	Encouraging the completion of tasks in the best way (achievement motivation)
3	Inspire others with the best achievements	Relational	Motivating members to always excel (affective work contract)
4	Inspire others to always achieve the best	Mutualism	Encouraging the implementation of tasks properly, directed, precise and complete (cognitive work contract)

3.2 Measurement Development

Leading is not only able to order subordinates to carry out orders but to lead needs to use the heart because humans are different from objects that have no response to refuse. Successful leadership is an action that is ordered to subordinates to complete but does not cause further negative impacts. Humans psychologically have a feeling of being detached from their duties in companies/organizations. Leaders will be successful when they are able to link psychological and non-psychological factors so that performance will increase. Psychological factors of place have a positive effect on employee performance. It contributes to the success of individuals, groups, organizations, industries, and countries.

The next psychological factor is related to a leader who empowers subordinates with the object of service to the community states that employees who are empowered by leaders feel valued and considered as part of the organization so that it is an achievement in itself so that it motivates them to improve their performance. Practically leaders can improve the performance of their members not only from members' financial factors which are stimulated but can use psychological factors such as a comfortable workplace and empower employees properly. Table 2 explains the definition of dimensions and indicators for psychological achievement leadership.

Table 2. Dimensions and indicators of psychological achievement leadership.

No. 1-4	Dimensions of PAL	Indicators
1	Achievement motivation is the ability to motivate members to achieve while still emphasizing the obedience and adherence of members to religious rules	1. Encourage achievement for the good of the organization; 2. Awards for achievements; 3. Motivate achievement 4. Encourage members to motivate each other to excel
2	Inspirational motivation is the ability to be inspired in completing tasks in a good way	1. Inspire others by being hardworking and honest; 2. Inspire others with extraordinary self-achievement; 3. Be a good example with commendable actions; 4. Complete orders with full sense of responsibility
3	Cognitive Work Contract is a psychological work contract that fosters personnel compliance with applicable rules and norms	1. Build relationships and work relationships that comply with regulations 2. Building member adherence to norms 3. Building an attitude of never giving up 4. Build altruistic behavior
4	Affective Work Contract is a psychological work contract that encourages good, directed, precise and thorough execution of tasks	1. Build a working relationship to complete the task well; 2. Build a working relationship to complete the task properly; 3. Build working relationships to complete tasks according to the direction of the leadership 4. Build working relationships to complete tasks thoroughly

4   Conclusions

This study proposes a new concept related to Transformational Leadership (intellectual stimulation and inspirational motivation) and psychological contract theory which forms a new concept called Psychological Achievement Leadership. The two dimensions of transformational leadership, namely intellectual stimulation and inspirational motivation, create change, formulate a vision that will pave the way for changes to be made and carry out the necessary plans for these changes to occur and bring the organization to a better direction. However, transformation towards this change only identifies mundane changes, not to mention the change in the individual who in his psychological contract must change towards a better direction. Psychological achievement leadership is a leadership style that encourages the transformation of change in the organization by inspiring organizational members to achieve according to the psychological work contract that is built between leaders and members. This psychological work contract is structured by referring to ethical relational values. We propose that psychological achievement leadership has four dimensions: achievement motivation, inspirational motivation, cognitive work contract and affective work contract. An organization that has leadership with the psychological achievement style is believed to have good organizational performance.

References

1. Kelloway, E.K., Weigand, H., Mckee, M.C., Das, H.: Positive leadership and employee well-being. J. Leadersh. Organ. Stud. **20**(1), 107–117 (2013). https://doi.org/10.1177/1548051812465892
2. Bass, B.M.: Two decades of research and development in transformational leadership. Eur. J. Work Org. Psychol. **8**(1), 9–32 (1999). https://doi.org/10.1080/135943299398410
3. Bass, B.M.: From transactional to determining for a group of loyal transformational leadership: learning to followers, the direction, pace, and share vision. Organ. Dyn. **18**(3), 19–32 (1990). https://www.mcgill.ca/engage/files/engage/transformational_leadership_bass_1990.pdf
4. Indriastuti, D., Fachrunnisa, O.: Achieving organizational change: preparing individuals to change and their impact on performance. Public Org. Rev. **21**(3), 377–391 (2021)
5. Middleton, J., Harvey, S., Esaki, N.: Transformational leadership and organizational change: how do leaders approach trauma-informed organizational change…twice?. Fam. Soc.: J. Contemp. Soc. Serv. **96**(3), 155–163 (2015). https://doi.org/10.1606/1044-3894.2015.96.21
6. Avolio, B.J., Bass, B.M.: Individual consideration viewed at multiple levels of analysis: a multi-level framework for examining the diffusion of transformational leadership. Leadersh. Q. **6**(2), 199–218 (1995). https://doi.org/10.1016/1048-9843(95)90035-7
7. Bass, B.M.: Leadership and performance beyond expectations. Acad. Manag. Rev. **12**(4), 756–757 (1987). https://doi.org/10.5465/amr.1987.4306754
8. Ngaithe, L.N., K'Aol, G.O., Lewa, P., Ndwiga, M.: Effect of idealized influence and inspirational motivation on staff performance in state owned enterprises in Kenya. Eur. J. Bus. Manag. **8**(30), 6–13 (2016)
9. Sudarti, K., Fachrunnisa, O.: Religious value co-creation: measurement scale and validation. J. Islamic Mark. (2023). https://doi.org/10.1108/JIMA-08-2022-0223
10. Sudarti, K., Fachrunnisa, O.: Religious personal value towards knowledge conversion process: the power of collaboration between sales team. Int. J. Knowl. Manag. Stud. **12**(2), 136–160 (2020)
11. Fachrunnisa, O., Assyilah, F.: Blockchain for Islamic HRM: potentials and challenges on psychological work contract. In: Barolli, L., Kulla, E., Ikeda, M. (eds.) EIDWT 2022. LNDECT, vol. 118, pp. 114–122. Springer, Cham (2022). https://doi.org/10.1007/978-3-030-95903-6_13

Semantic Wrap and Personalized Recommendations for Digital Archives

Alba Amato[1], Rocco Aversa[2], Dario Branco[2], and Salvatore Venticinque[2(✉)]

[1] Department of Political Science, University of Campania "Luigi Vanvitelli",
81031 Aversa, Italy
alba.amato@unicampania.it
[2] University of Campania "Luigi Vanvitelli", Aversa, Italy
{rocco.aversa,dario.branco,salvatore.venticinque}@unicampania.it

Abstract. Many efforts have been made in the Cultural Heritage field to build digital repositories based on standards for technological interoperability and for enriching the semantic content and the interconnection among the archived objects and with other repositories. The availability of standard APIs for linked information allows for the design and implementation of personalized recommendation systems, which are a relevant service for digital archives that aims at promoting the valorization of their content to visitors. In this paper, a prototype development of such a mechanism that is based on IIIF standards and IIIF-compliant open-source technologies is described.

1 Introduction

The delivery of cultural heritage through the web is continuously growing through the publication of digital archives by content providers such as museums or libraries and according to new kinds of service providers, which behave as aggregators that collect and enrich metadata allowing for advanced presentation models which exploit the interconnection of contents along several cultural dimensions. For example, Europeana, relying on the thousands of European galleries, libraries, archives, and museums that bring and share collections online, delivers cultural heritage material from a network of aggregating partners. Aggregators collect the data, check it thoroughly, verify the formal correctness, and enrich it with information like geo-location, or link it to other material or datasets through associated people, places, or topics. Web Documents are connected on the web through hypertext links, which allow the user to navigate and carry out research. These connections are semantically limited, as they are not characterized by explicit semantics. The Semantic Web aims at Order in an environment that has an amount of data that grow day by day exponentially extending the potential of the current web regarding the search operations of documents and information. In this context, Linked Data allow for publishing structured data that use metadata to be connected and enriched, so that different representations of the same content can be found, and links made between semantically related resources. The enrichment of metadata with linked data and the mappings of concepts belonging to open vocabularies allows for the building of interconnection of contents archived in distributed repositories, but

L. Barolli (Ed.): CISIS 2023, LNDECT 176, pp. 299–308, 2023.
https://doi.org/10.1007/978-3-031-35734-3_30

the construction of the semantic web must be accompanied by the need to preserve the expressiveness of the data, to avoid a dangerous semantic flattening. To address this issue ontologies are used. An ontology can be considered as a set of terms of knowledge, including vocabulary, semantic interconnections, and some rules of inference and logical rules for precise purposes. Enriching meta-data with ontology terms, or mapping linked data to ontology terms, allows for supporting advanced reasoning and for the development of smart applications.

Semantic wrapping concerns the practice of incorporating metadata or semantic information within a database, such as a document, a file, or an object, to describe its content in a more accurate way. In the case of media contents, semantic wrapping is implemented annotating parts of the content, such as a specific area of an image or the frame of a video.

In this paper, we present the design and the prototype implementation of techniques for automatic profiling and recommendation that exploits the combined utilization of semantic wrapping and ontologies to cultural heritage delivered by interoperable technologies.

2   Related Works

Several projects have been created in the field of knowledge organization, whose products have been ontologies and metadata systems, and thesauri, for the specific reference domain of digital archives. CIDOC CRM: ISO 21127 is a reference ontology for the interchange of cultural heritage information [3]. The CIDOC Conceptual Reference Model (CRM) provides definitions and a formal structure for describing the implicit and explicit concepts and relationships used in cultural heritage documentation. The CIDOC CRM is intended to promote a shared understanding of cultural heritage information by providing a common and extensible semantic framework to which any cultural heritage information can be mapped. In this way, it can provide the "semantic glue" needed to mediate between different sources of cultural heritage information, such as that published by museums, libraries, and archives [4]. CIDOC CRM is a domain ontology, whose main classes are those of space-time, events, and tangible and intangible assets. MACE (Metadata for Architectural Contents in Europe) [9] is a pan-European initiative for the interconnection and dissemination of digital information in the domain of architecture. Dublin Core Metadata Initiative (DCMI) is a project started in 1995 during a workshop held in the city of Dublin Ohio, in order to define a set of basic elements with which to describe digital resources. Following the launch of this initiative, in 1996 a first set of 15 basic elements was defined - the Dublin Core Metadata Element Set - which included, among others: the Title, the Author, the Subject, the Description, an Identifier, the Source, and so on. With the start of the first Semantic Web studies and the study of RDF, Dublin Core became one of the most popular and used vocabularies for describing RDF documents. In [8], to meet particular implementation requirements, application profiles are proposed to describe metadata records using Dublin Core together with other specialized vocabularies. During that time, the World Wide Web Consortium's work on a generic data model for metadata, the Resource Description Framework (RDF), was maturing. As part of an extended set of DCMI Metadata Terms, Dublin Core

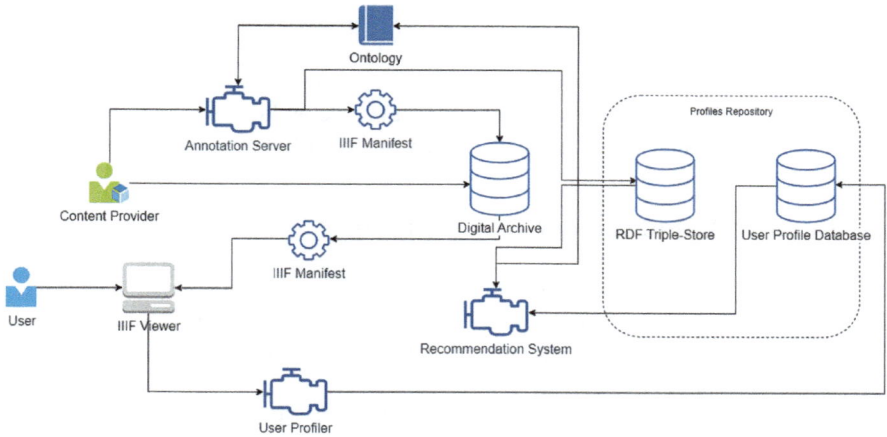

Fig. 1. Semantic wrapping of an IIIF image.

became one of the most popular vocabularies for use with RDF, more recently in the context of the Linked Data movement. Harmonise [2] enables seamless data exchange by following a semantic mapping approach. Instead of a fixed standard, Harmonise provides a reference model, the Harmonise ontology, which local data formats can be mapped against and in this way mapped to other local data formats or standards. In the sense of such a reference model, the Harmonise ontology defines all relevant concepts and relationships between concepts of the problem domain. A specific mapping mechanism and reconciliation engine then enable the translation of data from one local format to another and enable seamless data exchange and global interoperability. In order to solve the interoperability problem we have proposed an Ontology-mediated integration process based on a framework for data integration - instead of a new standard for the tourism domain. Increasingly these knowledge bases are being used in synergy with recommendation systems, in [5] a recommendation system is defined as a system that aims to generate meaningful recommendations for its users on articles that may be of interest to them. Recommendation systems, especially in the last period, are enjoying great success also in the field of cultural heritage and a large number of advancements are being made in this field. Another motivation behind recommendation systems is well described in [6] where it is noted that recommendation systems can be used to enhance those sites of interest that are often overlooked by tourists. Promoting lesser-known cultural sites is of fundamental importance for the preservation of a region's cultural heritage.

3   Software Architecture

The software architecture shown in Fig. 1 is composed of three main components. The annotation server that allow Content Providers for the semantic annotation by of IIIF contents using a domain ontology. Media contents are published in a digital archive IIIF compliant. Semantic annotations are indexed in a RDF triple-store for fast retrieval. The user profiler tracks the utilization of IIIF APIs by the client. In particular it records the list of viewed contents and, on each IIIF event, it eventually update a weighted list of ontology concepts which have been used to annotate the contents. The content recommendation system periodically uses the updated profile and the indexed annotations to compute a metric that evaluate the affinity with the available contents in the repository.

4   Semantic Annotation of IIIF Contents

The International Image Interoperability Framework (IIIF) defines, among others, an API for the presentation of digital content through the web. The semantic wrapping is defined by standard elements of the standard presentation APIs and supported by compliant IIIF technologies.

The semantic annotation of the IIIF manifest is supported by annotation servers and annotation tools, which respectively allow for the visual editing of the manifest and the retrieval of enriched metadata. In Fig. 2, it is shown how Simple Annotation Server allows for the creation of the semantic wrap of an area of the visualized image and its association to a free text or a tag. We extended the tool with the possibility to annotate the area with a concept of ontology. The button highlighted by the red circle opens a popup that allows for the selection and visualization of an imported ontology.

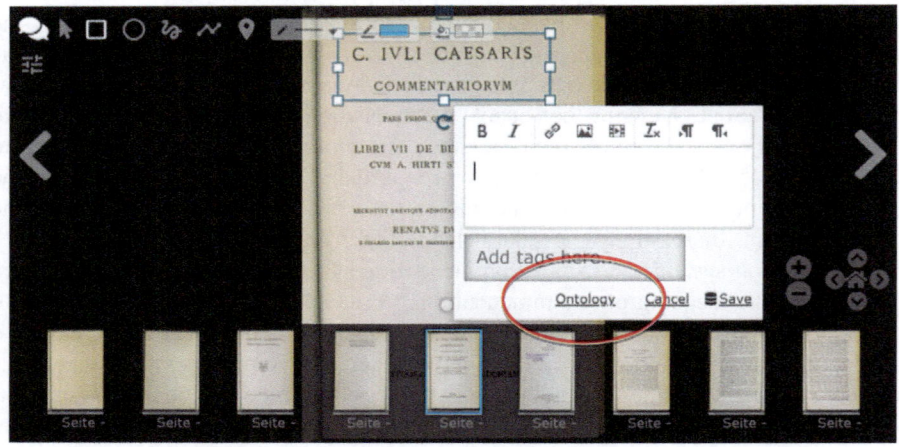

Fig. 2. A web dashboard of Simple Annotation Server.

At the left side of the popup shown in Fig. 3 the ontology tree is visualized. The selected nodes of the tree are automatically concatenated in the list shown on the right side of the form and used for the semantic annotation.

In Listing 4 the IIIF-generated annotation is shown. From line 4 to line 12 the annotation type and the annotation content are shown.

From line 13 it is declared the target of the annotation, which means the annotated manifest and the area of the image that should be selected by the annotation. In particular, the *selector* property defines an area to be highlighted, and at line 25 the SVG object for highlighting the selected part of the canvas. The canvas address is specified at 27.

```
1   { @id: "http://digitalarcheo.eu/annotation/11136",
2     @type: "oa:Annotation",
3     resource: [
4     { @type: "dctypes:Text",
5       {http://dev.llgc.org.uk/sas/full_text:} "",
6       format: "text/html",
7       chars: "The city gate of Norba"
8       },
9     { @type: "ca:Tag",
10      {http://dev.llgc.org.uk/sas/full_text:} "http://digitalarcheo.
                unicampania.it/ontology/rasta.owl#city_gate",
11      chars: "this_tag"
12      }],
13    on: [
14    { @type: "oa:SpecificResource",
15      within:
16      { @id: "https://digitalarcheo.eu/norba/roads/index_new.json",
17        @type: "sc:Manifest"},
18      selector:
19        { @type: "oa:Choice",
20          default:
21          { @type: "oa:FragmentSelector",
22            value: "xywh=1828,863,157,121" },
23            item: {
24              @type: "oa:SvgSelector",
25              value: "<svg xmlns="http://www.w3.org/2000/svg"><path
                    xmlns="http://www.w3.org/2000/svg" d="M=1828,863h78.
                    5v0h78.5v60.5v60.5h-78.5h-78.5v-60.5z" /></svg>"}
26            },
27            full: "https://digitalarcheo.eu/norba/roads/index.json/
                    canvas/0"
28        }]
29        [...]
30  }
```

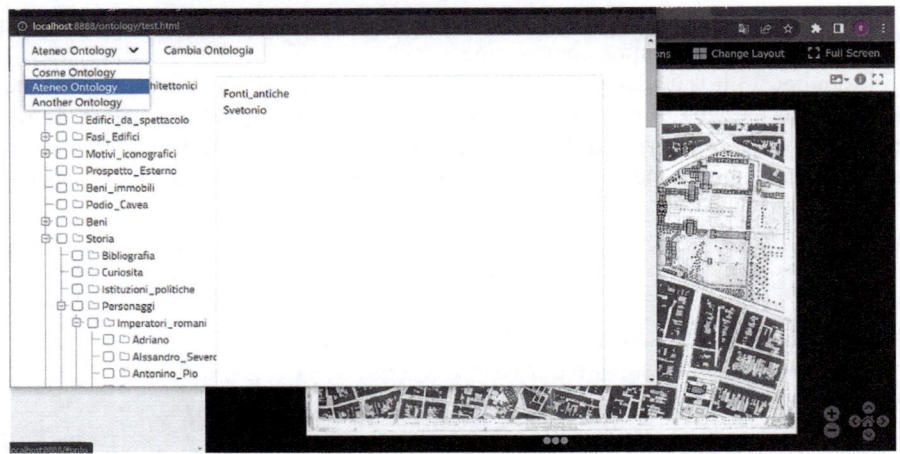

Fig. 3. Ontology visualization.

5   Optimal Recommendation

In this section, we will discuss how users' preferences can be taken into account. The basic idea is to evaluate each annotated content and to measure its relevance to the user's profile **p**.

The Model chosen to represent the user's profile is the Vector Space Model (VSM) which is a model created by G.Salton [7] utilised for semantic representations of contents as vectors of items.

In our definition, $\mathbf{a_j}$ describes the annotation of content j. It contains l_j concepts $c_{i,j}$ of the domain ontology, each one occurring $o_{k,j}$ into the annotation. The user profile contains p concepts of the domain ontology which are relevant to her own interests.

- $\mathbf{a_j} = <\{c_{1,j}, o_{1,j}\}, \{c_{2,j}, o_{2,j}\}, ..., \{c_{l_j,j}, o_{l,j}\}> \forall j = 1..N$
- $\mathbf{p} = <c_1, c_2, ..., c_m>$

Furthermore, we defined a score $\phi(\mathbf{a_j}, \mathbf{p})$ to measure the relevance of an annotation $\mathbf{a_j}$ to the profile $\mathbf{p}$ in Eq. 1.

$$\phi_j = \phi(\mathbf{a_j}, \mathbf{p}) = \frac{1}{l} \sum_{k=1}^{l} r_k$$

$$\text{where } r_k = o_{k,j} * \frac{1}{m} \sum_{i=1}^{m} \frac{1}{d_{k,i}+1} \tag{1}$$

where r_k is the relevance of the profile respect to the concept $c_{k,j}$ in the annotation. In order to calculate r_k, we established the correlation between a concept $c_{k,j}$ from the annotation and any other concept c_i from the profile. This was achieved by taking the inverse of $d_{k,i}$, which represents the minimum number of connections between the nodes in the ontology plus one. To obtain r_k, the occurrences $o_{k,j}$ of $c_{k,j}$ are multiplied by the affinity of each concept in the profile. Then, the resulting values are averaged across m elements of $\mathbf{a_j}$. As ϕ_j increases, the user's interest in the annotated content also increases. Furthermore, factors such as the user's time constraints for the visit and the device's ability to play certain types of content can be considered, as demonstrated in [1]. One can simplify this issue by framing it as a discrete optimization problem, which involves finding the maximum or minimum value of a function, denoted by f : $\mathbf{x} \in \mathscr{Z}^n \to \mathscr{R}$, where $\mathbf{x} = x_1, \ldots, x_n$ represents the solution in which the function's value is optimal. The function $f(\mathbf{x})$ is known as the cost function, and it is typically subject to a set of t constraints that limit its domain, such as time constraints or user device compliance. A collection of inequalities typically specifies the constraints, defining the feasible values for the x_i variables, which form the solution space for the problem. For our problem, the score ϕ_i and a set of constraints B comprising N integers for each $\mathbf{b_i}$ define the constraints $w_i \geq 0 : \forall i = 1..N$. Our objective is to compute the maximum value

$$max \sum_{k=1}^{N} \phi_i x_i$$

so that

$$\sum_{k=1}^{N} b_{i,j} x_i \leq b_{j,N+1}$$

with $x_i \in \{0, 1\} \ \forall i = 1..N$. The aim is to maximize the delivered value by selecting the set of recommended content without violating the constraints. The vector $\mathbf{x}$ represents a possible solution, where its components x_i are either 1 or 0, depending on whether the item is included or not in the optimal set. In the following subsection, the description of the prototype implementation is detailed.

6   Implementation

The preceding paragraphs discussed how an annotation software for IIIF standards was modified to introduce concepts of ontology for multimedia content annotation and the mathematical aspect governing the recommender system. In this section, however, we will focus on the user profiling task and on the recommendation system implementation. In principle, thanks to the annotator modifications we are able to profile multimedia content with respect to a domain ontology, there remains the need, however, for the profiling of the user himself with respect to his interests so that we can apply the methodology described in the Sect. 5. For the purpose of user profiling, a listener integrated with the IIIF viewer takes care of tracking the user's navigation through the digital repository. Each time the user selects a new image or triggers IIIF events because of the interactions with the viewer, the associated concepts are stored in a user's profile. The profile will be built as an array of concepts. In order to build the user profile, and semantically calculate what content is most relevant to the user, SPARQL queries are performed. Simple annotation server, stores annotations in a Jena RDF database. For each content displayed by the user, the list of ontological annotations is extracted by the SPARQL query shown in Listing 1, where g represents the annotation id o represent the ontology concept and t represent the annotated canvas.

```
SELECT ?g ?o ?t {
 GRAPH ?g {
   ?a <http://dev.llgc.org.uk/sas/full_text> ?o}
. GRAPH ?g {
   ?b <http://www.w3.org/ns/oa#hasSource> ?t}
 }
```

Listing 1. SPARQL query to retrieves annotations

An excerpt of returned results is listed in Table 1.

Table 1. Annotation list from the SPARQL query.

Annotation ID	Ontology concept	Canvas
11129	"bridle"	\<https://digitalarcheo.eu/norba/roads/index.json/canvas/2\>
11163	"sidewalk"	\<https://digitalarcheo.eu/norba/roads/index.json/canvas/8\>
11178	"spa"	\<https://digitalarcheo.eu/norba/roads/index.json/canvas/6\>
11179	"labrum_house"	\<https://digitalarcheo.eu/norba/roads/index.json/canvas/6\>
11182	"ninfina_gate"	\<https://digitalarcheo.eu/norba/roads/index.json/canvas/6\>
[...]	[...]	[...]

At the end of the queries, it is possible to calculate an ordered list of occurrences of the ontology concepts that recur the most times in the content displayed by the user. After calculating the number of occurrences of the ontology concepts, we proceed to invoke the recommendation system in order to find multimedia contents semantically related to those already visited and appreciated by the user through a second SPARQL query that calculates the semantic distance in terms of hops from the ontology concepts already visited. The distance between two concepts of the ontology in terms of hops along the inheritance relationship is computed by the following SPARQL shown in Listing 2 which generate the table like the one shown in Table 2.

```
select ?super ?sub (count(?mid) as ?distance) {
   ?super rdfs:subClassOf* ?mid .
   ?mid rdfs:subClassOf+ ?sub .
}
group by ?super ?sub
order by ?super ?sub
```

Listing 2. SPARQL query to compute concepts distance

Table 2. Distance from ontology concepts as a result from the SPARQL query.

Concept A	Concept B	Distance
Anfiteatri	Beni	4
Anfiteatri	Beni_immobili	3
Anfiteatri	Edifici_da_spettacolo	1
Anfiteatri	Edilizia_Pubblica	2
[...]	[...]	[...]

Once the distances for the various concepts in the ontology have been obtained, we proceed to select the concepts that have the least distance from those most viewed by the user and finally invoke the RDF Database to obtain the annotated multimedia content with that ontological concept to provide it to the user.

7   Conclusions

In this paper, a system making use of IIIF semantic content annotation through concepts of a Cultural Heritage domain ontology by updating an existing standard IIIF annotation tool with the ontology annotation feature, a user profiling system using a tracker of the user's own operations, and a content recommendation system using an expert system for the purpose of enhancing the enjoyment of digital museums has been proposed. Piquing the curiosity of users is one of the basic requirements of any digital tool that wants to present content concerning art and history and to incentivize users in their discovery of cultural heritage. It is now clear that modern tools can no longer be mere static mechanisms for presenting content but must somehow adapt to the needs and heterogeneity of users who can and should have different interests with respect to art and history.

Acknowledgement. This work is supported by the *Cleopatra Project*, funded by the University of Campania "Luigi Vanvitelli" through the VALERE 2019 research program.

References

1. Amato, A., Di Martino, B., Scialdone, M., Venticinque, S.: Adaptive recommendation to dynamically changing profiles for delivery of ubiquitous services. Int. J. Comput. Sci. Eng. **13**(4), 322–332 (2016). https://doi.org/10.1504/IJCSE.2016.080209
2. Dell'Erba, M., Fodor, O., Ricci, F., Werthner, H.: Harmonise: a solution for data interoperability. In: Monteiro, J.L., Swatman, P.M.C., Tavares, L.V. (eds.) Towards the Knowledge Society. ITIFIP, vol. 105, pp. 433–445. Springer, Boston, MA (2003). https://doi.org/10.1007/978-0-387-35617-4_28
3. Doerr, M.: The CIDOC conceptual reference model (CRM) (2014). https://www.cidoc-crm.org
4. ICCSI Group: Definition of the CIDOC conceptual reference model. Version 4.2 (2005). https://www.cidoc-crm.org/docs/cidoc_crm_version_4.2.pdf

5. Pavlidis, G.: Recommender systems, cultural heritage applications, and the way forward. J. Cult. Heritage **35**, 183–196 (2019). https://doi.org/10.1016/j.culher.2018.06.003. https://www.sciencedirect.com/science/article/pii/S1296207418302577. Modern and Contemporary Art
6. Rajaonarivo, L., et al.: Recommendation of heterogeneous cultural heritage objects for the promotion of tourism. ISPRS Int. J. Geo-Inf. **8**(5) (2019). https://doi.org/10.3390/ijgi8050230. https://www.mdpi.com/2220-9964/8/5/230
7. Salton, G., Lesk, M.E.: Computer evaluation of indexing and text processing. J. ACM **15**(1), 8–36 (1968)
8. Weibel, S.L., Koch, T.: The Dublin core metadata initiative. Mission, current activities, and future directions. D-Lib Mag. **6**(12) (2000)
9. Wolpers, M., Memmel, M., Stefaner, M.: Supporting architecture education using the mace system. Int. J. Technol. Enhanced Learn. **2**(1–2), 132–144 (2010). https://doi.org/10.1504/IJTEL.2010.031264

Towards the Interoperability of Metadata for Cultural Heritage

Alba Amato(✉)

University of Campania "Luigi Vanvitelli", Caserta, Italy
alba.amato@unicampania.it

Abstract. Archiving is essential for digital preservation that allows the conservation and access to the general heritage of humanity. This process is structured in a set of distinct activities which include scanning, inventory and the creation of special databases, accompanied by historical, artistic and aesthetic analysis of the property in question. The archiving activity is closely related to compliance with the principles of protection of cultural heritage. This paper describe several studies and a technological stack for the interoperability of metadata for archiving and for cataloging cultural heritage.

Keywords: Interoperability · Cultural Heritage · Digital Humanities · Metadata

1 Introduction

The term cultural heritage refers to a set of cultural assets, which, due to particular historical, cultural or aesthetic importance, are of public interest and specifically constitute the wealth of a place and its population. Cultural heritage represents the set of assets such as physical objects, places, traditions and knowledge of a society that are handed down and inherited from the past, maintained in the present and preserved for the benefit of future generations. The term heritage emphasizes the economic value attributed to the cultural assets that compose it in relation to their artistry and historicity [6]. Cultural heritage assets constitute wealth of towns, cities, nations or any legally circumscribed territorial sector. Cultural assets are of public interest as they are intended for collective use, in other words everyone must be able to freely enjoy the vision of the heritage and the knowledge connected to it [8]. The assets that constitute cultural heritage can be classified into three distinct areas [5]:

- tangible cultural assets, i.e. concrete objects such as works of art, books, buildings and in general physical objects made by man;
- intangible cultural assets, i.e. phenomena such as traditions, popular knowledge, languages, craft techniques and in general all knowledge handed down orally;
- natural cultural heritage, such as landscapes, biodiversity and geodiversity.

L. Barolli (Ed.): CISIS 2023, LNDECT 176, pp. 309–317, 2023.
https://doi.org/10.1007/978-3-031-35734-3_31

Tangible assets are mainly documented through the use of metadata, i.e. information that describes the properties and characteristics of the asset in question. Tangible goods, as well as the documentation associated with them and with other types of goods, are collected in different physical spaces and preserved by specific public or private bodies. The main places responsible for their conservation are libraries, archives and museums, the latter specializing in specific areas such as art, archaeology, natural history, science and so on. A small set of cultural assets is maintained and conserved by cultural associations and private citizens. The set of cultural assets conserved by an institution, body or private individual is called a collection. Archiving is essential for digital preservation that allows the conservation and access to the general heritage of humanity. This process is structured in a set of distinct activities which include scanning, inventory and the creation of special databases, accompanied by historical, artistic and aesthetic analysis of the property in question. The archiving activity is closely related to compliance with the principles of protection of cultural heritage. This paper describe several studies and a technological stack for the interoperability of metadata for archiving and for cataloging cultural heritage.

2   Related Works

In order to correctly define a conceptual model, it is necessary first of all to analyze the peculiar characteristics of the contents associated with cultural heritage. The contents are multi format as they are present in various forms: text documents, images, audio and video recordings, etc. Moreover the contents are multi-topic because they can refer to different topics: art, history, traditions, etc and multi-language. The contents are connected and interpreted differently based on the culture and traditions of a country, city or nation and can be aimed at different types of people: experts and non-experts in the sector, young and old, etc. The contents are collected, maintained and published by various institutions and associations operating in the cultural sector, as well as by private citizens. A common problem deriving from these characteristics and the heterogeneity of the contents is the difficulty of making the information mutually interoperable [9]. In general, a set of contents is interoperable if the data present can be searched, semantically linked and presented to the user in an organic and standardized way, overcoming the constraints imposed by ad hoc implementations of the services of the various participating institutions. The interoperability of the conceptual model must be guaranteed both at the syntactic level, through the linking of different data formats and different collections, and at the semantic level, through the use of common metadata schemas and shared vocabularies for describing the information. The publication of cultural heritage data on the Web is not just a technological problem. A good part of the contents could be protected by licenses and/or copyrights. Moreover not all available information can be freely made public and published information may not be reliable [4]. Since it is difficult to ignore or avoid these problems, the main objective becomes to maximize the amount of cultural content published and shared as Linked Data, in order to disseminate the largest possible volume of knowledge. During the last years archives and museums have worked to create websites used as places to display the goods belonging to their collections. These virtual exhibitions are usually accompanied by detailed text descriptions, images and possibly audio and video recordings relating to the property described. The main problem of these websites is that a user interested in looking for information relating to a specific cultural asset must know where it is exhibited and in which collection is found. Furthermore, each website uses its own interfaces with different search systems, which the user must learn to use. The process of searching for the desired information could therefore be very complex and tedious. To overcome these problems, cultural portals were born that collects information about cultural heritage from different data sources, combines them and presents them in a uniform way to the user. The purpose of these portals is to try to offer a global centralized service that allows users to find information relating to the cultural heritage preserved in the various collections present on national and international territory in a quicker, simpler and more organic way. The search is carried out by the system on the data sources of the institutions that have given permission to access their information. There are two types of information portals: Specific information related to a well-defined theme of cultural heritage. Generic portals, which provide search, navigation and visualization systems for large amounts of generic cultural information derived from various heterogeneous sources. Among these, Europeana [1] deserves a dutiful mention, a digital library that brings together millions of contributions already digitized in 30 different languages from various institutions of the member countries of the European Union. The search for information within these types of portals is facilitated by the presence of user-friendly interfaces that offer a limited number of filters, called facets, to narrow the list of results based on the characteristics of the cultural asset to be searched.

3   Metadata for Archiviation

Metadata is important for describing the content, context, and structure of an archive, and for facilitating access and discovery of the archive's resources. Here are some general steps for metadata creation for an archive:

Define the scope of the archive: Before creating metadata for an archive, it's important to define the scope of the archive, including the types of resources it contains, the intended audience, and the purpose of the archive.

Identify the metadata elements: The next step is to identify the metadata elements that will be used to describe the archive's resources. Some commonly used metadata elements for archives include title, creator, date, description, subject, and format.

Determine the metadata standards and schemas: It's important to determine which metadata standards and schemas will be used to describe the archive's resources. Some commonly used metadata standards for archives include Dublin Core, EAD (Encoded Archival Description), and MODS (Metadata Object Description Schema).

Create the metadata records: Once the metadata elements and standards have been identified, metadata records can be created for each resource in the archive. These records should include all relevant metadata elements, and should be consistent and accurate.

Organize the metadata: The metadata records should be organized in a consistent and logical way, such as by resource type, subject, or date. This can make it easier to search and browse the archive.

Provide access to the metadata: The metadata records should be made accessible to users, either through a catalog or database, or through a search interface. The metadata should be designed to facilitate discovery and access to the archive's resources.

Maintain and update the metadata: Metadata is not a one-time task. It's important to regularly review and update the metadata records as needed, to ensure that they remain accurate and up-to-date.

These are some general steps for metadata creation for an archive. However, the process can vary depending on the specific requirements of the archive, and may involve additional steps, such as quality control and preservation planning. Archives typically use metadata standards that are designed specifically for describing archival materials. These standards help ensure that the materials can be easily located, understood, and used by researchers, scholars, and other users.

The most widely used metadata standard for archives is Encoded Archival Description (EAD). EAD is an XML-based standard that provides a flexible framework for describing archival materials at various levels of granularity, including collection, series, folder, and item. EAD allows archivists to capture information about the scope and content of archival collections, their provenance and arrangement, and any intellectual or physical access restrictions that may apply. EAD schema is based on XML, which stands as the digital transposition of the standard for the description of ISAD(G) archival documents, of which it maintains the hierarchical structure. It is very complete, but presents some problems in terms of complexity and automatic management, as well as in the exchange of portions of information: an EAD file, in fact, describes the entire archive and it is not easy to separate individual portions, such as a single unit archival.

It consists of an header section describing the EAD file itself, the description of the entire collection, more detailed descriptions of archival series and units and possible links to digital objects corresponding to archival units. Many EAD elements have been or can be mapped to standards (such as MARC or Dublin Core), increasing the flexibility and interoperability of the data.

Another widely used standard for archival metadata is Metadata Encoding and Transmission Standard (METS). METS provides a framework for describing complex digital objects, including archival collections that may contain a mix of textual, audiovisual, and other materials. METS can be used in conjunction with other metadata standards, such as EAD, to provide a comprehensive description of archival materials in both physical and digital formats. XML schema created for the description and management of complex electronic bibliographic objects. The scheme is divided into seven main sections:

1. METS Header: contains data useful for describing the METS document itself and some information such as the author, publisher...
2. Descriptive Metadata: may contain descriptive metadata external to the METS document (e.g. a MARC record in a catalog) and/or descriptive metadata integrated with the object. Repeatable. There are no specific guidelines for completing this section for which, as well as for the following section, one of the three metadata standards recommended by METS can be used: DC, MARCXML and MODS
3. Administrative Metadata: contains information regarding the creation and storage of files, intellectual property rights, the original object from which the digital object is derived and the provenance of the files
4. File Section: list of all content files that make up the digital object
5. Structural Map: explains the hierarchical structure of the digital object and the links between these elements, the files and metadata pertinent to each element
6. Structural Links: allows you to register any hyperlinks between nodes of the hierarchical structure
7. Behavior: it can be used to associate certain behaviors with the contents of the object

Other standards are:

– Dublin Core (DC)
  The strengths of the Dublin Core profile are that it is concise and simple. In fact, the set is made up of only 15 elements, all optional and repeatable, which can be inserted in any order. The use of controlled vocabularies for the compilation of some fields (for example the Subject field) is recommended, but it is not mandatory. The Dublin Core is somewhat at the opposite extreme compared to some hyper-specific and complete but equally complex metadata schemas, since it turns out to be quite simple, not very rigid, usable even by non-experts and suitable for many types of resources. This makes it a popular metadata schema and this ensures interoperability with numerous platforms. On the other hand, however, it has the disadvantage that there may be a need to integrate it with other metadata to respond to the needs of specific reference communities or to describe particular resources. The elements of the Dublin Core can be divided into three groups: Content related: Coverage, Description, Type, Relation, Source, Subject, Title; Related to intellectual property: Contributor, Creator, Publisher, Rights; Related to the instance: Date, Format, Identifier, Language
  Precisely to try to overcome some limitations inherent in the scheme, a "richer" version of the Dublin Core was developed, defined as qualified Dublin Core, which includes an additional element (Audience) and a series of qualifiers which further specify the meaning of the various elements (refinement elements) or identify patterns that aid in the interpretation of the value of an

element, such as controlled vocabularies (encoding patterns). The use of the qualified Dublin Core could decrease the possibility of interoperability, since not all platforms recognize the qualified elements.

– MARC (Machine Readable Cataloging)

It is a family of metadata formats developed as variants of the MARC format, created in the late 1960s s by the Library of Congress from cataloging cards to facilitate the exchange of cataloging records between libraries. We speak of the MARC "family" since numerous national library organizations have developed their own variants of the format (e.g. ANNAMARC in Italy, UKMARC in the United Kingdom). There is also the UNIMARC format (UNIversal MAchine Readable Cataloguing), developed by IFLA to allow interoperability between the different MARC formats

The record consists of a header, followed by a series of control fields and fields containing data, divided into one or more sub-fields and identified by a system of numbers, letters and symbols, which codify the information making it interpretable by a car. The fields of a record of any MARC format are divided into 10 blocks, which, in the case of the UNIMARC format, are:

- 0xx: block of control information and record identification numbers
- 1xx: blocking of encrypted information
- 2xx: block of descriptive data, which follow the ISBD standard
- 3xx: Notepad
- 4xx: blocking of links between bibliographic records
- 5xx: title forms block
- 6xx: block of subjects, classifications, thesauri terms
- 7xx: blocking of intellectual liability links
- 8xx: blocking of international data (e.g. cataloging agency code, url of the electronic resource...)
- 9xx: undefined block, for fields for national use

– Text Encoding Initiative (TEI)

It is an international project dedicated to the markup of electronic texts, in particular to promote research in the humanities, based on the use of XML. A so-called "header portion" is integrated into the resource, which contains the metadata. It is used in digital libraries because, although it contains more data, it can be mapped to the MARC format, but in most cases the light version is used, since it is a rather complicated scheme.

– Metadata Object Description Schema (MODS)

Schema of descriptive metadata expressed in an XML language, it derives from MARC but uses textual rather than numeric tags. It is very flexible and can be used on its own or to complement other metadata schemas, especially METS, thanks to its structure designed for a granular description of complex objects, which are also the basis of the METS schema. Its purpose is a rich description of electronic resources and in this respect it has some advantages over other schemes that are mainly a richer set of elements than Dublin Core; elements more compatible with bibliographic data than other metadata schemas, such as the Dublin Core and easier to use than a full MARC schema.

The DC, MODS, EAD and METS metadata standards and schemas have common characteristics and some peculiarities [2] and are more apt to describe digital resources for the purposes of discovery, retrieval, presentation and interoperability [2].

Moreover the analysis of the results performed in [2] indicates that Dublin Core, MODS and EAD supported METS in detecting and documenting technical aspects of sites and proving their authenticity, context, and origin. METS can manage archived site while Dublin Core proved to be an exponent for Web archiving through its use in remarkable area initiatives.

Europeana Data Model (EDM) has been specifically designed to enable Linked Data integration and to solve the problem of cross-domain data interoperability [3]. The EDM builds on the reuse of existing standards from the SemanticWeb environment but does not specialize in any community standard acting as a top-level ontology.

In [3] and [7] authors demonstrate how an EAD-XML encoded archival can be modeled in a RDF-based representation elaborate on the EDM-RDF representation of a concrete EAD encoded finding aid.

EDM copes with the representation of hierarchical and sequential order in EAD. Representing elements in the hierarchy as fully-fledged, distinct resources allows the creation of explicit links between them. Together with the properties (types of links) that EDM re-uses or mints in order to reflect 'part-of' or 'next-in-sequence' relationships, this feature enables the precise representation needed [1].

In next section is shown a technology stack that perform a conversion between EAD and EDM in order to obtain the interoperability among the man standards in archiving.

4   Technology Stack

In Fig. 1 the software components of the interoperability platform are shown. AtoM[1] (Access to Memory) is primarily intended for describing and accessing archives, it acts as a public web-based catalog so that users can search and browse your archives. It uses templates to map Dublin Core and another standard to the EAD application profiles. AtoM represents the access point for the editor that needs to store and describe new items, eventually organized into collections and it is the technology that allows for describing metadata according to a chosen standard. After transformation of the original EAD native model to EDM and IIIF, the generated IIIF manifest can be automatically validated[2]. The IIIF artifact generated from the EAD metadata is semantically annotated by an extended version of Simple Annotation Server[3]. The IIIF annotations can be imported into the ATOM archive and exported in EAD by extended fields. As shown in Fig. 1 the IIIF representation automatically generated from the EAD guarantees the generation of a self-consistent artifact that can be used

---

[1] https://www.accesstomemory.or.
[2] https://github.com/IIIF/presentation-validator.
[3] https://github.com/glenrobson/SimpleAnnotationServer.

for the web delivery of images and other media contents and for the utilization of IIIF-compliant viewers. Moreover, as the IIIF standard does not restrict the kind o metadata which can be used to describe collections and single items, the utilization of AtoM as the origin of the IIIF manifest generation guarantees the utilization of standard schema for meta-data.

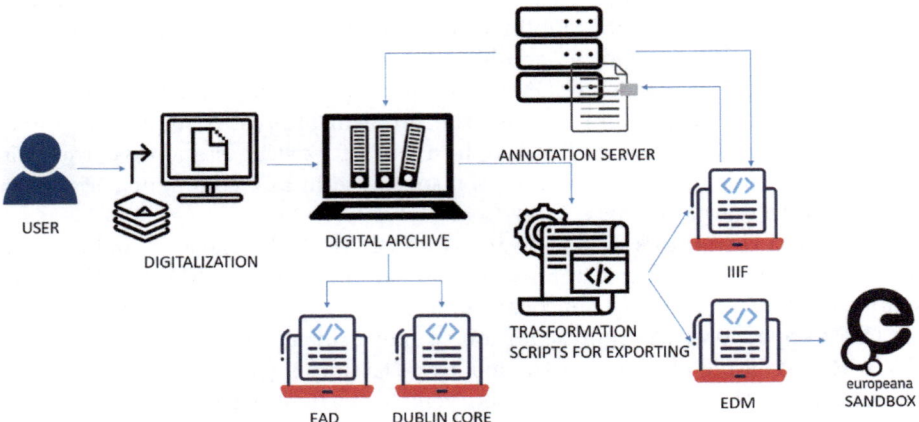

Fig. 1. Technology stack for the iteroperability

EAD to EDM transformation allows for sharing the exported metadata with aggregators, such as Italian and European aggregators which support such a standard. The transformation is based on a mapping schema and integrates, for the visualization of high-resolution images and of 3D objects, a reference to the IIIF manifest describing the exported item or a collection. The generation of the EDM metadata is validated by the Metis Sandbox[4], which is a functional application that enhances the overall data aggregation workflow from data providers to Europeana.

5   Conclusions

The existence of numerous metadata schemas has created and continues to create problems both in terms of interoperability and in terms of understanding the data, as well as compilation and completeness of the record. In fact, the use of a more or less rich schema, as well as the use of a schema that includes mandatory elements compared to one that instead has only optional elements, means that the level of completeness of the information conveyed by the metadata can be very variable and, moreover, the fact that sometimes metadata are created by people who are not experts in metadating has given rise to descriptions

---

[4] https://metis-sandbox.europeana.eu/.

that may lack elements considered indispensable by experts in the field. This paper describe several studies and a technological stack for the interoperability of metadata for archiving and for cataloging cultural heritage.

References

1. Cesare, C., Stefan, G., Sjoerd, S.: Not just another portal, not just another digital library: a portrait of Europeana as an application program interface. IFLA J. **36**, 61–69 (2010)
2. Formenton, D., de Souza Gracioso, L.: Metadata standards in web archiving: technological resources for ensuring the digital preservation of archived websites. RDBCI J. **20**, 1–28 (2022)
3. Hennicke, S., Olensky, M., Boer, V., Isaac, A., Wielemaker, J.: Conversion of EAD into EDM linked data 801, 15–22 (2011)
4. Janssens, M.C., Gorbatyuk, A., Rivas, S.P.: Chapter 10: Copyright issues on the use of images on the Internet, pp. 191–213. Edward Elgar Publishing, Cheltenham (2022). https://www.elgaronline.com/view/book/9781800376915/book-part-9781800376915-18.xml
5. Linaki, E., Serraos, K.: Recording and evaluating the tangible and intangible cultural assets of a place through a multicriteria decision-making system. Heritage **3**(4), 1483–1495 (2020). https://www.mdpi.com/2571-9408/3/4/82
6. Pérez-Hernández, E., Peña-Alonso, C., Hernández-Calvento, L.: Assessing lost cultural heritage. A case study of the eastern coast of Las Palmas de Gran Canaria city (Spain). Land Use Policy **96**, 104697 (2020). https://www.sciencedirect.com/science/article/pii/S0264837719317442
7. Sugimoto, G., van Dongen, W.: Technical report: archival digital object ingestion into Europeana (ESE-EAD Harmonisation) v1.0. Technical report, 34–67 (2016)
8. Theodora, Y.: Cultural heritage as a means for local development in mediterranean historic cities-the need for an urban policy. Heritage **3**(2), 152–175 (2020). https://www.mdpi.com/2571-9408/3/2/10
9. Wijesundara, C., Monika, W., Sugimoto, S.: A metadata model to organize cultural heritage resources in heterogeneous information environments. In: Choemprayong, S., Crestani, F., Cunningham, S.J. (eds.) ICADL 2017. LNCS, vol. 10647, pp. 81–94. Springer, Cham (2017). https://doi.org/10.1007/978-3-319-70232-2_7

A Methodology for Formal Modeling and Evaluation of the Judicial Process

Angelo Ambrisi, Rocco Aversa, Marta Maurino, and Salvatore Venticinque[✉]

University of Campania - Luigi Vanvitelli, Aversa, Italy
{angelo.ambrisi,rocco.aversa,marta.maurino,
salvatore.venticinque}@unicampania.it

Abstract. This paper proposes a methodology for the formal representation and the automatic evaluation of the organization model of the Italian Courts. This methodology is based on the implementation of BPMN models and their simulation by an open source technologies. In particular, the paper describes the application of the methodology to the activities of all phases of the Italian civil trial, focusing on a new Office denominated *Office for the trial*.

1 Introduction

The overall duration of Italian civil trials is much longer than the European average [3, 8]. However, public spending on justice is not low at all and Italy has a number of judges per inhabitant that is even higher than Countries that show better performance of their judicial system. Many actions have been taken to improve the performance of the Italian judicial system. In this context, it is relevant to design and develop mechanisms to model, estimate in advance and monitor the impact of organization choices on judicial performance. This activity must be specialized for the heterogeneous territorial realities, which can choose different organizational models according to their peculiarities and requirements. A well defined structured practice must be integrated into the management plan so that the investments implemented through growing computerization and use of additional resources do not have a limited impact, on new or consolidated practices, such as to prevent achievement of the results expected in terms of efficiency.

The use of formal models is ì proposed for the representation [7], analysis [1, 2] and optimization [5] of processes in, information flows and the effectiveness of existing procedures in public administration. A limited number of contributions focus on the analysis of laws and procedures [4]. Here we exploit Business Process Model and Notation (BPMN) as tool for formal representation of the organizational model of the Court to handle the jurisdictional process. Supporting technologies and tools are used to configure the created models with both estimated values and historical data. Simulation techniques allow to evaluate their performance in terms of average duration of the trials, or of other subprocesses, in terms of throughput or queue length and waiting time. Simulation

L. Barolli (Ed.): CISIS 2023, LNDECT 176, pp. 318–326, 2023.
https://doi.org/10.1007/978-3-031-35734-3_32

provides a decision making support allowing for the evaluation in alternative scenarios, in which it is possible to configure the resources available in the Courts or to replace manual activities with ICT services [6]. The present study is part of a large project, involving researchers from University and personnel of the Italian Courts. The main goal of the project are the analysis and evaluation of the impact of the choices made by different Courts about the organization of the UPP, taking into account the opportunities of advancing the level of computerization of existing activities. In this paper we present some examples of application of the proposed methodology to model and analyse the organizational choices made at the Italian Court of Santa Maria Capua Vetere to exploit a time-defined support (Office of the process, UPP), in terms of human resources, granted by the Italian Ministry of Justice to improve the efficiency of the judicial process at national level.

2 Methodology

Figure 1 represents the proposed methodology through a BPMN model. It is developed in four steps:

- *BPMN modeling*: representation of the document flow and of the organizational model through BPMN.
- *Structural check*: automatic verification of BPMN structural completeness and correctness;
- *Supervised validation*: supervised verification of process correctness.
- *Quantitative analysis*: configuration, simulation and quantitative evaluation of the model.

In the first phase a visual tool is used to build the BPMN model according to the user's knowledge of the application domain. The second steps is supported by technologies which are able to check the compliance of diagrams with well defined rules. It is a structural check, that assures completeness of the model and the satisfaction of requirements to perform the following phases. The third step consists of the execution of the diagram by simulation tools that allows for the triggering of start events. This inserts new tokens that will follow the modeled workflow. Such tools usually provide interactive controls of BPMN elements. At this step, the user can verify if the process semantic correctly addresses the requirements, looking at the token flowing in the BPMN network, monitoring the occurrence of events and matching the resulting sequence against the expected one. Last step allows for the quantitative evaluation of the model trough simulation. The supporting tools are used for a batch simulation of the concurrent model instances activated by a stochastic process of start events. At the end of this step it is possible to estimate the model performance in terms of throughput, statistical distribution of resource usage and completion times.

Of course, a general prerequisite for the application of the methodology in the jurisdictional field is a deep knowledge of the domain, but also about the specific organization of the processes to be modeled.

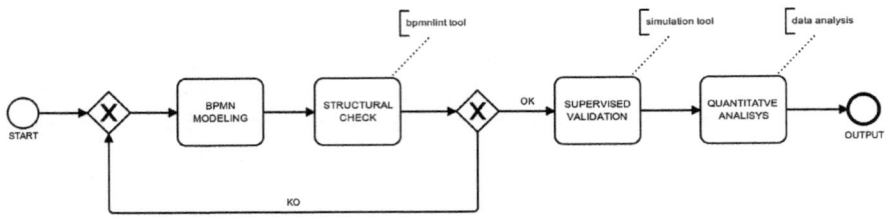

Fig. 1. BPMN based representation of the modeling and evaluation methodology.

3 The UPP Case Study

3.1 The UPP Organizational Model

The Court of Santa Maria Capua Vetere and the Court of Matera have been selected as case studies for the research activities. In order to acquire the necessary information to represent and analyze their organisational model interviews and surveys have been formulated as a grid of standardized questions for the UPP staff. The goal of this activity is to learn about the types of activities carry out by the individuals assigned to the UPP office as well as the operating procedures. The information collected is both quantitative and qualitative. In this paper we are going to focus on the assessment of UPP of the Court of Santa Maria Capua Vetere as it uses a more complex and articulated organizational model. The Officials assigned to the UPP of the Court of Santa Maria Capua Vetere are 99 and are distributed to several organizational units of the Court, which include the Central Office, Transversal Office (Statistics), Corpus Crimes Office, Court of Assizes, Single Civil or Penal Section and Registers Office. Each UPP personnel is required to cooperate with other officers of the same unit. The UPP personnel assigned to the Transversal Office (Statistics) is responsible for drafting ministerial statistics mainly relating to the penal sector, monitoring data on pending cases and carrying out inspections. UPP staff assigned to the Corrals Office is mainly responsible for backlog reduction. The UPP personnel assigned to the Court of Assizes also deal with the archiving and cataloging of folders. From the interviews carried out with the UPP officials assigned to other individual Sections in the civil and criminal field emerged the greatest heterogeneity of activities.

3.2 BPMN Modeling

On the basis of the classification of the types of ritual and the types of disputes, the research group started the mapping of the activities carried out by the UPP employees assigned to the individual civil Sections, describing their role during the civil trial of first instance[1].

---

[1] The data collected dates back to the period before the entry into force of Legislative Decree 10 October 2022, n. 149 (so-called Riforma Cartabia).

Following the provisions of the civil procedure code (henceforth "c.p.c."), the jurisdictional procedure before the Court was mapped distinguishing: preparatory/introductory phase; preliminary investigation phase, in turn divided into discussion and probative instruction; decision-making phase.

In each of the these phases were highlighted by those for which the law provides for broad powers of the judge and, consequently, the intervention of the UPP employees assigned to the single Section.

Figure 2 shows the complexity of the BPMN model and its complete structure. In particular, the model is composed of four BPMN pools. In each pool they are included the activities carried by a specific part of the jurisdictional process. In particular, the modeled roles are: parts in the Civil trial (plaintiff/defendant), the Software and the Court, which is divided in turn into three lanes: the Registry, the Judge and the UPP employees. According to what we described in Sect. 3.1, the UPP employees assigned to a single Section are then assigned either to a Judge or to the Chancellery of a Judge. Consequently, depending on the type of assignment, they carry out very different activities.

Firstly, the activities carried out by the UPP employees assigned to the single Judge within a section were analysed.

The research group analyzed the activities carried out by the UPP employees assigned to the Judge within a section. The interaction among the participants in the trial take place mainly through a centralized web service, provided by the Ministry of Justisce[2], that implements the access point to the telematic

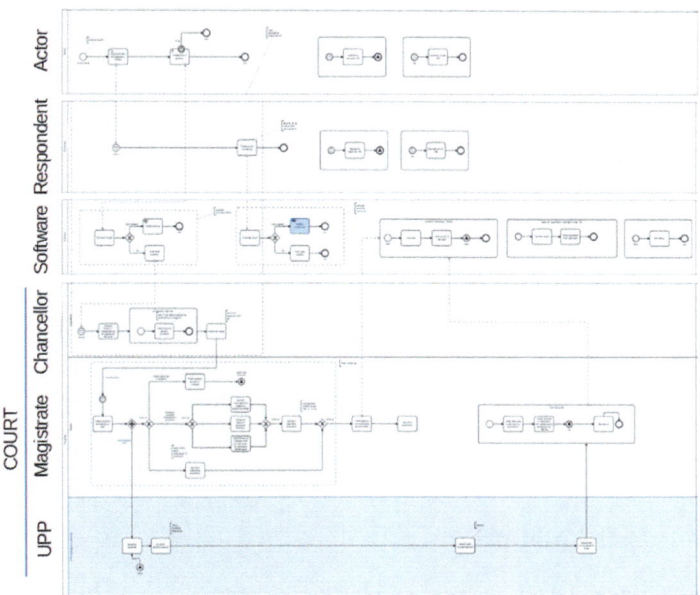

Fig. 2. BPMN model of the Italian civil trial

---

[2] https://www.accessogiustizia.it/.

justice for the lawyers. The Central Manager can be accessed trough proprietary clients, or by employees of the Ministry of Justice, according to the specific role, by specific web dashboards. The SICID ("District Civil Litigation IT System") manages the Registry of civil litigation voluntary jurisdiction and labor disputes. It is used in the Tribunals and Courts of Appeal by the Clerks of the Courts. Judges access to the System and to other services by a dedicated dashboard named the *Judge's Console*.

Computerized activities are represented in this specific BPMN model highlighting related processes and sub-processes and grouping them in a dedicated Software pool. For this BPMN scenario, activities are executed through specific tasks. In addition, to identifies those activities that the various actors carry out with the support of telematic systems, this pool also includes the management and verification of certified electronic communications. Each verification activity is modeled as an exclusive gateway (XOR). Each activity is then concluded with an *end event*.

With regard to the activities carried out by the Registry office, it should be noted that the main task of the Clerk of the Court is to assist the Judge in all activities that have to be documented through the drafting of a report. It follows that the activities carried out by the Judge through the "Judge's Console" appear on the SICID as" to be worked on" and are then defined after necessary intervention by the Registry office. Therefore, these activities are not represented in the Clerk of the Court pool, but in the pool of the Software. The interaction between the participants to the BPMN scenario are represented as message exchanges, which trigger new sub-processes. Exchanges of physical objects or of information between participants are represented through a dotted arrow.

3.3   Structural Check

The verification of the BPMN model is an operation that checks the structure of the model against a list or rules processing the XML code in the BPMN2.0 standard format. The check is an automatic process performed by specific technologies such as the *bpmnlint* tool[3], which currently defines a set of 16 structural checks. Addressing the notified violations, such as the lack of the termination event or disconnected link it is not a difficult task, but requires a sufficient level of technical knowledge about the BPMN theory. The definition of additional rules to be checked requires a multi-disciplinary skills, both technical and specific of the application domain.

3.4   Supervised Validation

A second level verification is used to check the semantic and functional compliance of the model with specification of requirements. An expert of the application

---

[3] https://github.com/bpmn-io/bpmnlint.

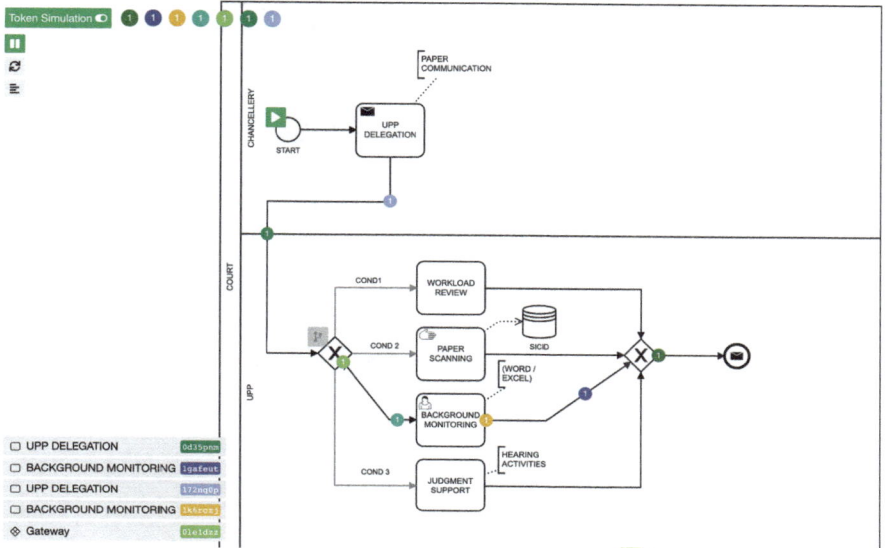

Fig. 3. Interactive execution of the model

domain (the jurist) must checks that the sequence of operations and the document flow respects the requirements. This operation is supported by visual tools that allow for the execution of the model, enabling the activation of several start events and the simulation of the consequent flow through the animation of tokens that follow a specific path through the network.

In Fig. 3 the BPMN details models the activities carried out by the UPP employee supporting the Registry office of the single Judge.

In Fig. 3 different coloured token are flowing trough the model. The start event into the Registry office pool represents the delegation of the activities by the Clerk of the Court to the UPP employee starts from. The activities delegated are the examination of the workload, the scanning of paper files with transmission of the scanned contents to the District Civil Litigation IT System and the monitoring of the backlog.

The XOR gateway used in split mode allows models the exclusive choice than different token can follow only one ongoing branch. The following OR gateway allows the rejoin of the alternative flows generated by previous gateways in the process. The process ends with a message-type end-event, which describes the communication of the completed activity by the delegated UPP employee to the delegating Registry office. The end event is combined with a signal which is followed by the conclusion of the process. As it is shown by the controls shown in Fig. 3 the operator can pause and restart the execution and he can control not only the generation of events. In fact, he can also change interactively the routing of the tokens along alternative paths acting on each gateway. The list of the events for the different tokens is dynamically updated and can be used as a trace of the completed execution.

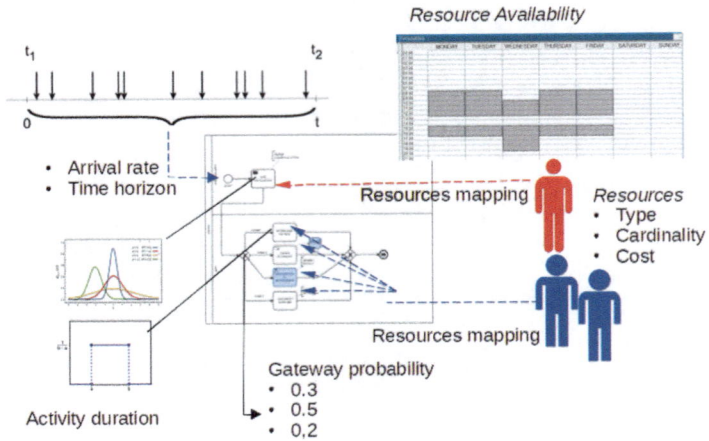

Fig. 4. Simulation-specific information

3.5 Quantitative Analysis

The execution of the model described in the previous section does not support batch simulations and the quantitative evaluation of model performance. The model, possibly reduced, is then simulated complementing its structure with the some additional information. *Resources* represent any kinds of human, software or physical infrastructure which are required to perform the activities composing the model. Each resource type is characterized by its cost, its availability and cardinality. In Fig. 4 resources are depicted as red and blue human icons and their availability as time-tables. For each activity *duration* and *resource* mapping must be defined. These parameters will keep busy the assigned resource type and will cause the pause of the workflow in the case of resource unavailability. As it is shown in Fig. 4, duration of an activities can be defined in terms of statistical distribution. The *simulation workload* is characterized in terms of statistical distribution of the arrival rate of start events, total number of events and eventually time horizon of the simulation. Finally, the probability of activation of the alternative routes must be defined for each gateway. Example of probability values are reported in Fig. 4. Simulation results are generated both in terms of process duration and usage (and cost) of the assigned resources until the completion of input load, and within the time horizon. To this aim, Scylla[4], an open source simulator has been used. Scylla is an extensible BPMN process simulator which relies on the building blocks of discrete event simulation. The process simulator offers an UI where BPMN models can be extended with the simulation-specific information defined before. Scylla is implemented in the Java programming language. It offers well-defined entry points for extensions based on a plug-in structure.

Simulation-specific information are grouped in three files.

---

[4] https://github.com/bptlab/scylla.

The *Current Global Configuration* file includes the resources and their time-table, which indicates how many hours the resources are available for (it could simply be the working hours of a employee or, if the resource is computational, not having a time limit in which it can be used).

In the previous example, the defined resources are the Court's employees, i.e. Chancellors and UPP employees. The configuration of the *Current Global Configuration* file is independent of the BPMN model, i.e. the modeled diagram file, which is needed for the definition of the *Current Simulation* file. The *Current Simulation* binds together the resources that are available and the activities of the workflow. Moreover, it defines the number of start events, which correspond to the tasks that the Chancellors will send to the UPP. If there are one or more XOR gateways, the probability must be set for each possible output condition.

Scylla provides in output an XML file and, in order to make the results easier to read, the Scylla-UI interface has been used. The Scylla-UI[5] application allow for the visualization of some specific Key Performance Indicators (KPIs) extracted from the simulation output. KPIs are used to compare different configurations of the simulation. Available KPIs are related to process, to individual activities and to resources utilization.

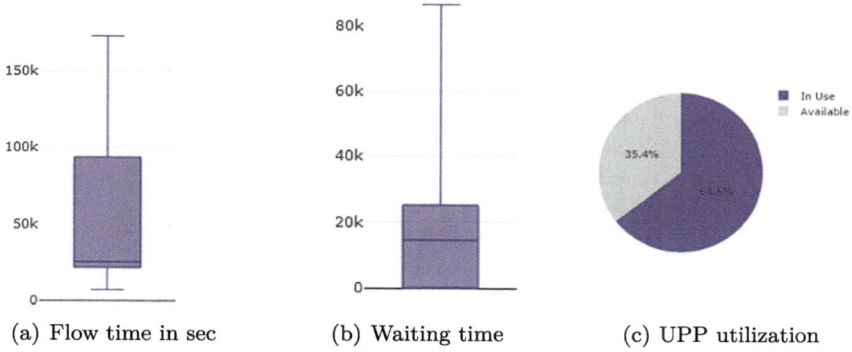

   (a) Flow time in sec   (b) Waiting time   (c) UPP utilization

Fig. 5. Scillya-UI visualization of evaluated KPIs

BoxPlots related to the flow time of tokens, which are the execution time of processes, and to the waiting time of a token because of resource unavailability are shown respectively in Fig. 5(a) and Fig. 5(b). In Fig. 5(c) the ratio between the busy and the idle time of UPP resources is shown. For each KPI the total value, the average and statistical parameter (q1, q3, median, max/min) are provided.

---

[5] https://github.com/bptlab/scylla-ui.

4   Conclusion

The presented study proposes an original methodology, based on the utilization of open source software tools, for the formal representation and evaluation of the information flow and of the organizational model through the BPMN with a specific application to the judicial process. The described approach allows the statistical estimation of some KPIs related to execution time of process, utilization and cost of consumed resources. It allows to identify critical tasks, such as human activities which should computerized and automized in order to improve the performance of the process. It allows to design new organizational models, or to apply changes to consolidated procedures, and evaluate their impact trough simulation.

Acknowledgments. This work is supported by the *Cleopatra Project*, funded by the University of Campania "Luigi Vanvitelli" through the VALERE 2019 research program, and by the Project PON_MDG_1.4.1_17, *Modelli Organizzativi e Digitalizzazione Ufficio Per il Processo (MOD-UPP)*, cofinanced by the European Union, Programma Operativo Nazionale Governance e CapacitaÍstituzionale 2014–2020 FSE/FESR.

References

1. Bhagya, T., Vasanthapriyan, S., Jayaweera, P.: Collaboration modelling framework for courts hearing workflow specification. Developing Country Stud. **4** (2014). https://doi.org/10.1007/s12652-021-03490-5
2. Capuzzimati, F., Violato, A., Baldoni, M., Boella, G., et al.: Business process management for legal domains: supporting execution and management of preliminary injunctions. In: JURIX, pp. 149–152 (2015)
3. Casamonti, M.: La giustizia civile italiana resta la più lenta d'europa, ma c'è qualche miglioramento. Osservatorio sui Conti Pubblici Italiani (2020). https://osservatoriocpi.unicatt.it. osservatoriocpi@unicatt.it
4. Ciaghi, A., Weldemariam, K., Villafiorita, A., Kessler, F.: Law modeling with ontological support and BPMN: a case study. In: Proceedings of the Fifth International Conference on Digital Society-CYBERLAWS, pp. 29–34 (2011)
5. Di Martino, B., Colucci Cante, L., Esposito, A., Graziano, M.: A tool for the semantic annotation, validation and optimization of business process models. Softw.: Pract. Experience (2023)
6. Di Martino, B., Esposito, A., Colucci Cante, L.: Multi agents simulation of justice trials to support control management and reduction of civil trials duration. J. Ambient Intell. Hum. Comput. (2021). https://doi.org/10.1007/s12652-021-03490-5
7. Di Martino, B., Graziano, M., Colucci Cante, L., Ferretti, G., De Oto, V.: A semantic representation for public calls domain and procedure: housing policies of Campania region case study. In: Barolli, L. (ed.) CISIS 2022. LNNS, vol. 497, pp. 414–424. Springer, Cham (2022). https://doi.org/10.1007/978-3-031-08812-4_40
8. Fronzetti Colladon, A., Figà-Talamanca, G.: Disegno, simulazione e business process re-engineering del processo civile di cognizione. Temi Romana (2008)

A Comparative Analysis of Formal Storytelling Representation Models

Luigi Colucci Cante[1(✉)], Beniamino Di Martino[1,2,3], and Mariangela Graziano[1]

[1] Department of Engineering, University of Campania "L. Vanvitelli", Aversa, Italy
{luigi.coluccicante,beniamino.dimartino,
mariangela.graziano}@unicampania.it
[2] Department of Computer Science and Information Engineering, Asia University,
Taichung, Taiwan
[3] Department of Computer Science, University of Vienna, Vienna, Austria

Abstract. This paper presents a comparative analysis of the various formal models that can be used to represent a story. The analysis focuses on two types of representation families: semantics-based representations, which use ontologies, and process-based representations. The aim is to provide a comparative overview of the models, analyzing their weaknesses and strengths, in order to determine the formal model that best lends itself to modeling a story by highlighting its main components in terms of the actors involved, events, actions, spatio-temporal relations, as well as cause and effect, in hopes of identifying the formal story representation model that can be used as the starting point for developing a framework that can perform automated storytelling generation. Finally, examples are given of the uses of these models to represent a mythological story.

1 Introduction

Storytelling is defined as the art of writing or telling stories while capturing the attention and interest of the audience. It consists of organizing selected content into a coherent system, governed by a narrative structure. The storytelling approach can be applied in various fields of human life, from sales to marketing, from public and political communication to medicine, and from education to corporate identity; in almost every human interaction in which information is transferred, storytelling can play a decisive role.

The steadily increasing use of storytelling in our times has encouraged research into techniques and approaches for generating storytelling automatically.

In order to achieve this, it is necessary to start by establishing a coherent narrative structure. This is the focus of our research, which proposes to analyze all possible methods by which a story can be represented, with the aim of finding the most suitable formal model to model a story in all the necessary elements, which are useful for identifying a basic structure for automated storytelling activities. A formal representation of story structure is defined as a set of structural axioms, e.g., causality, precedence, continuity, and time, which are followed to generate stories automatically.

The remainder of this paper is organized as follows: Sect. 2 presents an overview of the various models analyzed in the literature for modeling stories focusing in particular on those based on semantics described in Sect. 2.1, and those based on processes

L. Barolli (Ed.): CISIS 2023, LNDECT 176, pp. 327–336, 2023.
https://doi.org/10.1007/978-3-031-35734-3_33

described in Sect. 2.2; Sect. 2.3 provides a detailed comparison of the models described before; Sect. 3 reports the final model proposed for the representation of a story and a practical example of the application of the proposed model; Sect. 4 closes the paper with final remarks and addresses future research directions.

2   Storytelling Models: State of the Art

Several models have been proposed in the literature for the representation of a story. Work [8], for example, describes the use of a StoryFlow Model for modeling a story, using "storylines" to describe the evolution of the plot of each actor within the story. Another interesting model for representing a story is the Story Graph, described in paper [14], which allows a story to be modeled as a graph whose nodes can represent the events and/or main concepts of a story, while the arcs represent the causal relationships between events or the semantic relationships between concepts. A more in-depth analysis of the main semantic and process-based models for modeling a story follows in Subsects. 2.1 and 2.2.

2.1   Semantic-Based Models for Story Representation

This section describes the main semantic models for representing a story and its components.

The first ontology proposed in this comparison is the **RST Ontology** described in work [11], which proposes a generic ontology model of storytelling in OWL (Ontology Web Language) based on the organization of events using the relations proposed by the Rhetorical Structure Theory (RST). Figure 1 shows the structure of RST Ontology,

Some of the main classes present in Fig. 1 are described below:

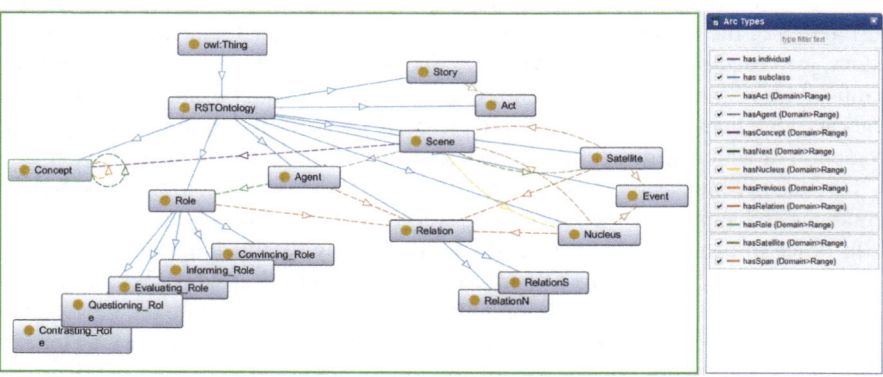

Fig. 1. RST Ontology

- **Act and Scene**: defines the acts into which a story is composed. A story is divided into several acts, and each act can be divided into several scenes. An act is a recursive structure in that one of its nuclei or satellites can contain another act. This recursive structure is of considerable benefit when we want to semantically describe stories that recall other stories within them.
- **Nucleus and Satellite**: in Rhetorical Structure Theory, a nucleus represent a central part of the story whose removal would result in the loss of information relevant to the correct continuation of the storyline. Satellites, on the other hand, represent secondary parts of the story that clarify or add information about a nucleus.
- **Event**: defines the events of a story. Each Nucleus is associated with a single event via the "*hasSpan*" relation.
- **Concept**: defines a specific topic that a story or part of it may refer to.
- **Agent**: defines the actors who participate in a Scene by performing or being part of one or more Events.
- **Role**: defines the parts that the agents play during a Scene.
- **Relation**: A Relation is a rhetorical binding between two entities, which refers to a specific rhetorical function. As specified in RST [10], entities in a Relation can be both Nucleuses (which is defined as a Multinuclear Relation Type) or a Nucleus - Satellite pair (which is defined as a Nucleus-Satellite Relation Type). It is also possible to use the concept of "Relation" to associate an actor and the role he plays in a specific scene.

The second ontology proposed in this comparison is the **ONAMA Ontology** described in work [7], which enables the description of the essential features of literary narratives, particularly medieval ones. ONAMA provides an OWL[1] model for a cross-media description of actions, actors, settings, and temporal structures, providing the possibility of linking these elements to narratives using layered semantic roles. Figure 2 shows the structure of ONAMA Ontology.

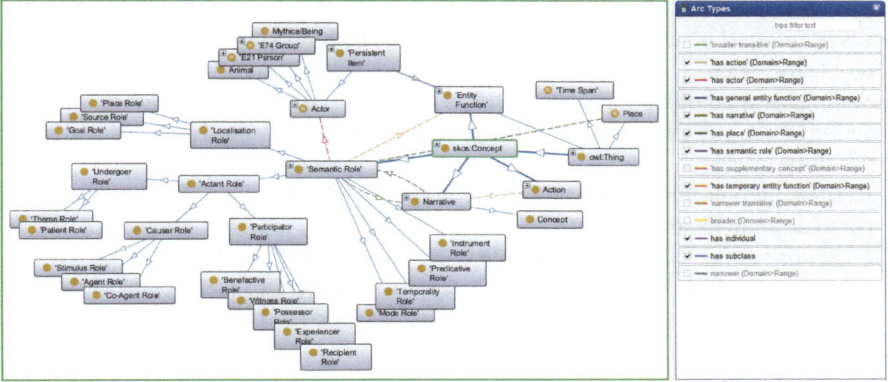

Fig. 2. ONAMA Ontology

---

[1] http://onama.sbg.ac.at/en/ontology-2/.

Some of the main classes present in Fig. 2 are described below:

- *Actor*: describes the protagonists of a narrative. Actors can be people, animals, groups of individuals, or even mythological beings.
- *Semantic Role*: describes the semantic role of an entity associated with the narrative. ONAMA distinguishes different roles: i) "*Causer Role*" to denote the role of an entity that is the author or instigator of an event; ii) "*Participator Role*" to denote the role of animated entities that participate in an event but have no control over it; iii) "*Benefactor Role*" to denote the role of entities that are advantaged or disadvantaged by an event.
- *Place*: defines the location where a narrative can take place;
- *Time Span*: defines the temporal extension in which a narrative may take place.

The third ontology proposed in this comparison is the **Common Narrative Model Ontology (CNMO)** described in work [2], which attempts to provide a narrative representation model for the development of a shared knowledge base that can be populated in the future with the outcomes of new storytelling systems without the need to modify it for each unique representation model-specific system. Figure 3 shows the structure of RST Ontology.

The model has been designed in a hierarchical structure, in which the root concept is the "**Story**". It is composed of two classic narratological components: the plot and the space. The "**Plot**" is represented as a sequence of scenes, while a "**Scene**" represents a single episode inside the plot, which consists of a sequence of events. An "**event**", on the other hand, may consist of one or more actions or happenings; unlike "**Actions**", which represent voluntary acts performed by one or more characters in the story, a "**happening**" represents involuntary acts that occur as a consequence of a previous action or event. Also related to the scene is the concept of "**Time**", which identifies the time span within which the scene is set. Space, which includes the entire universe in which the story takes place, is the other narrative element that is visible from this ontology. The most significant components of space are explained below:

- *Location*: can be considered the scenario in which every scene that composes the plot takes place.
- *Setting*: is a combination of a set of physical - or virtual - locations in which the action of the story takes place, and the set of cultural and physical rules that govern the story world.
- *Existent*: represents the whole set of actors that take part in the story. They can be persons, animals, or objects.
- *Function*: defines the functional role of a character in the plot.
- *Cognition*: defines the psychological - cognitive sphere of a character in the story, which can be expressed through a BDI (Belief - Desire - Intention) model [4, 6].

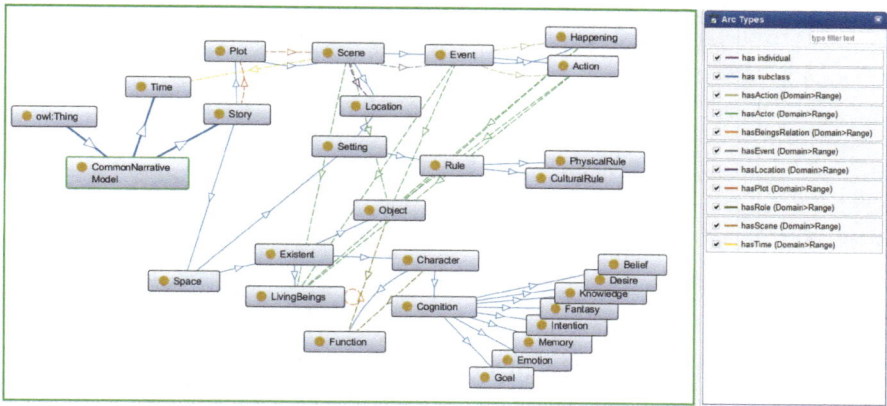

Fig. 3. Common Narrative Model Ontology

2.2   Process-Based Models for Story Representation

Several process-based models have been investigated, but one of the most interesting is the BPMN (Business Process Model and Notation), which is one of the most successful standards for modeling business processes in the literature. Some works, such as [13] and [1], propose the idea of using the BPMN's graphical notation to describe a story. The standard provides several useful features for story modeling, the most important of which are listed below:

- **Actions and Events Representation**: the constituent elements of a scene can be represented through *BPMN Activities* and *BPMN Events*.
- **Actors**: it is possible to specify who the actor performing an activity by placing them in "*pools*" and "*lanes*".
- **Flows**: using "*sequenceFlows*", the BPMN allows the causal sequence between actions and events within a scene, or between different scenes, to be represented very well. By using "*Parallel Gateways*", it is possible to indicate that the plot splits into several flows executed in parallel, while with "*Exclusive Gateways*", it is possible to indicate that a flow of the plot will only be executed if an appropriate condition is satisfied.
- **Time**: the BPMN standard does not offer an explicit way of representing the time in which a scene or event takes place. However, it is possible to use a "*Timer Start Event*" or a "*Timer Intermediate Catch Event*" to indicate the time between two activities/events, or the time in which an event is to occur.
- **Recursive Scene Modelling**: it sometimes happens that an action or event in the story leads to the recursive initiation of a new scene or act. The BPMN offers the construct "*SubProcess*" [12] for the definition of subprocesses that can call up a new scene, a new act, or even a new story.

Figure 4 shows a fictitious example of modeling a story using BPMN notation.

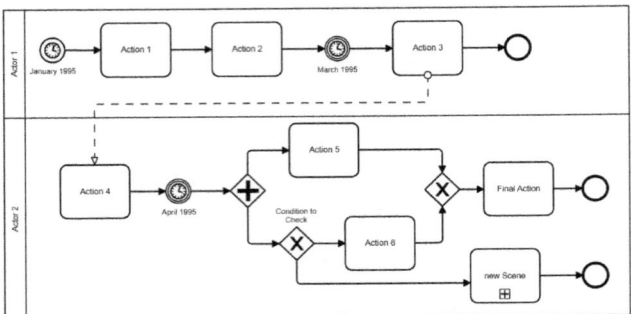

Fig. 4. Example of BPMN notation for modeling a story

2.3　Model Comparision

This subsection provides a comparison of the models described in the previous sections. The aim is to highlight each model describing the peculiarities and expressive potential in order to select the most suitable one for the representation of a story. Table 1 shows the comparison of the models described in Sects. 2.1 and 2.2.

By way of comparison, it is clear that the Common Narrative Model Ontology is the most complete model for the representation of stories. It allows for a good description of the flow and organization of a story into acts, scenes, events, and actions; it models the concepts of actors, the setting, and the timing of a scene; and it models the role and psychology of an actor. The idea of "Connection", which connects an actor's many roles in various scenes, as well as a mechanism to specify recursion in a story's structure, are two of the most glaringly absent aspects, nevertheless.

On the other hand, the RST Ontology includes these elements, but the lack of the notions of location and time connected to a scene might significantly constrain how tales are represented. On the other hand, the ONAMA ontology has many shortcomings, the most glaring of which is its inability to specify the organization or flow of a tale. To

Table 1. Comparison of Models for Storytelling

Feature	RST	ONAMA	CNMO	BPMN
Representation of Actors	X	X	X	X
Representation of Locations		X	X	
Representation of Times		X	X	X
Description of Story Structure (Acts, Scenes, Events)	X		X	
Modelling Story Flow	X		X	X
Modelling of "Recursive Stories"	X			X
Defining an Actor's Role	X	X	X	
Associating different Roles to the same Actor in different Scenes	X			
Representation of an actor's psychology			X	

simulate the actors in a story, ONAMA defines a set of fascinating roles as well as some helpful concepts (groups, animals, mythological beings, etc.).

The BPMN offers an interesting way to represent a story due to the comprehensive graphical notation of its elements; it excels at modeling the actors and flow of a story and also supports modeling recursive stories by defining "Sub-Processes", but it has limitations when it comes to modeling the setting of a scene as well as the roles of the actors. By semantically annotating the BPMN's structural parts with concepts from domain ontologies, as defined in [3,5], these limitations can be solved. The expressiveness of BPMN's graphical notation might be combined with all the benefits each of the semantic models has to provide if one or more of ONAMA, RST, or CNM were adopted as the domain ontology properly.

3   Proposed Framework for Story Representation and Illustrative Example

As a result of the analyses discussed in Sect. 2.3, we concluded on a final framework that we presume is the most appropriate for formally and comprehensively representing a story for storytelling. The final framework is just an ontology that combines the RST, ONAMA, and CNM ontologies. One example of how the framework may be applied is to semantically represent the mythological story *"Teseo and the Minotaur"* [9]. Figure 5 demonstrates the story's modeling.

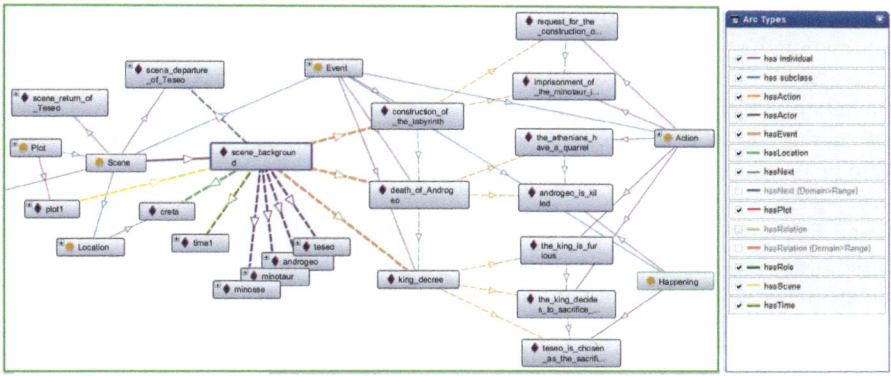

Fig. 5. Representation of Story "Teseo and the Minotaur" with the framework proposed

As illustrated in Fig. 5, the story is modeled with a single plot of 4 scenes: i) Background; ii) Departure of Teseo; iii) Labyrinth; iv) Return of Teseo. Each scene is an instance of the CNM Ontology class *"Scene"*, and the scenes are connected together by the *"hasNext"* relationship to indicate their temporal order of them. Each scene is associated with a Location, a Time, and a set of actors participating in it. Moreover, each scene is composed of an ordered sequence of Events that are linked together, connected by them through the *"hasNext"* relation.

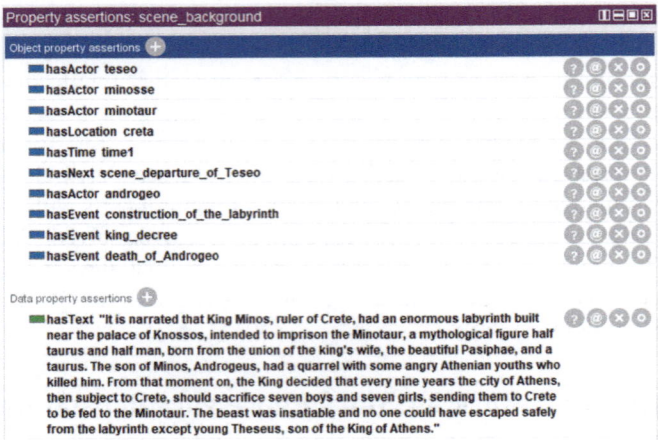

Fig. 6. Semantic Description of the Scene "Background"

The first scene of the story consists of 3 events: i) the Construction of the Labyrinth; ii) the Death of Androgeo, brother of King Minosse; iii) the Decree of the King for the selection of seven boys and seven girls each year. This scene has "*Creta*" as its Location and "*Minosse*", "*Androgeo*", "*Minotaur*" and "*Teseo*" as its actors.

Furthermore, each event is linked to its constituent actions. For example, the event "Construction of the Labyrinth" consists of two temporally consecutive actions: i) Request for the Construction of the Labyrinth; ii) Imprisonment of the Minotaur in the Labyrinth. This structured organization of the story was achieved solely through the application of CNM ontology concepts. The story is represented as a graph, with nodes representing the actions, events, scenes, and plots of the story and edges representing the logical ordering of the story's pieces. Each node in the graph is a "piece" of the story, whose combination yields the plot. Figure 6 shows a comprehensive view of the "Scene Background" node; the text related to the scene is associated with it through the data property "*hasText*". It is possible to relate a scene, event, or action to media contents inherent to the portion of the narrative by defining appropriate properties.

The ONAMA ontology serves to specify the roles of actors in the story: the Minotaur, for example, is an instance of the "*Mythological Being*" class of ONAMA.

RST ontology, on the other hand, is extremely useful in defining the roles that each actor in the story may play in each scene. Figure 7 shows the roles that the actor "Teseo" has in the first two scenes.

Even during narration, an actor can play a variety of roles. Teseo, for example, interprets the role of "sacrificial victim" in the Background scene (Fig. 7a), in which he is one of the seven children selected to access the labyrinth and fight the minotaur, and he interprets the role of "hero" in the later scene (Fig. 7b).

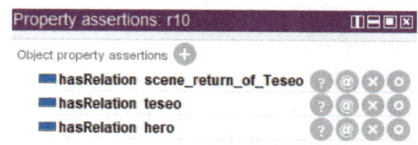
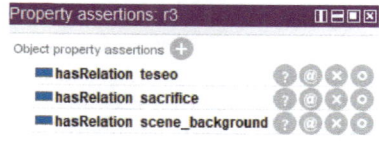

(a) Teseo' role in the Scene "Return of Teseo"   (b) Teseo' role in the Scene "Background"

Fig. 7. Use of Relation to Model the Role of an Actor for each Scene

4   Conclusion and Future Works

The paper compares different semantic and process-based models, emphasizing that neither of the models proposed in the literature completely models all of the main components and features of a story. A new framework for story modeling is proposed, which combines several existing semantic models. Finally, an example of a story description using the final framework is shown to demonstrate the framework's potential. This comparative study offers an intriguing starting point for constructing a knowledge base for an expert system capable of automated storytelling.

Acknowledgments. The work described in this paper has been supported by the research project RASTA: Realtà Aumentata e Story-Telling Automatizzato per la valorizzazione di Beni Culturali ed Itinerari; Italian MUR PON Proj. ARS01 00540.

References

1. Antunes, P., Simões, D., Carriço, L., Pino, J.A.: An end-user approach to business process modeling. J. Netw. Comput. Appl. **36**(6), 1466–1479 (2013)
2. Concepción, E., Gervás, P., Méndez, G.: A common model for representing stories in automatic storytelling. In: 6th International Workshop on Computational Creativity, Concept Invention, and General Intelligence, C3GI 2017, Madrid, Spain (2017)
3. Di Martino, B., Cante, L.C., Esposito, A., Graziano, M.: A tool for the semantic annotation, validation and optimization of business process models. Softw.: Pract. Experience (2023)
4. Di Martino, B., Esposito, A., Cante, L.C.: Multi agents simulation of justice trials to support control management and reduction of civil trials duration. J. Ambient Intell. Hum. Comput. 1–13 (2021)
5. Di Martino, B., Graziano, M., Colucci Cante, L., Esposito, A., Epifania, M.: Application of business process semantic annotation techniques to perform pattern recognition activities applied to the generalized civic access. In: Barolli, L. (ed.) CISIS 2022. LNNS, vol. 497, pp. 404–413. Springer, Cham (2022). https://doi.org/10.1007/978-3-031-08812-4_39
6. Guerra-Hernández, A., El Fallah-Seghrouchni, A., Soldano, H.: Learning in BDI multi-agent systems. In: Dix, J., Leite, J. (eds.) CLIMA 2004. LNCS (LNAI), vol. 3259, pp. 218–233. Springer, Heidelberg (2004). https://doi.org/10.1007/978-3-540-30200-1_12
7. Landkammer, M., Hinkelmanns, P., Schwembacher, M., Zeppezauer-Wachauer, K., Nicka, I.: ONAMA. Ontology of Narratives of the Middle Ages: Ontology 1.5. University of Salzburg. Salzburg (2020). https://doi.org/10.5281/zenodo.4285987. Accessed 23 Nov 2020
8. Liu, S., Yingcai, W., Wei, E., Liu, M., Liu, Y.: StoryFlow: tracking the evolution of stories. IEEE Trans. Vis. Comput. Graph. **19**(12), 2436–2445 (2013)

9. Lutterbach, B.: Theseus and the minotaur (2016)
10. Mann, W.C., Thompson, S.A.: Rhetorical structure theory: a theory of text organization. University of Southern California, Information Sciences Institute Los Angeles (1987)
11. Nakasone, A., Ishizuka, M.: Storytelling ontology model using RST. In: 2006 IEEE/WIC/ACM International Conference on Intelligent Agent Technology, pp. 163–169 (2006)
12. Natschläger, C.: Towards a BPMN 2.0 ontology. In: Dijkman, R., Hofstetter, J., Koehler, J. (eds.) BPMN 2011. LNBIP, vol. 95, pp. 1–15. Springer, Heidelberg (2011). https://doi.org/10.1007/978-3-642-25160-3_1
13. Simões, D., Antunes, P., Cranefield, J.: Enriching knowledge in business process modelling: a storytelling approach. In: Razmerita, L., Phillips-Wren, G., Jain, L.C. (eds.) Innovations in Knowledge Management. ISRL, vol. 95, pp. 241–267. Springer, Heidelberg (2016). https://doi.org/10.1007/978-3-662-47827-1_10
14. Valls-Vargas, J., Zhu, J., Ontañón, S.: Towards automatically extracting story graphs from natural language stories. In: AAAI Workshops (2017)

Towards the Reconstruction of the Evolutionary Behaviour of Finite State Machines in the Juridical Domain

Dario Branco[1], Luigi Colucci Cante[1], Beniamino di Martino[1,2,3],
Antonio Esposito[1(✉)], and Vincenzo De Lisi[4]

[1] Department of Engineering, University of Campania "Luigi Vanvitelli", Caserta, Italy
{dario.branco,luigi.coluccicante,beniamino.dimartino,
antonio.esposito}@unicampania.it
[2] Department of Computer Science and Information Engineering, Asia University,
Taichung, Taiwan
[3] Department of Computer Science, University of Vienna, Vienna, Austria
[4] Direzione generale per i sistemi informativi automatizzati - Ministero della Giustizia,
Rome, Italy
vincenzo.delisi@giustizia.it

Abstract. Juridical systems are inherently complex, since they address numerous and critical processes that are regulated by several laws and regulations. Furthermore, such regulations change with more or less frequency, thus provoking the evolution of the entire Juridical system. In order to adequately regulate a sophisticated system, it is imperative to establish and utilize models as well as to monitor the system's progression. In this paper, historical logs of Trials are used to build model that can be used to keep track of the regulation changes, and their effect on the Juridical system, through the years. In particular, the formalism provided by Finite State Machines will be used to represent the information extracted from the logs, and to provide a homogeneous model.

1 Introduction

Complex systems require simplified models to be correctly analysed and managed: the more complex the system, the more useful modelling them becomes. Choosing a model is not a simple task either: one needs to take into consideration the nature of the domain to be represented, to spot characteristics that can be used to identify a methodology and a formalism for modelling. The Italian Juridical System, that in recent years has rapidly evolved towards an almost completely digitalised framework, offers the ideal case study as it has a very complex structure and, at the same time, it collects information regarding all the activities carried our within Courts. In particular, the Telematic Civil Process offers a good view on the history of Trials, allowing the retrieval of temporal information regarding the different processes taking place in Courts.

Previous works conducted on the analysis of Italian Civil Trials, both on Documents involved in them [2] and on the whole workflow [3] have provided useful information

L. Barolli (Ed.): CISIS 2023, LNDECT 176, pp. 337–347, 2023.
https://doi.org/10.1007/978-3-031-35734-3_34

regarding characteristics and properties of the Italian Telematic Civil Process. In particular, such studies enabled the identification of time consuming events [6] and made it possible to build multi-agent simulation tools to understand the impact of organisational changes in Courts [4].

However, a model to detect and monitor juridical changes and their effect on the system that enables the whole Telematic Process has not been defined yet.

For this reason, methodology and tools to reconstruct the processes underlying the current Juridical System from execution logs are needed. Process Mining techniques can be exploited to analyse the historical logs, and to determine the best FSM model to represent the underlying system [1]. However, such techniques need to take also in consideration the evolution and changes that have influenced the Juridical System. In particular, we are interested in observing the evolution of the model through the years, through the comparison of several FSMs extracted from the historical logs referring to different operating years.

The remainder of this paper is organised as follows: Sect. 2 describes the generic models that can be applied to this case study, that is Finite State Automata; Sect. 3 provides a first reconstruction of the model, through the analysis of historical logs from the Court of Livorno; Sect. 4 provides a high granularity view on the evolution of the FSMs over the years, using a graphical depiction of the extracted models; Sect. 5 refines the evolutionary view on the system, considering the changes happening to the model during different years; Sect. 6 closes the paper with final remarks and future directions.

2 Use of Finite State Automata to Model the Juridical System

A Finite State Automaton (FSA) or Finite State Machine (FSM) [5] [7], is defined as an Event State Machine, i.e. a mathematical model used to describe the behavior of a system, in a clear and simple way. A Finite State Machine is made up of a set of **States**, which define the situation in which the system finds itself in a certain instant, and a set of **Inputs**, which represent what causes, or may cause, the change of State, i.e. the evolution of the system. The Inputs can be of various types: in the case of an Event States Machine, an Input corresponds to a particular Event that has occurred in the system. Each triple **Current State-Input-Next State** is called a **Transition**. A Finite State Machine has three main characteristics:

- The ability to evolve over time, changing one's State.
- The possibility to represent States and Inputs of the system in a discrete way.
- The property of being able to enumerate in a Finite way States and Inputs of the system.

There are two specific types of Finite State Automata. **Deterministic Automata** are such that each Input to the Automaton determines, based on the current State, one and only one subsequent State. **Non-deterministic Automata**, on the other hand, can have, even with the same Input, multiple destination States for each current State.

Both types of Automa have a finite number of possible **Initial States**, and a finite number of **Admitted Final States**. Furthermore, no State must be "isolated", i.e. there must always be at least one path in the Automaton that allows one to reach a State, starting from an initial State.

There are two main representations for a Finite State Automaton:

- A table or Transition matrix, which shows in tabular form all the starting States and all the target States for each possible Input, i.e. all possible Transitions for each starting State.
- A Directed (Directed) graph, where the nodes represent the States and the edges the Transitions, labeled with the Input symbol that generates the Transition. One can mark the initial State with an arc entering from nowhere and the final States with a double circle.

In the case of the Italian Ministry of Justice, an FSM represents the ideal model to describes all the possible evolutions of a Judicial Trials, identifying all the States in which each Trial can be found, and all the Events (Input of the Machine) that generate Transitions from one State to another.

There are potentially several FSMs, associated with different Rituals/Rites, which follow a separate evolution. Each ongoing process in the Justice System is indeed characterized by a unique code (process number **numpro**) and by a code that characterizes its subject. The object code is a six-digit code, made up of two digits corresponding to the Role, two digits referring to the Matter and two specific digits of the Object. The association between the Process and the specific triad of Role-Material-Object takes place during the registration phase of the same.

The Object Code determines the Rituality, which is expected to be modelled through a different FSM to follow.

Different Transitions are foreseen for each of the existing Machines. At the moment, we know of 68 different Rituals, characterized by a total of 155 States and 878 events.

For each FSM it is important to know:

- The Initial States, i.e. the entry points of the Machine.
- The Definitive States, i.e. the final States in which the Machine can be.
- The Definitive Events, i.e. those events that bring the Machine into a Definitory State.
- The Transitions, i.e. the triple Current State-Event-Next State that define its evolution.

Just theoretically speaking, the high number of Events compared to the States determines the existence of numerous Transitions in which the Current State coincides with the next one. In other words, the Directed Graph that represents each Machine is expected to expose numerous Self Loops, i.e. Transitions that fall back to the same starting State.

Managing such a complex set of FSMs requires much effort, as it is necessary to verify and check all the paths that can be followed by trials, ensuring that such paths fulfill the current legislation requirements, and that no ambiguity nor incomplete or erroneous transitions have been added over time. Considering that, over the years, changes in the legislation have triggered modifications to the Juridical System, having instruments and tools to support the analysis of the different Machines and Rituals can help in identifying possible inconsistencies.

3   Finite State Machine Model Reconstruction

This section reports the studies that have been carried out to reconstruct the Finite State Machine model that all Trials are subject to, by exploiting well known techniques described in literature [1]. Indeed, as already mentioned, the various types of judicial Trials are divided according to different rituals, which in turn include several objects, each object taking charge of a specific type of judicial Trial. The study portrayed in this Section focuses on a particular object pertaining to a specific ritual. The reconstruction was done by analysing the historical logs of the operations performed on about 700 different judicial Trials by the court of Livorno. The number of Trials taken in consideration per year is reported in Table 1.

Table 1. Number of Trials examined per year

Year	Number of Trials
2005	100
2006	73
2007	87
2008	54
2009	61
2010	48
2011	44

Each historical log file, describing a single Trial, has been pseudonimized first, and then provided in .csv format, with each row of the files containing the following relevant information: the object code to which the process refers, the process identification number, the current state identifier, the next state identifier, the event identifier that produced the transition, and a date indicating the exact moment the event was triggered. With this information, it was possible to reconstruct a Finite State Machine that all the Trials, during the reference year, followed. An example of a reconstructed Finite State Machine is provided in Fig. 1. Given the large number of self-edges affecting the various nodes of the machine itself, the self-edges themselves have been collapsed into a single self-edge. From this initial analysis, it was noted that different self-edges with the same identifier occur on several different nodes. After a discussion with domain experts it became clear that different events, such as the ones referring to document annotations or access to records, can be triggered in different States and at any time in the Trial regardless of the state in which the Trial itself is, without altering the workflow of the machine. For this reason, these events were identified and cut from subsequent analysis.

In order to identify recurrent self-edges spread over all nodes, the event-state machine-generated from all historical logs was reconstructed regardless of the year in which the processes were initiated. The resulting Graph is provided in Fig. 2.

The machine's complexity primarily stems from its events, as demonstrated in this Figure. Therefore, it is necessary to identify and eliminate events that can be triggered regardless of the machine's current state. To this end, an analysis of the percentage of occurrence, with respect to states, of the self-edges of the Finite state Machine was carried out from the total reconstructed machine. The analysis yielded the results presented in Fig. 3. From the figure, we can see how a good number of self-edges are reported on more than eighty percent of the machine. This behavior, therefore, suggests that we should eliminate from future analyses all events that have as their identifier one of those that, in the total reconstructed Finite State Machine, has an occurrence rate greater than eighty percent. This choice turns out to be necessary because, the goal of the work is to identify substantial years of change in the Finite State Machine and that they may reflect the regulatory and legislative change, having self-edges spread over multiple states does not benefit the calculation, identification, and treatment so it was decided simply to ignore them is that they did not contribute any information content with respect to the goal of the work.

4   FSM Models Comparison

The analysis continued by examining the evolution of the Finite State Machine over time, year by year, comparing the percentage of change from the previous year, after removing recurring self-edges in the graph. An initial analysis was made by comparing graphs generated from two consecutive log years, essentially identifying three main subgraphs:

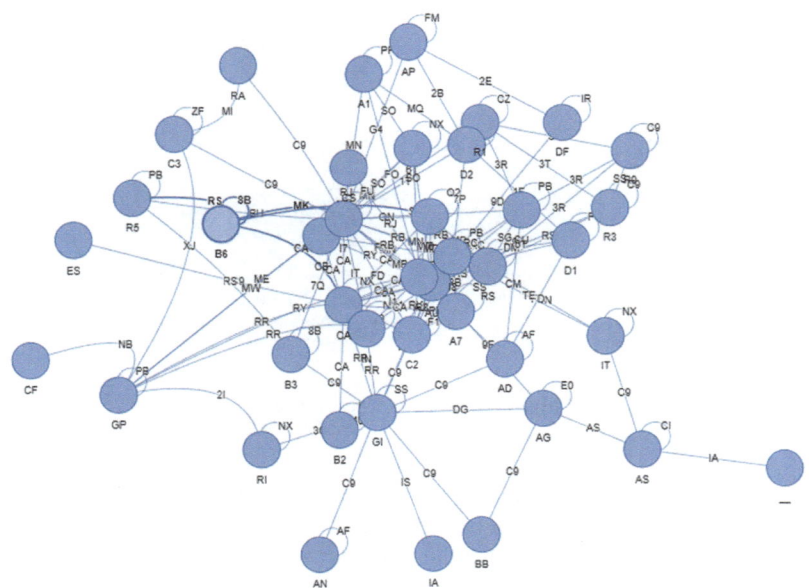

Fig. 1. FSM Reconstruction - Year 2005

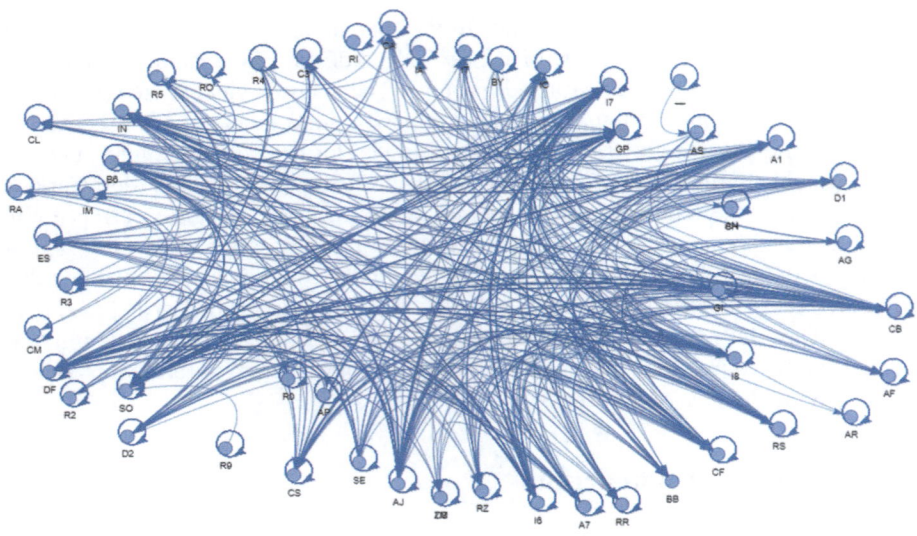

Fig. 2. Total FSM Reconstruction

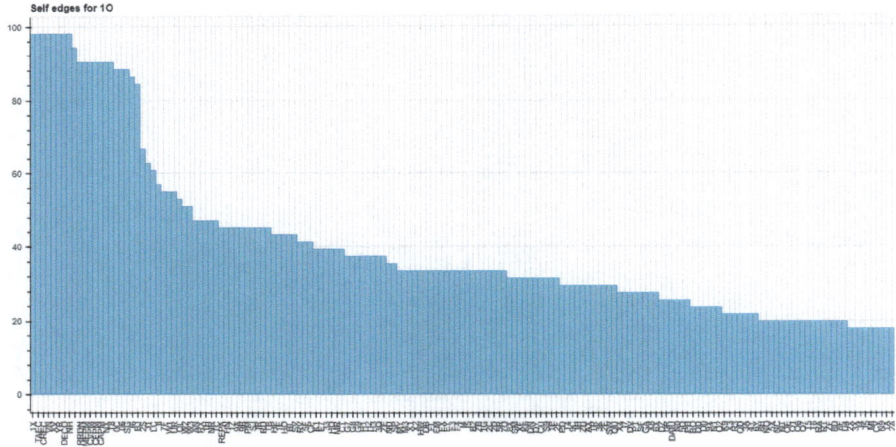

Fig. 3. Self-Edges Occurrences

- Nodes and edges that are common to both Finite State machines;
- Nodes and edges that are present in the previous year but not in the following year;
- Nodes and edges that are present in the next year but not in the previous year.

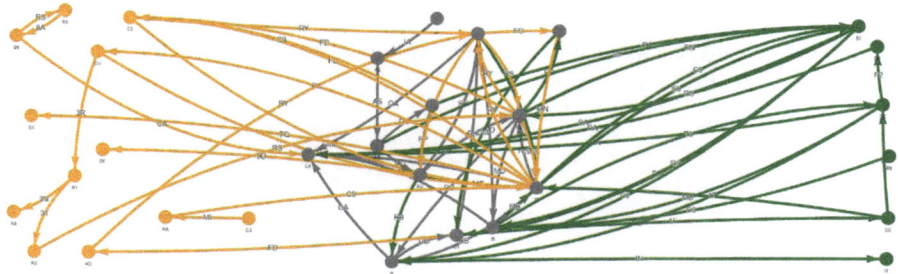

Fig. 4. FSM Comparison - 2005 vs 2006

In this regard, we provide the image in Fig. 4 which portrays in gray the nodes and edges that are common to both machines, in green the states and edges that are present only in the following year, and in orange the nodes and edges that are present only in the previous year. This graphical visualization is the first prototype of a tool that can be extremely useful to domain experts in analyzing, comparing, and validating regulatory and legislative changes. Thanks to the differentiation with colors applied to the graph representation of the Finite State Machine, in fact, it is possible to have an immediate knowledge of the changes made to the Finite State Machine and it is much easier to identify any discrepancies or inconsistencies in order to intervene promptly to restore a functioning and consistent process model. In addition, from this analysis, year-by-year percentage changes in terms of events and states of the Finite State Machine were calculated. Table 2 provides the experimental results obtained.

Table 2. Percentage variations along years

Transition	Percentage Variation for edges	Percentage Variation for states
2005–2006	22.80%	24.00%
2006–2007	11.43%	0%
2007–2008	10.23%	16.19%
2008–2009	4.28%	2.77%
2009–2010	2.73%	0%
2010–2011	2.0%	2.70%

5   Focus on States and Events Evolution

The generic information on FSM variations reported in Fig. 4 can be refined to better observe the behaviour of the system, in particular considering States that have been added and/or removed over the years.

Figure 5 shows FSM States that are used only until a specific year, taking in consideration a single object and not an entire Rite. This limited perspective suggests that numerous states may have been eliminated from the system over time, highlighting the importance of developing a comprehensive model that accounts for this evolving dynamic and effectively depicts it.

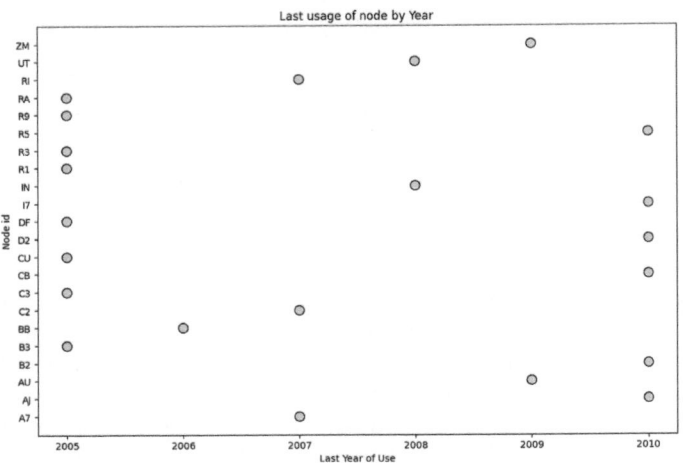

Fig. 5. State deprecation per year

A refined view is proposed by Fig. 6, where the blue bars represent the years in which a State is present. It is already evident by this still limited view that some State tend to disappear through the years, while others are added, modifying the Trial models. Despite the limits offered by this kind of visualisation, it still can be used to draw some very preliminary results. Indeed, states with codes R1, R3, R9, RA, DF, CU seem to be used in 2005 and then disappear the following years. If we look at the States' descriptions reported in the Ministry documents, we can obtain some possible hints on what actually happened:

- R1: ATTESA ESITO UDIENZA ex Art. 274 DAVANTI AL PRESIDENTE DEL TRIBUNALE/CORTE
- R3: ATTESA ESITO UDIENZA ex Art. 274 DAVANTI AL GIUDICE
- R9: IN CORSO DI RIUNIONE
- RA: ATTESA ESITO UDIENZA ex Art. 274 DAVANTI AL PRESIDENTE DEL TRIBUNALE
- DF: PROCEDIMENTO DEFINITO
- CU: RIMESSO AL PRESIDENTE DELLA CORTE/DEL TRIBUNALE

While some of these States seem completely unrelated to each other, there are some (R1, R3 and RA) that are specifically connected to Art. 274: according the Italian regulation, this specific article has been repealed in 2005, with effect starting from 2006. So, this would justify the absence of the aforementioned states starting from 2006. However, support from domain experts is needed in order to accept or reject this hypothesis.

Fig. 6. State variation over the years

Despite this, the analysis has pointed out to States that need to be further analysed in the future, thus demonstrating again the need of such an approach and its efficacy.

A better understanding of what is actually happening to the underlying model can be obtained through differential graphs, such as the ones reported in Fig. 7. The applied rationale is the following:

- Cancelled States/Event: States and Events that were present in Year X, but not from Year X+1 onward, portrayed in Red;
- Added States/Events: States and Events that were not present until Year X, but that appear in Year X+1, portrayed in Green;
- Stable States/Events: States and Events that were in Years [X−1; X+1], portrayed in Gray;

These graphs offer a comprehensive overview of the FSM's evolution over time, considering cumulative changes over the years, starting from 2005.

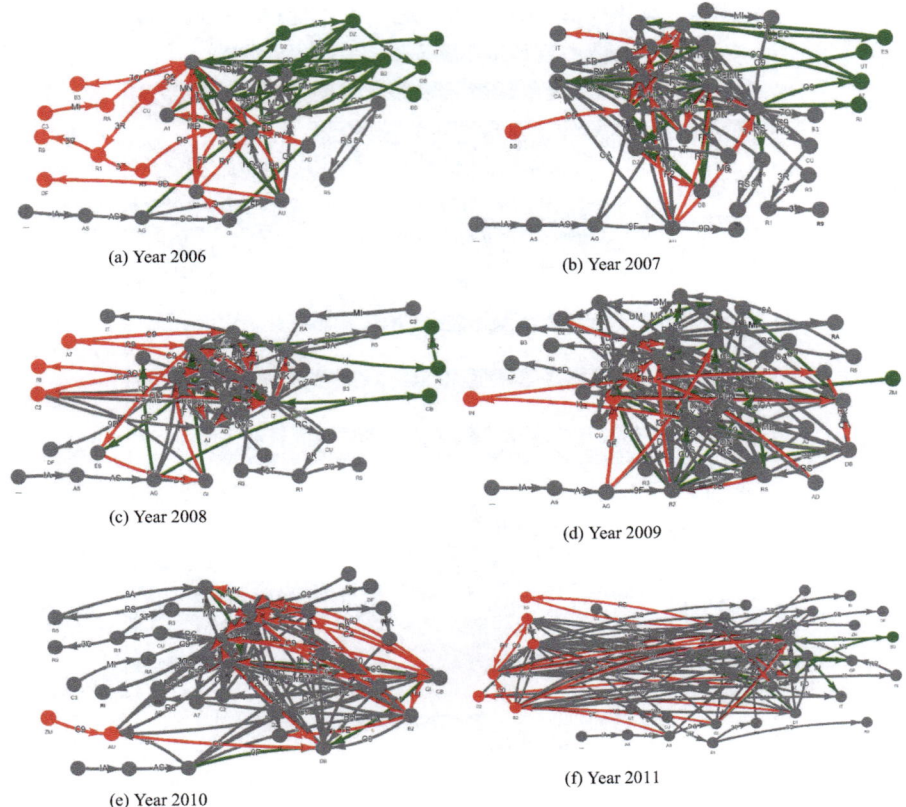

(a) Year 2006

(b) Year 2007

(c) Year 2008

(d) Year 2009

(e) Year 2010

(f) Year 2011

Fig. 7. Comparison of FSAs: cumulative changes from 2005 to 2011

6   Conclusion and Future Works

This paper emphasizes the significance of having an uncomplicated model to comprehend the actions of Civil Trials and its alterations over the years. To achieve this, a Finite
State Machine approach was employed to portray the workflow of the Trials. Actual historical log data, obtained from the Court of Livorno, were used to create the FSA model,
which was tailored to distinct years. Comparisons of the output data revealed variations
in the States and Events, providing insight into the changes over time. By analysing
such results, it is not only evident that changes happen from year to year, but it is also
possible to determine such changes with a fine granularity, and to provide material for
more informed analysis of the model, which can be carried out with the assistance of
domain experts, that can help in recognising the norms and regulation that have brought
to the Machine alteration.

In Future Works, the objective will be to focus on the identification of variable States
and Events, and on their connection to such legislative changes, in order to.

References

1. Agostinelli, S., Chiariello, F., Maggi, F.M., Marrella, A., Patrizi, F.: Process mining meets model learning: discovering deterministic finite state automata from event logs for business process analysis. Inf. Syst. **114**, 102180 (2023)
2. Di Martino, B., et al.: A big data pipeline and machine learning for uniform semantic representation of data and documents from it systems of the Italian ministry of justice. Int. J. Grid High Perform. Comput. (IJGHPC) **14**(1), 1–31 (2022)
3. Di Martino, B., Colucci Cante, L., Esposito, A., Lupi, P., Orlando, M.: Supporting the optimization of temporal key performance indicators of Italian courts of justice with OLAP techniques. In: Barolli, L., Yim, K., Enokido, T. (eds.) CISIS 2021. LNNS, vol. 278, pp. 646–656. Springer, Cham (2021). https://doi.org/10.1007/978-3-030-79725-6_65
4. Di Martino, B., Esposito, A., Cante, L.C.: Multi agents simulation of justice trials to support control management and reduction of civil trials duration. J. Ambient Intell. Humaniz. Comput. 1–13 (2021)
5. Hopcroft, J.E., Motwani, R., Ullman, J.D.: Introduction to automata theory, languages, and computation. ACM SIGACT News **32**(1), 60–65 (2001)
6. Di Martino, B., Cante, L.C., Esposito, A., Lupi, P., Orlando, M.: Temporal outlier analysis of online civil trial cases based on graph and process mining techniques. Int. J. Big Data Intell. **8**(1), 31–46 (2021)
7. Sipser, M.: Introduction to the theory of computation. ACM SIGACT News **27**(1), 27–29 (1996)

Reinforcement Learning-Based Root Planner for Electric Vehicle

Pietro Fusco[✉], Dario Branco, and Salvatore Venticinque

University of Campania "Luigi Vanvitelli", Aversa, Italy
{pietro.fusco,dario.branco,salvatore.venticinque}@unicampania.it

Abstract. Electric Vehicles (EVs) have been identified as the current innovation for sustainable mobility that reduces carbon emissions and pollution. The transition to electric mobility is accelerating, and this means that services and infrastructures must be ready to support the impact of such a change. Smart applications can leverage this transition contributing to a seamless integration of the expected increase of new energy loads into the electric grid assisting users' behaviour. At the same time, they must comply with the strict regulations in terms of privacy and deal with limited users' acceptance. In this context, we propose a Policy-Based Reinforcement Learning agent-based route planner that is able to suggest a route to a driver who starts from a location A traveling to a location B minimizing the number of re-charge sessions along the journey.

1 Introduction

Generally, an electric vehicle root planner is a crucial component in the adoption of electric vehicles (EVs) as a sustainable mode of transportation. The planner is responsible for selecting the best route for an EV driver, considering various factors such as battery range, charging infrastructure, traffic conditions, and driving patterns. The main challenge in developing an EV route planner is to ensure that the vehicle has sufficient charge to complete the journey or to arrive at the next charge station while minimizing the re-charge sessions and avoiding range anxiety for the driver. To achieve this goal, the planner needs to accurately predict the energy consumption of the EV for the given route and to identify the charging stations along the way, and the re-charge plan, to reach the destination. Electric vehicle route planning can be formulated as a combinatorial optimization problem [4], where the goal is to find the optimal route for a given set of locations and charging stations. Combinatorial problems are challenging optimization problems that involve finding the best combination or arrangement of a set of discrete elements. Reinforcement learning is a powerful technique for solving such problems [2], as it allows an agent to learn from its environment by trial and error. In this context, a Reinforcement Learning methodology is proposed in order to guide route planners for selecting the best number of re-charge

L. Barolli (Ed.): CISIS 2023, LNDECT 176, pp. 348–357, 2023.
https://doi.org/10.1007/978-3-031-35734-3_35

sessions along the planned itinerary. In particular, a Policy-Based algorithm is used as a Reinforcement Learning algorithm for equipping the agent with a brain.

2   Related Work

In recent years, researchers have explored the application of Reinforcement Learning (RL) to solve various combinatorial problems, such as the traveling salesman problem, the graph coloring problem, and the bin packing problem [1,8,11,16]. One possible RL approach is to use a *Value-Based* algorithm as Deep Q-network (DQN) [3,5,6,10] to *indirectly* learn an optimal policy that maps the current state, where the agent is located at time t, to the next action through the calculation of a value state function. The DQN is trained using experience replay, where the agent stores its experiences in a memory buffer and samples them randomly to update its policy. Another approach is to use a *Policy-Based* algorithm [7,12,15], such as the REINFORCE algorithm, to *directly* learn a stochastic policy that outputs the action to put in place. The policy is optimized using a variant of the Monte Carlo method, where the returns are estimated by simulating the entire episode from the current state. In both approaches, the policy that the agent learns can achieve near-optimal solutions to the problem and can generalize to unseen problem instances. Here the EV routing problem is modeled as a combinatorial problem and the usage of RL is investigated, focusing on the efficiency of the learning algorithm and scalability.

Approaches to develop an EV route planner use RL techniques for nightly offline rebalancing operations in free-floating electric vehicle sharing systems [1]. For example, researchers have used deep learning models for making predictions of both the expected travel time as well as energy use in long-distance route planning [9]. In the following approach, the policy of the proposed RL algorithm is *directly* approximated with a neural network called *Policy Network*. Another network called *Value Network* acts as a *baseline* and corrects the variance issue in the gradient updates improving the original algorithm [13].

3   Problem Formulation

In the electric vehicle routing problem (EVRP), with intermediary charging stations, a driver asks the route planner for an optimal itinerary to reach her destination, minimizing recharge sessions along the journey. A simple scenario is shown in Fig. 1, where icons represent different actors. The EV is identified through 5 parameters, namely $ev = \{\lambda, \phi, SoC, C, \eta\}$. The position is defined by λ and ϕ parameters, which are latitude and longitude respectively. Status of charge (SoC), capacity (C), and energy efficiency ($\eta, [Wh/Km]$) are necessary to estimate how long it is still possible to drive and the amount of energy it can be charged, once the user arrived at target station. On the other side, charging stations $S = \{cs_1, ..., cs_m\}$, are characterized by their position: $s_j = \{\lambda_j, \phi_j\}$.

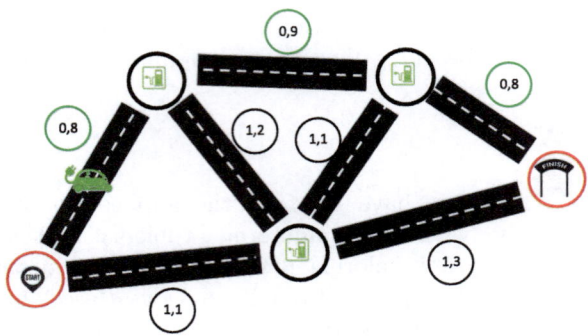

Fig. 1. Conceptual model

In particular, it is mandatory to take into account the constraint on the distance which can be covered by each car.

$$\eta * d(s_t, s_{t+1}) \leq SoC * C \tag{1}$$

The maximum distance that can be covered by a car is formulated in Eq. (1).

$$d_{Max} = \frac{SoC * C}{\eta} \tag{2}$$

The RL-based system tries to find out the best path from the start point to the end point minimizing the re-charge sessions. To evaluate how good the current action is, the system needs to calculate a reward value for an action a moving the EV from a state (position) s_t to the next state (position) s_{t+1}. Figure 2 shows the reward function used by the RL system to find the best path, which consists of a negative second-order polynomial function. The red vertical line represents d_{Max}.

The distance between two subsequent states (positions) s_t and s_{t+1} needs to be as equal to the d_{Max} as much as possible, in order to maximize the usage of the current level of charge. As an example, if $SoC = 0.4\%$, $\eta = 16.0 \left[\text{Wh/km}\right]$ and $C = 40 \left[\text{Wh}\right]$, d_{Max} is equal to 1. By keeping the latter in mind, the more the distance between two subsequent states (positions) is equal to 1, the more the reward value for the action a that has determined the change from state s_t to state s_{t+1} is high. As another example, in Fig. 1 the best path consists of all stretches of the road marked with a green circle.

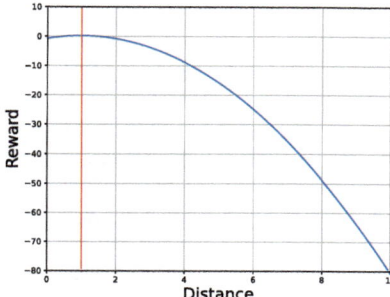

Fig. 2. Reward function

4 Reinforcement Learning Based Solution

Here we investigate a solution to the formulated problem by exploiting a Reinforcement Learning method. RL is about learning the optimal behavior (policy) in an environment to obtain the maximum reward. This optimal behavior is learned through interactions with the environment and observations of how it responds, similar to children exploring the world around them and learning the actions that help them achieve a goal. The goal of RL is to learn the optimal policy, which is the policy that provides actions to take in a particular state of the environment and that permits the maximum return to the agent.

4.1 Policy Gradient Approach

We applied a stochastic Policy-Based RL method. Policy Gradient (PG) is one of the most popular approaches in RL. It is a Policy-Based method by which the user can find the optimal policy without computing the $Q(state, action)$ function which states what the value of a state s and an action a under the policy π is. It uses a neural network as a function approximator of the optimal policy π parameterized by $\boldsymbol{\theta}$, where $\boldsymbol{\theta}$ is the parameters vector of the neural network (NN). The PG method uses a stochastic policy, then it is possible to select an action based on the probability distribution over the action space, as shown in Fig. 3, where is depicted a state which is fed to the neural network, then the latter elaborates the input state and returns the probability distribution over all actions in the action space.

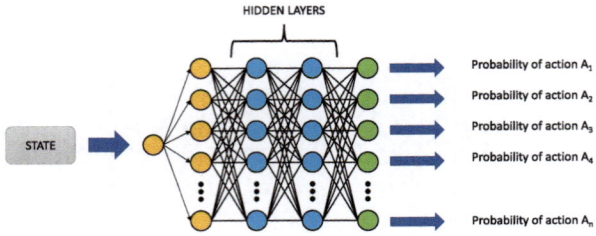

Fig. 3. A policy network

The objective of our NN is to assign high probabilities to actions that maximize the expected return of the trajectory. It is possible to write objective function **J** as follows:

$$\mathbf{J}(\boldsymbol{\theta}) = E_{\tau \sim \pi_{\boldsymbol{\theta}}(\tau)}\left[R(\tau)\right] \tag{3}$$

where:

- τ is the trajectory of the agent;
- $\tau \sim \pi_\theta(\tau)$ means that the trajectory followed by the agent is drawn using the policy π_θ;
- $R(\tau)$ is the return of the trajectory τ.

It should be noticed that maximizing objective function **J** is equivalent to maximizing the return of the trajectory τ. In order to obtain a practical form for the objective function **J**, it is necessary to start by observing that the probability distribution of a generic trajectory τ can be written as follows:

$$\pi_{\boldsymbol{\theta}}(\tau) = p(s_0) \prod_{k=0}^{K-1} \pi_{\boldsymbol{\theta}}\left(a_k \mid s_k\right) p\left(s_{k+1} \mid s_k, a_k\right) \tag{4}$$

where $p(s_0)$ is the initial state, $\pi_{\boldsymbol{\theta}}\left(a_k \mid s_k\right)$ represents the probability to take the action a_k at state s_k and $p\left(s_{k+1} \mid s_k, a_k\right)$ is the probability the agent ends up in the state s_{k+a} from state s_k taking action a_k. Applying the *log* function to both sides of the Eq. (4), it is possible to rewrite the same equation as follows:

$$\log\left[\pi_{\boldsymbol{\theta}}(\tau)\right] = \log\left[p(s_0) \prod_{k=0}^{K-1} \pi_{\boldsymbol{\theta}}\left(a_k \mid s_k\right) p\left(s_{k+1} \mid s_k, a_k\right)\right] \tag{5}$$

Applying *log* function rules, Eq. (5) becomes:

$$\log\left[\pi_{\boldsymbol{\theta}}(\tau)\right] = \log\left[p(s_0)\right] + \sum_{k=0}^{K-1} \log\left[\pi_{\boldsymbol{\theta}}\left(a_k \mid s_k\right)\right] + \log\left[p\left(s_{k+1} \mid s_k, a_k\right)\right] \tag{6}$$

Applying the gradient operator to the Eq. (6) you obtain:

$$\nabla_{\boldsymbol{\theta}}\Big\{\log\left[\pi_{\boldsymbol{\theta}}(\tau)\right]\Big\} = \nabla_{\boldsymbol{\theta}}\Big\{\log\left[p\left(s_0\right)\right] + \sum_{k=0}^{K-1}\log\left[\pi_{\boldsymbol{\theta}}\left(a_k\mid s_k\right)\right] + \log\left[p\left(s_{k+1}\mid s_k, a_k\right)\right]\Big\}$$
(7)

Since the gradient operator is performed with respect to the parameters vector $\boldsymbol{\theta}$, as a result, it is obtained the following:

$$\nabla_{\boldsymbol{\theta}}\Big\{\log\left[\pi_{\boldsymbol{\theta}}(\tau)\right]\Big\} = \sum_{k=0}^{K-1}\nabla_{\boldsymbol{\theta}}\Big\{\log\left[\pi_{\boldsymbol{\theta}}\left(a_k\mid s_k\right)\right]\Big\}$$
(8)

By keeping Eq. (8) in mind and taking into account that the expected value of a random variable in Eq. (3) can be explicited as follows:

$$\mathbf{J}(\boldsymbol{\theta}) = E_{\tau \sim \pi_{\boldsymbol{\theta}}(\tau)}\left[R(\tau)\right] = \int \pi_{\boldsymbol{\theta}}(\tau)R(\tau)d\tau$$
(9)

Applying the gradient operator to the Eq. (9), it is possible to calculate the derivative of an objective function $\mathbf{J}$ with respect to the parameters vector $\boldsymbol{\theta}$:

$$\nabla_{\boldsymbol{\theta}}\Big\{\mathbf{J}(\boldsymbol{\theta})\Big\} = \nabla_{\boldsymbol{\theta}}\Big\{\int \pi_{\boldsymbol{\theta}}(\tau)R(\tau)d\tau\Big\} = \int \nabla_{\boldsymbol{\theta}}\Big\{\pi_{\boldsymbol{\theta}}(\tau)R(\tau)d\tau\Big\}$$
(10)

From Eq. (10), considering that $R(\tau)$ is not a function of the parameters vector $\boldsymbol{\theta}$ and that, by a simple application of the chain rule, it is possible to get:

$$\nabla_{\boldsymbol{\theta}}\Big\{\pi_{\boldsymbol{\theta}}(\tau)\Big\} = \nabla_{\boldsymbol{\theta}}\Big\{\pi_{\boldsymbol{\theta}}(\tau)\Big\} \cdot \left[\frac{\pi_{\boldsymbol{\theta}}(\tau)}{\pi_{\boldsymbol{\theta}}(\tau)}\right]$$

$$= \left[\frac{\nabla_{\boldsymbol{\theta}}\Big\{\pi_{\boldsymbol{\theta}}(\tau)\Big\}}{\pi_{\boldsymbol{\theta}}(\tau)}\right] \cdot \pi_{\boldsymbol{\theta}}(\tau)$$
(11)

$$= \nabla_{\boldsymbol{\theta}}\Big\{\log\left[\pi_{\boldsymbol{\theta}}(\tau)\right]\Big\} \cdot \pi_{\boldsymbol{\theta}}(\tau)$$

Equation (10) can be rewritten as follows:

$$\nabla_{\boldsymbol{\theta}}\Big\{\mathbf{J}(\boldsymbol{\theta})\Big\} = \int \pi_{\boldsymbol{\theta}}(\tau)\nabla_{\boldsymbol{\theta}}\Big\{\log\left[\pi_{\boldsymbol{\theta}}(\tau)\right]\Big\}R(\tau)d\tau$$

$$= E_{\tau \sim \pi_{\boldsymbol{\theta}}(\tau)}\left[\nabla_{\boldsymbol{\theta}}\Big\{\log\left[\pi_{\boldsymbol{\theta}}(\tau)\right]\Big\}R(\tau)\right]$$
(12)

Finally, after computing the gradient, it is possible to update the NN parameters vector $\boldsymbol{\theta}$ as follows:

$$\boldsymbol{\theta}^{k+1} = \boldsymbol{\theta}^k + \eta\nabla_{\boldsymbol{\theta}}\Big\{\mathbf{J}(\boldsymbol{\theta}^k)\Big\}$$
(13)

where η represents the *Learning Rate*. In order to use the Eq. (13) in a practical code implementation, it is possible to get rid of the expectation value in Eq. (12)

by using a Monte Carlo approximation method and change the expectation to the sum over N trajectories as follows:

$$\nabla_{\boldsymbol{\theta}}\Big\{\mathbf{J}(\boldsymbol{\theta})\Big\} \approx \frac{1}{N}\sum_{i=1}^{N}\sum_{k=0}^{K-1} \nabla_{\boldsymbol{\theta}}\Big\{\log\big[\pi_{\boldsymbol{\theta}}(a_k \mid s_k)\big]\Big\}R(\tau_i) \tag{14}$$

4.2   Variance Reduction Method

In the Sect. 4.1, a PG method REINFORCE method was introduced. One major issue with the PG REINFORCE method is that the gradient, $\nabla_{\boldsymbol{\theta}}\big\{\mathbf{J}(\boldsymbol{\theta})\big\}$, shows high variance in each update. Basically, high variance is due to the major difference in the episodic returns. Generally, the PG algorithm is an *on-policy* method, namely the policy with which episodes are generated is the same policy whose vector parameters $\boldsymbol{\theta}$ are updated in every iteration using Eq. (13). Since the policy is getting improved on every iteration, the return can vary greatly in each episode [14] and it introduces a high variance in the gradient updates. When the gradients have high variance, then it will take a lot of time to attain convergence. To reduce variance, a new function called *baseline* function is introduced. Subtracting the baseline b from the return (reward-to-go) R_t reduces the variance, so the gradient can be rewritten as follows:

$$\nabla_{\boldsymbol{\theta}}\Big\{\mathbf{J}(\boldsymbol{\theta})\Big\} \approx \frac{1}{N}\sum_{i=1}^{N}\sum_{k=0}^{K-1} \nabla_{\boldsymbol{\theta}}\Big\{\log\big[\pi_{\boldsymbol{\theta}}(a_k \mid s_k)\big]\Big\}\Big(R_t - b\Big) \tag{15}$$

where R_t is the return *reward-to-go* $R_t = \sum_{t=k}^{T-1} r_t$, which is the return of the trajectory starting from the state s_t. In this paper one of the most popular functions of the baseline will be used, that is the Value function which represents the value of a state, in other words, it is the expected return an agent would obtain starting from that state following the policy π.

5   Experimental Activities

We compare results obtained in three different scenarios of growing complexity. The first root planner problem layout is depicted in Fig. 4(a). The START and END blue points represent the departure and destination of the EV. Red points represent the available re-charge stations. The root planner's goal is to identify a path, for an EV, from the START point towards the END point minimizing re-charge sessions. Green points represent recharge stations actually used along the itinerary.

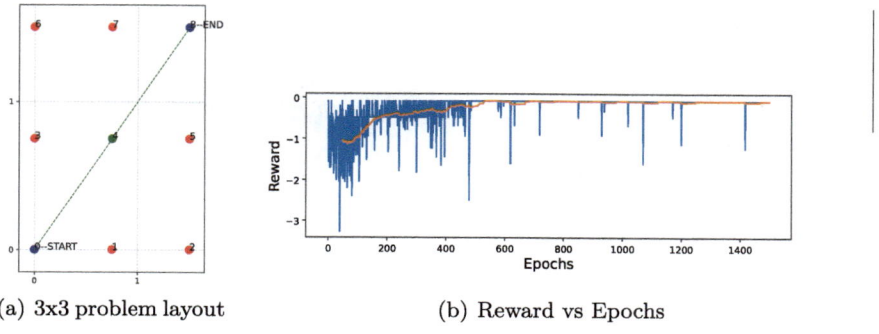

(a) 3x3 problem layout   (b) Reward vs Epochs

Fig. 4. 3×3 EV root planner problem

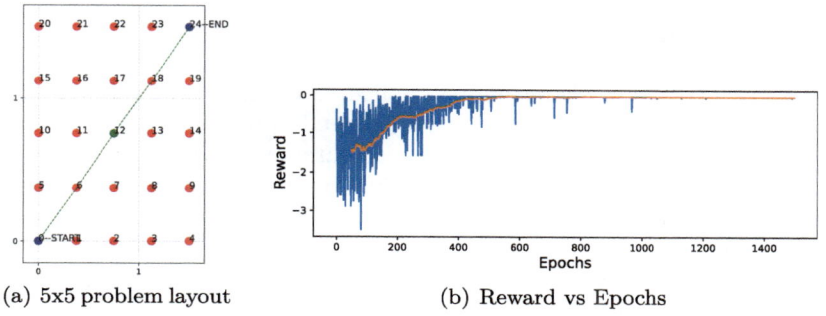

(a) 5x5 problem layout   (b) Reward vs Epochs

Fig. 5. 5×5 EV root planner problem

Figure 4(b) shows the value of reward against epochs over the training stage. It is possible to notice that the algorithm converges after about 700 epochs ranging over a state space of dimension 127. The REINFORCE algorithm is able to identify the best path among those it has computed. The reward value is quite constant without any swings. Analogously, Fig. 5(a) shows the same problem with a different configuration. In this case, the number of available recharge stations is bigger. Also in this case the REINFORCE algorithm is capable to identify the best path. The reward reaches the best value after 600 epochs maintaining a quite constant trend ranging over a state space of dimension 8388607. Finally, in the last configuration shown in Fig. 6(b) it is possible to notice a less clear convergence of the reward value due to the state space dimension which is of 172325161239.

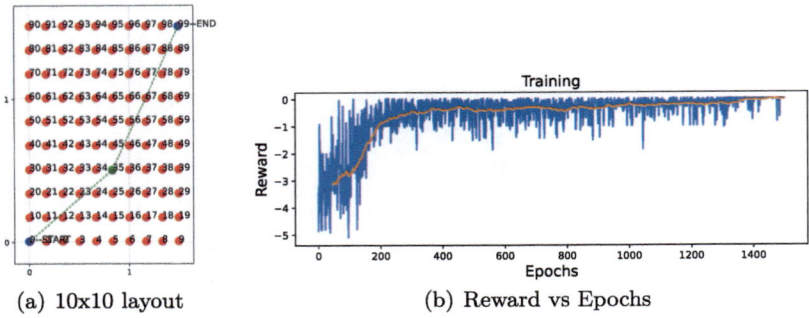

(a) 10x10 layout   (b) Reward vs Epochs

Fig. 6. 10×10 EV root planner problem

6   Conclusion

This paper proposed a RL technique applied to the EVPR. Using a stochastic Policy-Based method we demonstrated how it is possible to train the model for finding the most efficient route that is characterized by minimal re-charge sessions and the shortest distance. Even if experimental results demonstrated the feasibility of the proposed approach, other algorithms must be compared using real data and the same problem must be solved in a multi-user scenario, where multiple EVs must collaborate, among them and with charge stations, to compute the best itinerary and charge plan.

References

1. Bogyrbayeva, A., Yoon, T., Ko, H., Lim, S., Yun, H., Kwon, C.: A deep reinforcement learning approach for solving the traveling salesman problem with drone. Transp. Res. Part C Emerging Technol. **148**, 103981 (2023)
2. Dorokhova, M., Ballif, C., Wyrsch, N.: Routing of electric vehicles with intermediary charging stations: a reinforcement learning approach. Front. Big Data **4**, 586481 (2021)
3. Guo, H.: Generating text with deep reinforcement learning (2015)
4. James, J., Yu, W., Gu, J.: Online vehicle routing with neural combinatorial optimization and deep reinforcement learning. IEEE Trans. Intell. Transp. Syst. **20**(10), 3806–3817 (2019)
5. Lai, M.: Giraffe: using deep reinforcement learning to play chess (2015)
6. Mnih, V., et al.: Playing Atari with deep reinforcement learning (2013)
7. Nachum, O., Norouzi, M., Xu, K., Schuurmans, D.: Bridging the gap between value and policy based reinforcement learning (2017)
8. Ottoni, A.L.C., Nepomuceno, E.G., de Oliveira, M.S., de Oliveira, D.C.R.: Reinforcement learning for the traveling salesman problem with refueling. Complex Intell. Syst. **8**, 2001–2015 (2022)
9. Petkevicius, L., Saltenis, S., Civilis, A., Torp, K.: Probabilistic deep learning for electric-vehicle energy-use prediction. In: 17th International Symposium on Spatial and Temporal Databases, SSTD 2021, New York, NY, USA, pp. 85–95. Association for Computing Machinery (2021)

10. Schmidhuber, J.: Deep learning in neural networks: an overview. Neural Netw. **61**, 85–117 (2015). https://doi.org/10.1016/j.neunet.2014.09.003
11. Verma, R., et al.: A generalized reinforcement learning algorithm for online 3d bin-packing (2020)
12. Weng, J., Jiang, X., Zheng, W.L., Yuan, J.: Early action recognition with category exclusion using policy-based reinforcement learning. IEEE Trans. Circuits Syst. Video Technol. **30**(12), 4626–4638 (2020)
13. Williams, R.J.: Toward a theory of reinforcement-learning connectionist systems. Technical report, NU-CCS-88-3, Northeastern University, College of Computer Science (1988)
14. Williams, R.J.: Simple statistical gradient-following algorithms for connectionist reinforcement learning. Mach. Learn. **8**(3–4), 229–256 (1992)
15. Yu, M., Sun, S.: Policy-based reinforcement learning for time series anomaly detection. Eng. Appl. Artif. Intell. **95**, 103919 (2020)
16. Zhang, J., Zi, B., Ge, X.: Attend2Pack: bin packing through deep reinforcement learning with attention (2021)

A Study of Visualization System for Learning QoS Control

Kazuaki Yoshihara[1](✉), Katsuhisa Fujii[2], and Nobukazu Iguchi[1,3]

[1] Department of Informatics, Faculty of Informatics, Kindai University,
Higashiosaka 577-8502, Osaka, Japan
{yoshiharak,iguchi}@info.kindai.ac.jp
[2] Graduate School of Science and Engineering Research, Kindai University,
Higashiosaka 577-8502, Osaka, Japan
2133340417t@kindai.ac.jp
[3] Cyber Informatics Research Institute, Kindai University, Higashiosaka 577-8502, Osaka,
Japan

Abstract. As network traffic increases, congestion and delays that can degrade network service quality and cause network downtime to become a problem. QoS control is one way to solve this problem. Network engineers need to learn about the mechanism of QoS control and use it effectively. However, QoS control is difficult to understand intuitively because it operates inside network devices. Therefore, visualization of QoS control behavior is considered effective for verifying its operation and for efficient learning. In this paper, we implemented a system that visualizes the QoS control process inside a router using animation for learning.

1 Introduction

Network traffic continues to grow as networks continue to evolve and web services diversify [1]. And with projections that the total number of Internet users will reach 5.3 billion by 2023, or 66% of the world's population [2], the trend of increasing network traffic is likely to accelerate in the future. The issue that arises from increased network traffic is network congestion. Network congestion can cause serious failures such as poor quality of service and network downtime. The solution to this problem is Quality of Service (QoS) control, which includes bandwidth and priority control. QoS control enables the delivery of communication quality on demand and addresses issues such as poor quality of internet services. Therefore, network engineers need to understand how QoS control works and learn how to use it effectively.

Book learning and hands-on network construction are two methods for learning QoS control mechanisms. Information presented through diagrams and text in books can teach the individual processes involved in QoS control, but not their combined behavior. Hands-on network construction practice allows for the acquisition of practical knowledge as you can freely configure the equipment and learn how it works. However, it can be challenging to intuitively understand the operation of QoS control, as network devices operate and cannot be visually inspected.

L. Barolli (Ed.): CISIS 2023, LNDECT 176, pp. 358–366, 2023.
https://doi.org/10.1007/978-3-031-35734-3_36

It is necessary to perform network verification and reflect the results to make effective use of QoS control and achieve high quality communication networks. One way to verify the operation of QoS control is to analyze the statistical information generated by network devices. However, the analysis of statistical information is burdensome for network administrators, who must collect and analyze various types of information and verify that the network is working properly. To easily understand the QoS control process in each network device, it is useful to visualize the information.

Therefore, we developed a visualization system for learning QoS control. In this report, we describe the details of the developed system and the learning content using the system.

2   Support System for IP Network Construction Practice

We have been developing a support system for IP network construction [3, 4]. In the previous version of the system, users could practice building a virtual network by connecting virtual network devices and host machines on a single PC using User-Mode-Linux (UML) [5]. With the system, users could practice placing and wiring virtual devices and configuring them with a GUI on a web page. The server managed clients in groups to create a virtual network, allowing multiple users to simulate network construction at the same time.

Since the previous system lacked the ability to set QoS controls for virtual devices and draw animations of virtual device processes, our newly developed system has redefined QoS control settings for virtual routers and implemented a function for animating internal processes. This system is designed as a learning tool, where a user can learn by comparing their own QoS controls with the animated behavior. As such, this system is intended for individual use.

3   Related Work

Arai et al. have developed a tool for monitoring data and visually checking data structures and communication procedures for each application protocol, as well as a simulation tool for learning TCP control schemes [6]. Related our research, Tateiwa et al. have developed a system that visualizes the behavior of a network freely constructed by A learner using virtual devices in UML [7]. Although these studies can visualize the processing of network devices, they cannot visualize the operation of QoS control.

4   Development of the System

4.1   Outline of the System

Figure 1 shows a diagram of this system. This system is a client-server system. We implemented the client side in Vue.js and Typescript, and the server side in Java. The server has four functions: supporting network construction, traffic-generating, collecting traffic information, and generating animation. The system can also configure virtual routers to QoS control and create animations based on packet information in the virtual network. The QoS controls whose operation can be visualized in this system include packet marking, queuing, scheduling, and shaping.

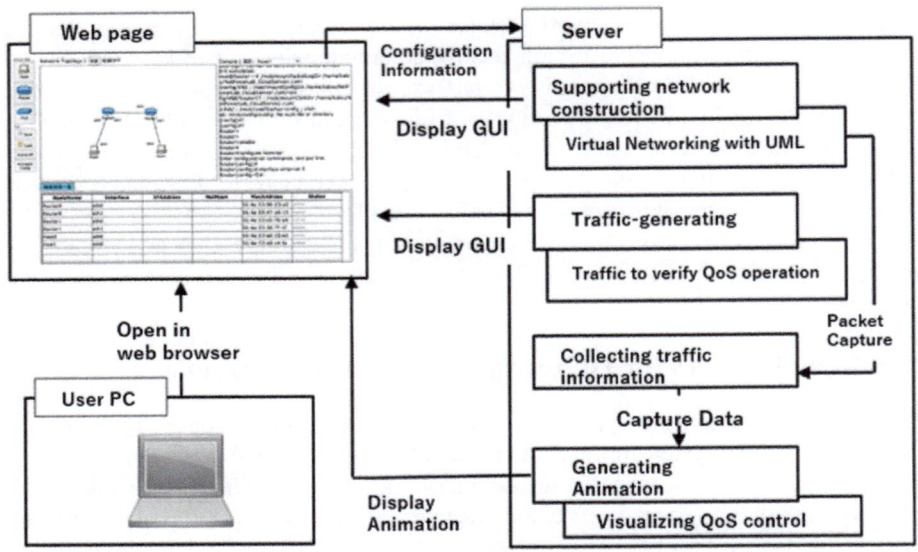

Fig. 1. System configuration diagram

4.2 Supporting Network Construction

This function provides a GUI for the client to operate on a web page and constructs a virtual network based on the client's operations. Figure 2 shows the GUI.

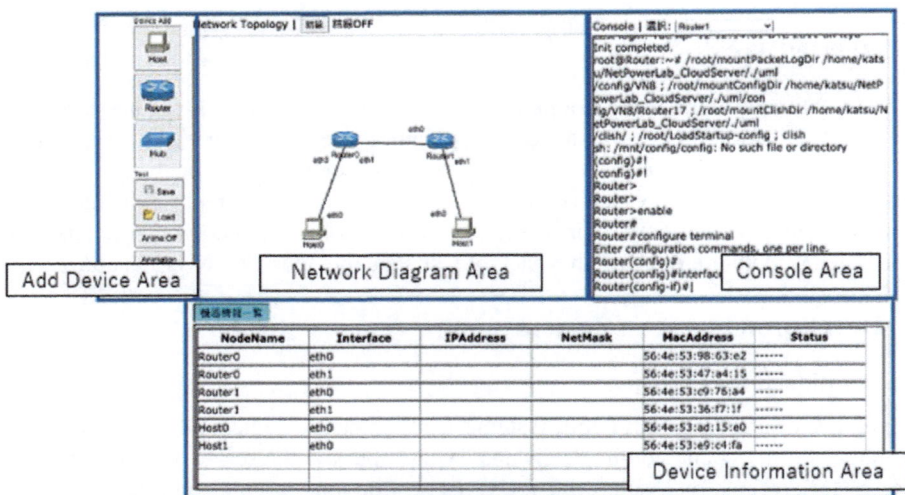

Fig. 2. Network construction support GUI screen

The GUI consists of "Network Diagram Area", "Add Device Area", "Device Information Area", and "Console Area". The network diagram display area shows the network topology of the constructed virtual network. The network diagram display area shows the network topology of the constructed virtual network. Add new devices to the network topology and enable virtual devices by dragging and dropping icons of hosts and routers from the area to the network diagram display area. The console area shows the terminal for each virtual device and allows configuration to the device using a CLI. The CLI implementation used CLISH [8], a framework for realizing a CLI like that of Cisco devices. Commands related to QoS control were defined using CLISH. QoS control can be activated by setting the defined commands in the router.

4.3   Traffic-Generating

This function provides a GUI to generate traffic on the network construction support GUI and generates traffic to the virtual network based on the information entered in the new GUI. Figure 3 shows the GUI for generating traffic.

Traffic is generated using hping3. This is an open-source packet generator. The fields the user enters in the GUI are the number of packets, the interval at which packets are sent, the protocol type, the packet size, and the source and destination IP addresses and port numbers. This function allows the system to generate a various packet and collect data to verify QoS control.

4.4   Collecting Traffic Information

This function captures and records traffic data in the virtual network. To draw packets flow as an animation in this system, information about input/output packets at each virtual device is required. Therefore, this system captures input and output packets from all network interfaces of each virtual device. Therefore, this system captures input and output packets from all network interfaces of each virtual device. Run the packet capture program on each virtual device to capture all network interfaces simultaneously. The packets captured by this function have five elements: the name of the device being captured, the name of the network interface, the direction of the packet (input or output), the capture time, and the packet data.

4.5   Generating Animation

This feature creates animations of packet flow and QoS control based on recorded traffic data. The animation is drawn in two sections, one between virtual devices and the other at the router. Animations are rendered using Pixi.js, a JavaScript 2D rendering library. The packet is drawn as a rectangular object with the protocol name.

Animation between virtual devices first generates an object from two packages with different input/output directions and all data matching. Based on the device names recorded in the two packets, the system obtains the coordinates of the source and destination virtual devices on the GUI. The system sets them as the starting and ending points of the animation. Then, the flow of packets is displayed as an animation by moving the

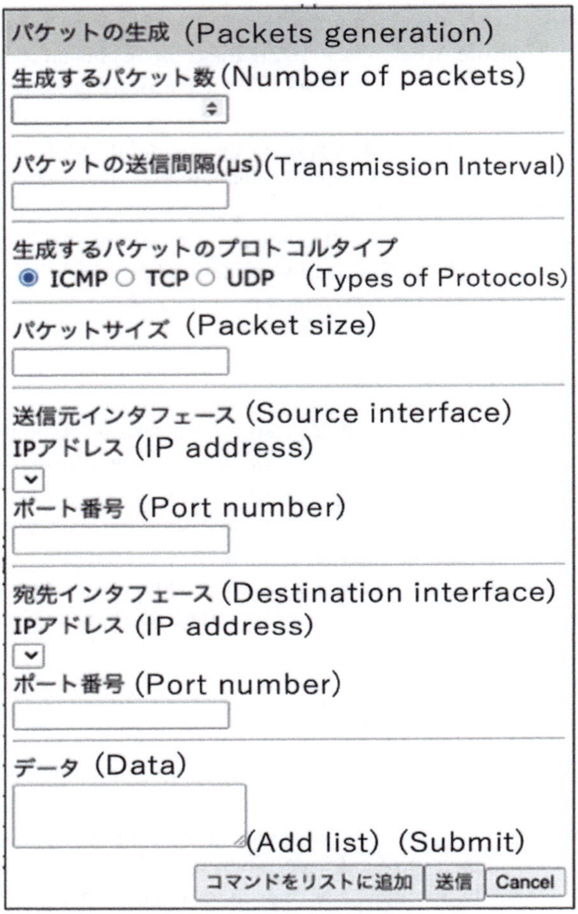

Fig. 3. Traffic generation GUI in Japanese

generated object from the start point to the end point. In the animation drawing of a router, the system obtains configuration information about the QoS of the virtual device and draws it based on that configuration as follows.

- Marking

Marking provides the capability to partition network traffic into multiple priority levels or classes of service. In this feature, the system changes the color of the object when the ToS (Type of Service) field value of the IP header changes, to visually verify the assigned priority. Figure 4 shows how the marking Animation of packets.

- Queuing and Scheduling

Queues are buffers in network devices that are used to store packets based on traffic classes when the interface is busy, or the bandwidth is full. The traffic can then be

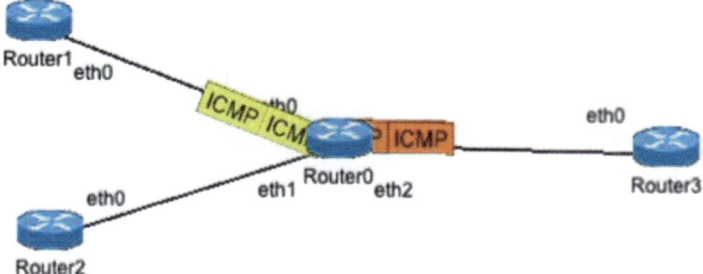

Fig. 4. Drawing marking

processed later when the interface or bandwidth becomes available again. In this function, when a packet object arrives at the router, the system determines the queue to store it by referring to the packet filter rules for each class in the animation drawing between virtual devices. Then, the system displays an image of the queue corresponding to each class at the top of the router, representing the packet objects being stored.

Scheduling is the process of assigning a packet to an internal forwarding queue based on its QoS information and servicing the queues according to a queuing scheme. This function expresses animation by changing the order in which objects are retrieved from the queue. First, the first object stored in each queue is retrieved and the transmission time information is compared. The object with the earliest transmission time is deleted and the drawing of the packet transmission begins. At the same time, all other objects in the queue are repositioned. Figure 5 shows how the queuing and Scheduling Animation.

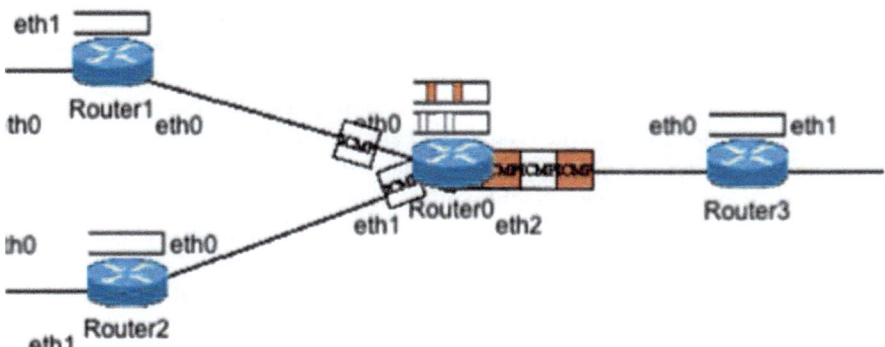

Fig. 5. Queuing and scheduling drawing

- Shaping

Traffic shaping adjusts the rate of outbound traffic to reduce traffic bursts so that outbound packets can be transmitted at a steady rate. In this function, the system expresses the suppression of the transmission rate by changing the height of the object to a lower level. It obtains the time information of input and output packets from the packet object and calculates the delay value by taking the difference between them. The system changes the height of the object according to the value obtained by dividing the packet size by the calculated delay value. The height of the packet object is predefined as a maximum value and is determined as a percentage of the maximum value. Figure 6 shows how the shaping Animation.

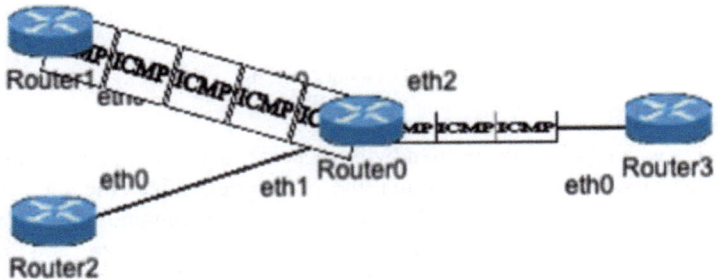

Fig. 6. Shaping drawing

5   Discussion

This chapter discusses the content for learning using the QoS control visualization system. Unlike the previous system, which could be learned by multiple people, the QoS Control Visualization System is designed for individual learners. To use the system, it must first be activated on the server side. Once activated, a learner can access the server and display the GUI for the system using a web browser, as shown in Fig. 2.

A learner constructs a virtual network by dragging and dropping virtual devices from the "Add Device Area" to the "Network Diagram Area" and connecting them. The learner can then configure routers on the virtual network by selecting the target router and entering and executing commands in the console area. The virtual router's IP addresses, routing settings, and QoS control settings can all be configured by executing the defined commands for QoS control.

A learner enters information about the packets into the traffic generation GUI, as shown in Fig. 3, and sends the input information to the server to generate traffic within the virtual network.

Finally, A learner can intuitively understand the virtual network they have built and the QoS controls based on their configurations by viewing the QoS control animation. In the system, we describe each of the QoS controls and how learners can learn them.

- Marking

A learner can learn the effects of the marking they themselves configured on the virtual router by watching the marking animations. Since the color of the object varies depending on the priority in that system, setting the marking to the virtual router will change the color of the packets that pass through the router as shown in Fig. 4. By seeing this change, A learner can visually confirm that the marking has changed the priority of packets passing through the router.

- Queuing and Scheduling

A learner can see how packets are stored in the queue by observing the animation of the queue object above the router object in "Network Diagram Area". The system generates a queue object for each class defined by a learner. The multiple packet objects are then stored in the queue object, allowing A learner to visually understand queuing. Then, packet objects reach the right side of the queue object in order, and packet objects move between virtual device objects in order as packets arrive as shown in Fig. 5. A learner can learn scheduling visually by watching these animations.

- Shaping

A learner can learn the effects of the shaping he configured on virtual router by observing the shaping animation. A learner can configure the virtual router to suppress the output rate. The system then adjusts the height of the packet object based on the configured output rate, so that when the packet object passes through a virtual router with a suppressed output rate, the height of the output packet object changes as shown in Fig. 6.

6 Conclusion

In this study, we developed a QoS control visualization system within a router, which allows for the visualization of QoS control using animation. We also explained how learners can use the system to learn about QoS control in an intuitive manner. To measure the system's learning effectiveness, we need to conduct evaluation experiments.

Furthermore, we plan to visualize QoS controls such as policing and Random Early Detection (RED), which were not included in this study.

References

1. Ministry of Internal Affairs and Communications: WHITE PAPER Information and Communications in Japan (2021). https://www.soumu.go.jp/johotsusintokei/whitepaper/ja/r03/pdf/index.html. Accessed 13 Mar 2023
2. Cisco Systems: Cisco Annual Internet Report (2018–2023). https://www.cisco.com/c/ja_jp/solutions/collateral/executive-perspectives/annual-internet-report/white-paper-c11-741490.html. Accessed 13 Mar 2023
3. Nobukazu, I.: Development of a system to support computer network construction practice using virtual router. IPSJ J. **52**(3), 1412–1423 (2011)
4. Nobukazu, I.: Development of a self-study and testing function for NetPowerLab, an IP networking practice system. Int. J. Space Based Situated Comput. **4**, 175–183 (2014)
5. User Mode Linux: The User-mode Linux Kernel Home Page. https://user-mode-linux.sourceforge.net/. Accessed 13 Mar 2023
6. Arai, M., Tamura, N., Watanabe, H., Ogiso, C., Takei, S.: Development and evaluation of TCP/IP protocol learning tools. IPSJ J. **44**(12), 3242–3251 (2003)

7. Tateiwa, Y., Yasuda, T., Yokoi, S.: Development of a system to visualize computer network behavior for learning to associate LAN construction skills with TCP/IP theory, based on virtual environment software. IPSJ J. **48**(4), 1684–1694 (2007)
8. CLISH: CLISH (Command Line Interface SHell). https://clish.sourceforge.net/. Accessed 13 Mar 2023

A Study on Changing Consciousness of Post Coronavirus Pandemic in Fashion Society and Use of Digital Technology

Momoko Sakaguchi[1]([⊠]), Eiji Aoki[1], and Koichi Nagamatsu[2]

[1] Institute for Hyper Network Society, Oita City, Japan
`{sakaguchi,blue}@hyper.or.jp`
[2] OEC Co., Ltd., Oita City, Japan
`knagamat@oec.co.jp`

Abstract. In recent years, the demand for EC has increased in the apparel industry due to the coronavirus pandemic. Previous researches have focused on trends in the fashion industry and researched consumer purchasing consciousness in the coronavirus pandemic. In response to the current situation of becoming post coronavirus pandemic, we conducted an ongoing research of consumer purchasing consciousness, and compared and study the trends. This discussion refers to sales trends in the apparel industry during the coronavirus pandemic. And the websites at EC are increasingly using digital services that utilize the latest technologies such as AI, VR, augmented reality, and Metaverse. Previous researches have also shown that the use of VR through smartphone applications is effective. One of the most obvious changes is virtual fashion service by smartphone, but at the same time, the hassle of trying on clothes in store is becoming apparent. Based on the status of those domestic and overseas digital services that utilize fashion tech, we tried to develop a smartphone application. By extension, we will consider whether the increased attention world of Metaverse will have an impact on the fashion industry. Digital Transformation in the apparel industry at reality would increase efficiency in design, production, distribution, sales, and inventory control. Therefore, from the perspective of Environmental, Social and Governance investment in light of the Sustainable Development Goals, we will study the information society of the future fashion tech.

1 Research Background

In our previous research, we focused on trends in the fashion industry and researched consumer purchasing consciousness in the coronavirus pandemic [1]. In 2021, based on the sales situation of the apparel industry before and after the coronavirus pandemic, we hypothesized that "for apparel companies to survive, it is important to combine real store and EC." To verify this, we conducted a consumer consciousness survey about real store and EC. We also considered the effectiveness of VR as an example of the use of new technologies in the post corona era. A survey of consumer consciousness revealed that real store and EC have their own merits, and that they complement each other's demerits. We also found that fashion services using VR are effective for consumers. What can be said from these two research studies is that the importance of digital strategy in the fashion industry is increasing. We hypothesized that the combined use of real store and EC and the further evolution of the use of VR will bring new innovations through new digital technologies in the future.

2 Changing Consumer Consciousness Toward Fashion

Now that the coronavirus is converging, how has consumer consciousness of fashion changed? During the three years of the coronavirus pandemic, the demand for EC has increased, and fashion services on the Internet have increased. Services have been created to solve problems that have been pointed out as problems at EC, such as the problem of trying on clothes, the feeling of size, and the problem of not being able to consult with a stores clerk. In recent years, ecological awareness has spread around the world, and as with the food loss problem, environmentally friendly consumer behavior has come to be demanded for clothes loss. Fast fashion is popular because you can buy trendy products at low prices. However, an increasing number of consumers are buying clothes that can be worn for a long time even if they are a little expensive, rather than clothes that are immediately discarded after purchase. In this way, consumers are entering an era in which they choose clothes in consideration of not only design-rich fashion trends, but also the environment.

2.1 Three Years with Coronavirus Pandemic

The textile industry, including the apparel industry, has seen a downward trend in sales since 2018. Especially in 2020, due to the global spread of the coronavirus, demand for industrial fibers as well as clothing decreased. In 2021, the textile industry's sales volume increased, led by clothing as the economy reopened. In 2022, sports and outdoor clothing will perform well. However, overall costs such as raw material prices will continue to rise, and profitability is expected to decline. Can apparel companies pass this on to sales prices? This may have an impact on capital investment in future digital technology. On the other hand, the Chinese brand "SHEIN", which is called ultra-fast fashion, expanded its sales force with the ultimate tax saving and the lowest cost price. Sales have reached 2.5 trillion yen, and most of the world's Z generation is enclosed. However, fast fashion brands such as "SHEIN" are viewed as a problem of environmental destruction due to the system of mass-producing and selling in a short period of time. It has gained popularity due to its overwhelmingly cheap price, but at reality, many of the items are low quality, do not last long, and are disposable. For example, 185,000 tons of used clothes were shipped to Kenya in 2019, but 55,500–74,000 tons became textile waste, as about 30–40% of the used clothes had no market value. This is just the tip of the iceberg, and it is clear that reducing clothing loss will become a global issue in the fast-changing apparel industry.

2.2 Research Activities in 2022

In 2021, we conducted a survey to find out which method of purchasing clothes, real stores or EC. As a result, the majority of purchases were made at real store, but now that services on the Internet have been enhanced, we thought that the ratio of purchase methods may change. This time, as a consumer consciousness survey at real store and EC, we conducted a questionnaire on four items to about 50 people between the ages of 16 and 22. The first question is, "When you buy clothes, at real store or EC". The result is 38% at real store, 42% at EC, and 20% at the same rate. The second question is, "What kind of problems do you have at EC". The most common answer is that they didn't know the size, fabric, and color (Fig. 1).

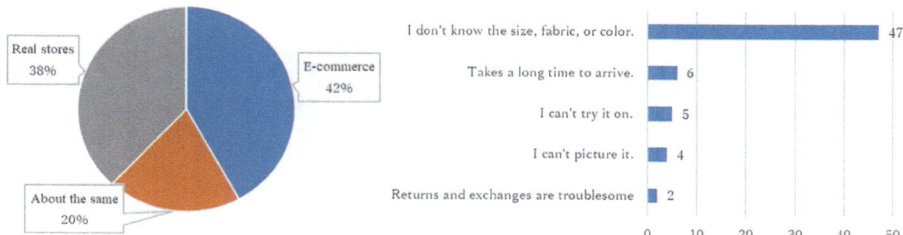

Fig. 1. Percentage of purchases at real store and EC. The problems at EC.

The third question is, "What kind of problems do you have at a real store". As a result, the most common answer is that they do not want to talk with store clerks. The fourth question is, "What kind of services and applications would you like to at EC". As a result, the most popular answer is wearing image by many photos of each height and body type (Fig. 2).

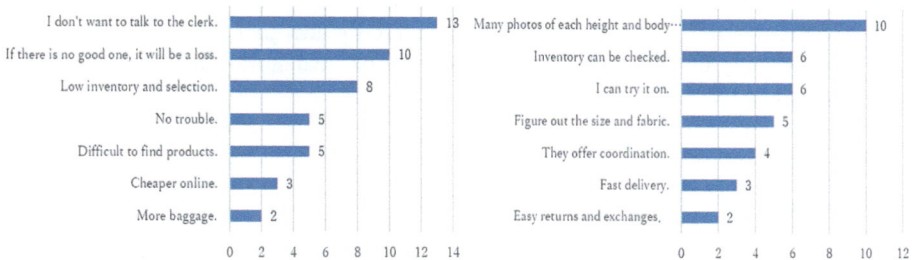

Fig. 2. The problems at real stores. New services and applications preferences.

2.3 Comparative Considerations of 2021 and 2022

Shows the latest apparel EC market size and EC rate data for the past four years announced by the Ministry of Economy, Trade and Industry. As you can see from the graph, the apparel EC market is on the rise, and the EC rate increased by more than 5% in 2020 due to the coronavirus pandemic (Table 1 and Fig. 3).

Comparing our questionnaire survey results for 2021 and 2022, regarding purchasing methods, the decrease at real stores and the increase at EC indicate that there has been

Table 1. Apparel EC market size and EC conversion rate.

Year	EC market size	EC conversion rate
2018	1,772.8 billion yen	12.96%
2019	1,910.0 billion yen	13.87%
2020	2,220.3 billion yen	19.44%
2021	2,427.9 billion yen	21.15%

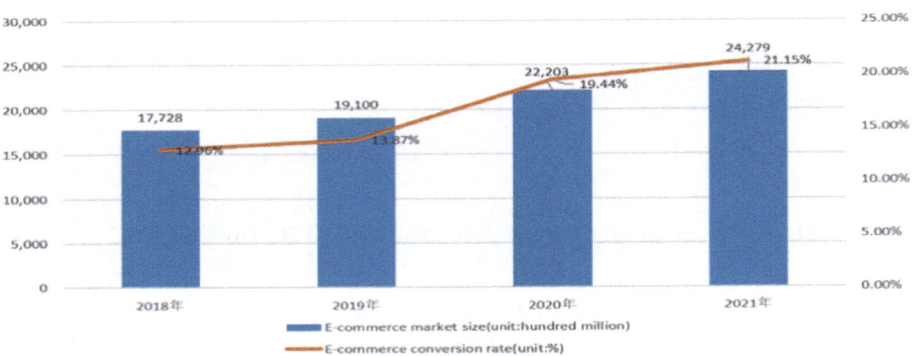

Fig. 3. Apparel EC market size and EC conversion rate.

a change in consumer purchasing consciousness. It means that this change is largely related to the fact that EC issues are being resolved. Still, there are many people who purchase at real stores or use both. This is related to the fact that there is an experience value that can only be obtained at real stores, and that real stores and EC complement each other's disadvantages (Fig. 4).

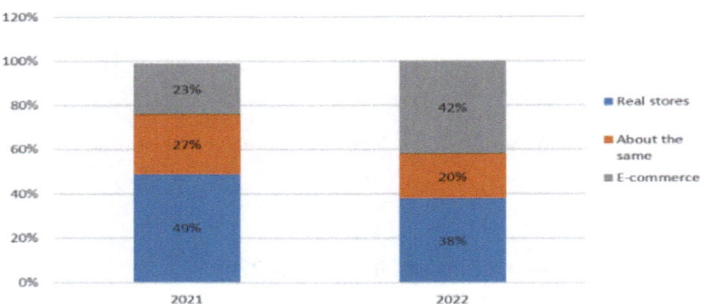

Fig. 4. Purchase method changing.

Comparing the problem at EC, in both 2021 and 2022, the problem of not knowing the actual product such as size and color is overwhelming. Since it is related to try-on, what is highly expected as a solution is what is called virtual fashion. A lot of services using VR have been born in the past few years. However, at present, it is less familiar and lesser known (Fig. 5).

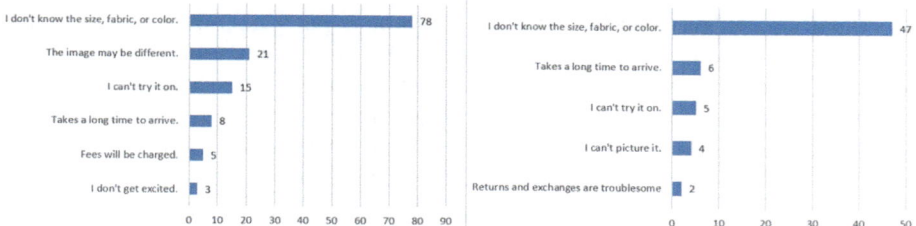

Fig. 5. The problems at EC in 2021 and 2022.

Looking at the results of comparing items such as what services and apps are good to have, the most common in 2022 was the service with wearing images it for each height and body type. Recently, the demand for pictures of people wearing clothes is increasing, and more and more people are looking at pictures of clothes they are interested in wearing and coordination on Instagram and TikTok. In addition, services using VR have been introduced at real stores. While there is a reputation for not having to worry about trying on clothes at real stores, it was pointed out that it was embarrassing for others to see how they moved in front of the digital signage. Due to these facts, it can be said that virtual fashion has produced a certain effect as a solution to the problem of fitting and size (Fig. 6).

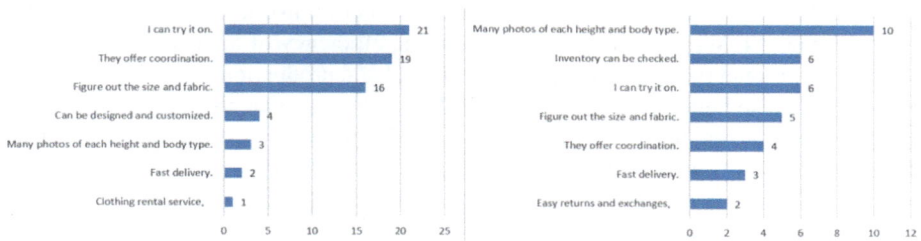

Fig. 6. New services and applications preferences in 2021 and 2022

2.4 Directions Post Coronavirus Pandemic

Among young people, there is a trend for people to purchase clothes by searching for images of clothes they are interested in and collections of outfits they are interested in on SNS. Consumers search for clothes online, try them on at real stores, and see the actual product. And it is a flow to purchase at EC cheaper than real stores. This trend will not change even after the post corona era. In addition, as a new purchasing experience at EC, the use of VR and AR is spreading on SNS. Snapchat, a photo and video sharing app, and TikTok, a video sharing app, are strengthening their position as shopping platforms. In January 2022, they added a function that displays the price, content, purchase link, etc. of the product in real time according to the image of the item that the user tries. In a two-week test of the new feature, U.S. cosmetics retailer Alta Beauty's products saw 30 million tries on Snapchat, driving $6 million in purchases. Based on these, the company came up with a new business model: enhancing information accessibility on the Internet and linking services with real stores. To make it easier to select clothes that interest you or that you want to try on online. Then, information about the selected clothes is transmitted to the real store, and when you go to real store, it will be prepared in advance in the fitting room, and the decision can be made on the store or via EC. Going to the store means that the desire to buy is quite high, so they try on several clothes instead of one. Purchasing efficiency in trying on products will increase. In addition, the utilization of EC data can also be used for inventory management and production sites at real stores. By digitizing consumer purchase intentions, AI-based demand forecasting will also contribute to solving the clothing industry's problem of closing losses. It will lead to the realization of a sustainable world that society demands.

3   Circumstances Surrounding Fashion Tech

The apparel industry has been rapidly digitized in recent years, and technology plays an important role. EC, in particular, has brought about significant changes in the fashion industry. It has become possible to provide personalized products by analyzing consumer preferences and past purchase history. Technologies such as VR and AR are being used in the fashion industry to improve the customer's shopping experience. It is possible to virtually try on products on your own smartphone or tablet terminal, so you can try on products without going to real stores. Like other industries, the fashion industry is powered by AI. We are working to improve the efficiency of data analysis, demand forecasting, product personalization, quality control and production control. 3D printing technology is also being used. Utilization of this technology can reduce product manufacturing costs and enable flexible production and the provision of customized products. Technology has become an integral part of the fashion industry.

3.1   Case Study of Fashion Service

Fashion services have been increasing in the domestic apparel industry. Some examples are shown below (Table 2).

Fitting services in virtual fashion are beginning to be introduced into real stores to create a digital experience. At Shibuya Parco, one of major department store, they have introduced a system called "FXMirror" that automatically measures your body just by standing in front of the virtual fitting mirror. Looking at the overseas situation, the demand is increasing, so we anticipate this trend to continue. In addition, consumers in Europe and the United States are increasingly seeking convenience and environmentally friendly styles rather than desire to own, and rental services are more popular than in Japan. Furthermore, fashion communities and apps using SNS are gaining popularity.

Table 2. Some apps technology and feature

Apps	Technology	Feature
PASHALY	AI clothing recognition	Similar product recommendation
Riko	AI chatbot	virtual stylist
unisize	Optimal size determination	Wearing silhouette display
kitemiru	VR	Try-on service

3.2 Demand for Fashion Services Using VR, AR, and Metaverse

With the development of VR and rapid onlineization, many companies have begun to use VR for promotion and entertainment. VR, which is not restricted by time or place and can deliver experiences just by looking at it, is mainly chosen as an advertising method in the fashion industry. For example, by holding a fashion show in VR or holding a virtual exhibition, you can approach a wide range of people. Well-financed apparel companies are developing their own VR and AR services. Similar to these technologies, Metaverse is attracting attention in the apparel industry. There is an image that it is nothing more than a mere item for avatars. However, there is a movement to acquire customers who do not usually visit the brand's real stores for another purpose. Just as the apparel industry's marketing channel has shifted to SNS, we anticipate that it will grow into one of the leading customer channels in the future. As environmental awareness of sustainability increases, the apparel industry is being forced to review its supply chain. Among them, Metaverse is attracting attention as a means of solving these problems. For instance, BEAMS, an apparel company, sells real products on Metaverse and induces customers to real stores. Overseas, RTFKT, a company under Nike, is developing a service that allows avatars to wear sneakers and clothes in Metaverse and use AR to experience as if they were actually wearing sneakers. It has potential as a service that increases sales at real stores and creates new value.

3.3 Smartphone Application Development Experiments

The development language adopts Unity, which is good at developing graphical applications that are good at 2D/3D games. And it adopts AR Foundation, an official Unity package for developing AR applications common to all platforms, as a method for realizing XR. Conventional AR application development often uses AR libraries prepared for each platform, and it was necessary to develop the same application for each platform. AR Foundation provides platform-independent AR libraries, so you can easily develop AR applications that are common to all platforms.

As for non-smartphone apps, there are web apps, which can be considered in the same way as normal web development. However, since it is an application that runs through a browser, it is necessary to think about how to store information that does not allow the camera to perform to its full potential (Table 3).

For XR. Use the terminal position as the origin and position the clothes object at eye level with the center displayed 1.2 m above ground. Use Face Tracking to place the

Table 3. Development environment

Category	Content
OS	Windows 11 Home
CPU	11th Gen Intel(R) Core(TM) i5-11400F @ 2.60 GHz
RAM	16.0 GB
ROM	500 GB SSD
Software	Unity 2023.1.0b1/Visual Studio Code 1.76.2
Debug equipment	Google Pixel 5 (Android 13)

clothes object under the user's face and control its movement based on the position and orientation of the face. Use the position (xyz coordinates) and orientation (up, down, left, and right) of the face to control the movement of the clothes object. This will ensure that the clothes object moves in a realistic way based on the user's head movements. Consider using 3D modeling software to create realistic clothes objects for the application (Fig. 7).

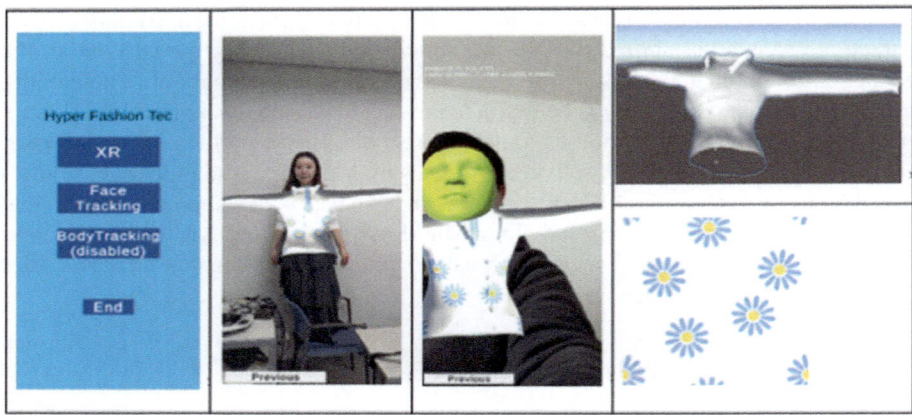

Fig. 7. The screen design and using data, top screen, XR, face tracking, shirt 3D object and clothes pattern.

In general, 3D data is comprised of two elements, with one file containing information about the object's shape, including lines and vertices, and the other file containing data about the object's appearance, such as color and texture. However, creating large amounts of clothing data efficiently is a challenge for real stores, which may have thousands to tens of thousands of items. Although a 3D scanner can be used to obtain exhaustive 3D data, it is difficult to do so in practice. Presently, smartphone apps use AI-generated images based on clothing images provided by companies to allow users to try on clothes virtually. Additionally, designers utilize 3D CAD during the design phase.

Nevertheless, accurately depicting fine details such as clothing thickness and texture remains an ongoing challenge.

3.4   Create Possibilities by Metaverse Utilization

The future possibilities of Metaverse, which is attracting attention in the apparel industry, so actually used an application that utilizes Metaverse (Table 4).

Table 4. Results of evaluation of smartphone applications utilizing Metaverse.

Category	Evaluation
Interest, fun	4
Number of items/edit	5
Number of users	5
Application Effectiveness	5

We utilized ZEPETO, a service that enables 300 million users worldwide to create avatars and interact with one another. With over 4 million items available for editing, including facial features, hairstyle, skin tone, clothing, and other attributes, users can customize every aspect of their avatars. The vast array of items and users results in a diverse range of fashion avatars, inspiring and stimulating creativity among the community. The Metaverse has created a worldwide desire for socializing and matching outfits with other users, generating excitement about the potential for virtual fashion to become reality. ZEPETO's popularity lies in its freedom to customize and design clothing, raising the question of whether Metaverse's strengths could be translated into the real world, such as by commercializing and delivering clothing customized in Metaverse. By integrating the Metaverse with real stores and EC, clothing can become more personalized, durable, and sustainable, reducing clothing waste and promoting attachment to garments. The apparel industry's potential unlocked by the Metaverse could lead to both fun and a more sustainable future (Fig. 8).

Fig. 8. Avatar of ZEPETO application

4 Conclusion

In the survey of changes in purchasing consciousness this time, we can see that there is a clear shift to the Internet society due to the impact of the coronavirus pandemic and consideration for the environment. The utilization of digital technology is inevitable for the efficiency of the Internet society. One of the keywords is virtual fashion, so we actually developed an app. It was surprisingly easy to create a demo version mockup in a few days, but it turned out that there were many challenges in providing it as a business. From this point of view, the question arises whether it is not always necessary to use technology to solve various problems. It is an analog concept that emphasizes real stores, but as a result, depending on the function of the app, we felt that the possibility of trying on clothes expanded. Many existing virtual fashion fitting services utilize avatars. In addition, the wearing image posted does not show the fabric texture and shrinkage rate of the clothes. This is as much of an annoyance as the size issue, as evidenced by surveys. In other words, the fashion service demanded by users is to understand the size, fabric texture, and shrinkage rate of clothes, and to create an image. Since application development evolves along with technological progress, epoch-making services will be provided sooner or later. On the other hand, in this research study, we considered the possibility of a new business model. With the fusion of Metaverse + EC + real stores, it is a mechanism in which the clothes that you customize or design are immediately commercialized and delivered to your hands. It is the realization of a sustainable world where the apparel industry contributes to society on a global scale in response to the issues of attachment to clothes, product life, and clothing loss.

5 Future Work

In the apparel industry, various changes are occurring due to technological innovation and trends after the coronavirus pandemic. With the advent of "online customer service," which allows customers to request styling over the Internet, and "live commerce," which allows customers to purchase products distributed in real time, EC market is expected to expand significantly in the future. Consumers are also spending more time in virtual communities due to technological advances. On the other hand, there is an attempt to strengthen the sales staff who coordinate at real stores. In any case, the formation of a fashion community is an important measure for apparel companies, along with capital investment in IT. The omni-channel, which considers real stores and EC as a consistent channel, will be promoted. We also anticipate that consumer environmental awareness will increase the need for long-lasting, reusable and recyclable products. Companies and brands that make clothing with sustainability in mind are one of the deciding factors in purchasing. We will continue to conduct research and surveys on what role fashion tech will play in the changing times and technological trends, taking into account the SDGs.

References

1. Sakaguchi, M., Aoki, E.: Study on the comparison of consumer impression of EC and real store in the fashion tech era, and the effectiveness of VR utilization. In: Proceedings of the 16th International Conference on Complex, Intelligent, and Software Intensive Systems (CISIS-2022), pp. 528–537, 29 Jun–1 Jul 2022 (2022)

Spatial Interpolation of Room Impulse Responses Using Information on Early Reflection Directions

Ichizo Furusho[1][✉], Keisuke Nishijima[2], and Ken'ichi Furuya[2]

[1] Graduate School of Engineering, Oita University, 700, Dannoharu, Oita-shi, Oita, Japan
`chainsmokers3383@yahoo.co.jp`
[2] Faculty of Science and Technology, Oita University, 700, Dannoharu, Oita-shi, Oita, Japan
`{k-nisijima,furuya-kenichi}@oita-u.ac.jp`

Abstract. In recent years, with the widespread advent of 3D movies and VR (virtual reality), studies have been advanced. In particular, opera and orchestral concerts are preferred in spaces with unique reverberations, e.g., concert halls. However, we cannot enjoy the recorded concerts in a homely. To obtain resonance, the room impulse response is measured; however, the impulse response is a measurement value at a certain point, and the measurement process is time-consuming and costly. Therefore, an interpolation method has been proposed to reduce the number of impulse response measurement points. Interpolation is a method of estimating the impulse response of the position between the microphones from acoustic models and actual measurement data, and performing numerical calculations. A problem with the conventional method is that the accuracy of the plane wave of the impulse response, i.e., early reflections portion is low. In this study, we further perform the direct wave/plane wave splitting method related to the conventional impulse response, splitting the plane wave into two parts, the early reflected wave and reverberation. We propose a method to determine a suitable model and range for early reflections that, which may help improve accuracy. In the first experiment, we determine the cutout position to split the plane wave; in the second experiment, we propose a method to compare the accuracies of early reflected and synthesized plane waves with the conventional results. Based on the results of the aforementioned methods, we confirm the effectiveness of the proposed method.

1 Introduction

In recent times, the measurement of room impulse responses using interpolation has attracted considerable attention [1, 2]. The use of interpolation can help reduce time and cost while measuring impulse responses [3]. The interpolation of an impulse response is formulated as an inverse problem and is conventionally modeled using the plane wave decomposition method (PWDM) or time-domain equivalent source method (TESM) [4]. However, one of the challenges of interpolation is the poor accuracy of the reflected wave, which is the latter part of the impulse response. Reflected waves are complicatedly reflected by floors and walls while reaching the sound-receiving point, making it difficult to weight them accurately.

To improve the accuracy of the interpolation of reflected waves, the reflected wave is divided into two parts: early reflected wave and late reverberation. As the early reflected

L. Barolli (Ed.): CISIS 2023, LNDECT 176, pp. 377–383, 2023.
https://doi.org/10.1007/978-3-031-35734-3_38

wave possess characteristics similar to the direct wave, e.g., the time it takes to reach the sound-receiving point and bias in the direction of arrival, we propose an interpolation method by applying the model that is used for the direct wave.

2   Previous Studies

2.1   PWDM

The PWDM is a model that reconstructs the impulse response with a finite weighted sum of plane waves [5], which is defined in the following equation:

$$p(x,t) = \sum_{l=0}^{N_\omega-1} \delta(t) * \phi_{l,x}(t) * \omega_l(t). \tag{1}$$

where N_ω denotes the number of virtual sources, δ denotes the Dirac delta function, ϕ denotes the plane wave, ω denotes the weight signal, and $*$ denotes linear convolution. The PWDM is used to measure reflected waves.

2.2   TESM

The TESM is a model that reconstructs the impulse response with a finite weighted sum of spherical waves [5], which is defined in the following equation:

$$p(x,t) = \sum_{l=0}^{N_\omega-1} \delta(t) * \psi_{l,x}(t) * \omega_l(t). \tag{2}$$

where ψ denotes the spherical wave. The TESM model is used for direct wave measurements.

2.3   Optimization Problem

We solve an optimization problem to determine weight signals needed to reconstruct the impulse response [6]. The optimization problem is defined in the following equation:

$$W^\star = \arg\min_W f(W) = \frac{1}{2}\|D(W) - \tilde{P}\|_F^2. \tag{3}$$

where f denotes the cost function, D denotes the model equation for impulse response, $\tilde{P}$ denotes the measured impulse response, and $\|\ \|_F$ denotes the Frobenius norm. The objective of this optimization problem is to minimize the difference between the measured impulse response and sound pressure of the acoustic model.

2.4   Ambisonics

To improve the accuracy of reflected waves with low interpolation accuracy, interpolation is performed using a first-order ambisonic microphone [7,8]. Figure 1 shows the characteristics of ambisonics; the zero-order W characteristic captures the surrounding sounds with the same sensitivity, and the first-order X, Y, and Z characteristics strongly capture sounds reflected by vertical and horizontal sides. These characteristics can be used to accurately weight reflected waves.

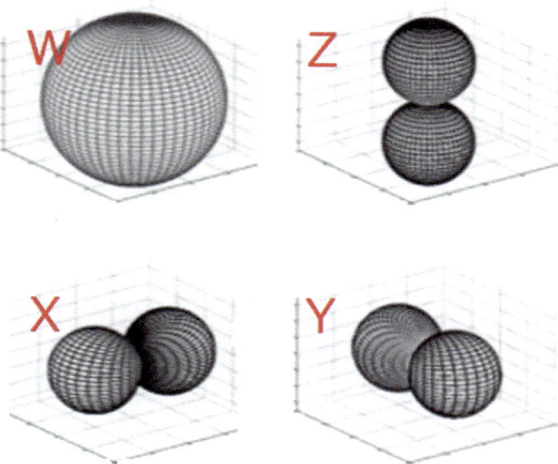

Fig. 1. Characteristics of Ambisonics.

3   Proposed Method

To improve the interpolation accuracy of the reflected wave, the reflected wave is divided into two parts: the early reflected wave and late reverberation. The early reflected wave possesses characteristics similar to the direct wave, e.g., the time of arrival at the sound-receiving point and bias in the direction of arrival. Therefore, the direct wave model is applied to the early reflected wave, and interpolation is performed.

3.1   Early Reflected Wave and Model

Figure 2 shows the wavefront when sound is emitted from a sound source. The red circle denotes the sound source, the right-pointing arrow indicates the direction of sound travel, and the light blue semicircles from A to F indicate wavefronts. As indicated by wavefronts A-C, sound emitted from the source distributes on a spherical surface, which is known as a spherical wave. This spherical wave model is used to reconstruct the direct wave portion of the impulse response. If the sound is distant from the source, the sound wavefront transforms into a plane wave, which can be assumed as a plane perpendicular to the travel direction. This plane wave model is used to reconstruct the reflected wave portion. As the initial reflected wave is relatively close to the sound source, the TESM model used for the direct wave is used to interpolate it. Then, it is combined with the separately interpolated late reverberation to form a single reflected wave, which is finally compared with the conventional reflected wave to confirm the effectiveness of the proposed method.

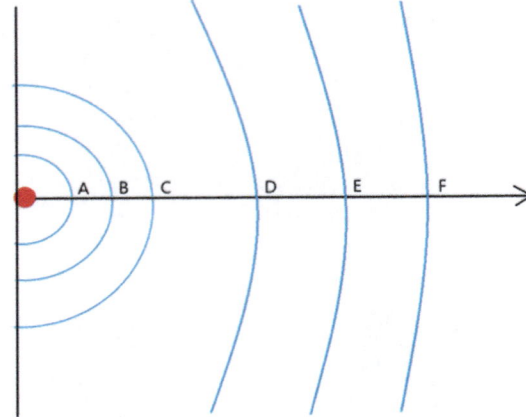

Fig. 2. Wavefront of the sound emitted from the sound source.

4   Evaluation Experiment

4.1   Early Reflected Wave Range

Figure 3 shows the reflected wave portion of the impulse response. As the range of the early reflected wave is significantly large, multiple experiments were conducted within the split location, and the range was taken to the point of greatest accuracy.

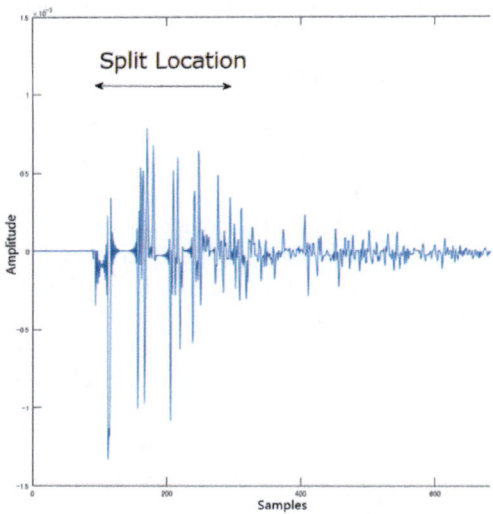

Fig. 3. Reflected wave portion of the impulse response.

4.2   Experimental Conditions

Table 1 lists the environment details of the simulation experiment to be conducted in this study. The RIR was created using the mirror image method [9, 10]. The sampling frequency was 8 kHz, and the size of the room was [7.34, 8.09, 2.87] m. The microphone array was an orthogonal array with its origin at [4.4, 3.1, 1.5] (Fig. 4). The number of microphones is 21, of which 15 are learning microphones (denoted by blue) and 6 are interpolation microphones (denoted by red). The range of the early reflected wave is up to where the horizontal axis is 296, which was the most accurate range obtained after multiple experiments.

Table 1. Environment of the simulation experiment.

Sampling frequencies	8000 Hz
Size of the room	[7.3, 8.1, 2.9] m
Room reverberation time	0.097 s
Source coordinate	[2.0, 6.5, 1.4] m
Microphone array coordinates	[4.4, 3.1, 1.5] m
Microphone spacing	5 cm
Learning microphones	15 pieces
Evaluation microphones	6 pieces
Number of virtual sound sources	700 pieces
Placement of virtual sound sources	On a spherical Fibonacci lattice
Evaluation methodology	Normalized Mean Squared Error
Directional characteristics used	X characteristic
Location of the early reflected wave	Horizontal axis = up to 296

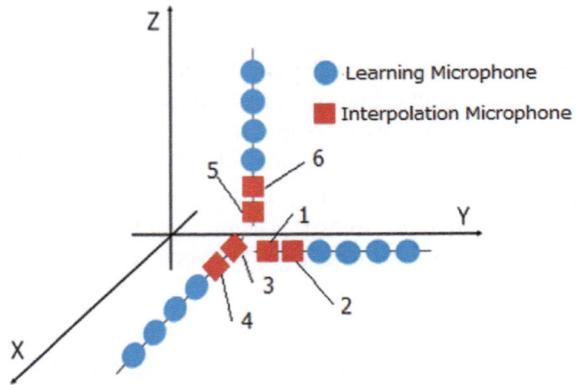

Fig. 4. Microphone layout.

4.3   Experimental Results

Table 2 summarizes the accuracies of the reflected waves interpolated using the conventional method and the proposed method, interpolation accuracies are expressed in terms of normalized mean upper error. It can be seen that the accuracy has slightly improved by around 0.08 dB. According to the results obtained for each microphone, the accuracy of microphone number 2 has improved by around 2.00 dB. However, the accuracy of microphone number 6 has decreased by around 1.50 dB. The areas where accuracy improved confirmed the effectiveness of the proposed method. However, the areas where the accuracy decreased, i.e., microphone numbers 5 and 6 were assumed to be related to the experimental environment. As listed in Table 2, the height of the room considered in this study is lower than the heights of the vertical and horizontal sides. Therefore, for the two microphones placed at the height, the sound was inconsistently reflected before it reached the sound-receiving point. When the reflected sound arrived, it was included in the late reverberation stage, and the TESM was not the appropriate model.

Table 2. Accuracies of the reflected waves interpolated using the conventional method and the proposed method.

Microphone	1	2	3	4	5	6	average
NMSE (dB): Conventional method	−11.11	−12.62	−10.31	−12.49	−9.38	−12.27	−11.37
NMSE (dB): Proposed method	−11.61	−14.59	−10.46	−12.19	−9.14	−10.71	−11.45

5   Conclusion

To enable the realistic viewing of concerts and other events, it is necessary to measure the room impulse response. However, measuring large areas is expensive and time-consuming. Therefore, spatial interpolation was used in this study the accuracy of reflected waves was poor. To solve this problem, the reflected wave was divided into two parts, early reflected wave and late reverberation, and interpolation was performed for the early reflected wave using the TESM, which was considered suitable. Finally, the divided waves were synthesized into a single reflected wave, and the interpolation accuracy was compared with the results of the conventional reflected wave. As a result, the average value improved by around 0.08 dB; however, but the value decreased by around 1.50 dB in some places depending on the microphone position, suggesting that several points need to be improved, e.g., the position of the divided early reflected wave and microphone placement.

Acknowledgements. This work was supported by JSPS KAKENHI grant numbers 19K12044.

References

1. Sato, F.: Measurement techniques of room impulse response. J. Acoust. Soc. Jpn. **58**(10), 669–676 (2002)
2. Kaneda, Y.: Notes for impulse response measurement. J. Acoust. Soc. Jpn. **55**(5), 364–369 (1999)
3. Ajdler, T., Sbaiz, L., Vetterli, M.: Dynamic measurement of room impulse responses using a moving microphone. J. Acoust. Soc. Am. **122**(3), 1636–1645 (2007)
4. Antonello, N., et al.: Room impulse response interpolation using a sparse spatiotemporal representation of the sound field. IEEE/ACM Trans. Audio Speech Lang. Process. **25**(10), 1929–1941 (2017)
5. Ota, M., et al.: Fundamental Physical Acoustics Engineering. Asakura Bookstore (1990)
6. Stella, L., et al.: Forward-backward quasi-Newton methods for nonsmooth optimization problems. Comput. Optim. Appl. **67**(3), 443–487 (2017)
7. Nishimura, R.: Ambisonics. J. Inst. Image Inf. Telev. Eng. **68**(8), 616–620 (2014)
8. Kato, M., Furuya, K.: Kyushu branch of acoustical society of Japan. In: The 15th Research Presentation for Students (2021)
9. Allen, J.B., et al.: Image method for efficiently simulating small room acoustics. J. Acoust. Soc. Am. **65**(4), 943–950 (1979)
10. De Sena, E., Antonello, N., Moonen, M., Van Waterschoot, T.: On the modeling of rectangular geometries in room acoustic simulations. IEEE Trans. Audio Speech Lang. Process. **23**(4), 774–786 (2015)

Co-browsing Cubic Gantt Charts with VR Goggles for Collaborative Immersive Visual Data Analytics

Shohei Nakamura[1]([✉]) and Yoshihiro Okada[2]

[1] Graduate School of Information Science and Electrical Engineering, Kyushu University, Motooka, 744, Nishi-ku, Fukuoka 819-0395, Japan
nakamura.shohei.386@s.kyushu-u.ac.jp
[2] Innovation Center for Educational Resources (ICER), Kyushu University Library, Graduate School of Information Science and Electrical Engineering, Kyushu University, Motooka 744, Nishi-ku, Fukuoka 819-0395, Japan
Okada.yoshihiro.520@m.kyushu-u.ac.jp

Abstract. This paper proposes a new 3D visualization tool based on Cubic Gantt Charts. This new tool has co-browsing functionality that allows multiple users to simultaneously see the visualization results of Cubic Gantt Charts and supports VR goggles. Therefore, it makes possible collaborative immersive visual data analytics. In addition, this new visualization tool has a recording function that stores users' visualization activities such as translation, rotation and zooming, and replay function that shows past activities to current users later. These recording and replaying functions are significant for multiple users to analyze data collaboratively. This paper introduces these functions and explains how they work for collaborative immersive visual data analytics.

Keywords: VR · Web-based Visualization tool · Co-Browsing · Interaction log

1 Introduction

Although there are many methods to analyze data statistically, visualization is one of the most effective methods for users to grasp features of target data interactively without any statistical process. In previous work [1], we proposed a new 3D visualization tool based on Gantt Charts called "Cubic Gantt Charts" and showed how the tool can be used for visualizing students' learning activity data. In Cubic Gantt Charts, the visualized charts could be observed from any direction. We speculate that this function could increase the potential for deeper, more collaborative data analysis by providing the framework for multiple users to watch the charts at the same time. Co-browsing, short for "Collaborative browsing", means multiple users watch the same webpage and interact with it simultaneously. The purpose of co-browsing is to enable experts to contribute to data analysis by viewing the visualization tool together in real time. Potentially, a user could provide detailed feedback while other users look at and manipulate the charts. The

L. Barolli (Ed.): CISIS 2023, LNDECT 176, pp. 384–394, 2023.
https://doi.org/10.1007/978-3-031-35734-3_39

new 3D visualization tool based on Cubic Gantt Charts proposed in this paper, developed by JavaScript, is a web-based VR visualization tool that has co-browsing functionality. This tool is more accessible than traditional systems, as it can be used with any browser and without any specific software.

The VR world has much more flexibility compared to normal 2D screens. It allows users to watch and manipulate the visualization tool freely, unlike traditional screens which inhibit flexibility of certain actions. If users find data points of interest while manipulating the charts, they may want to discuss them with others. Our previous Cubic Gantt Chart visualization tool was incapable of doing so. Contrarily, our new co-browsing Cubic Gantt Chart visualization tool has a function that records past users' visualization process and replays them to current users later. In addition, our co-browsing Cubic Gantt Chart visualization tool supports VR goggles. Therefore, it enables collaborative immersive visual data analytics.

This paper is organized as follows: there are related works in Sect. 2. We introduce our previous visualization tool based on Cubic Gantt Charts and explain how users can analyze the data with them in Sect. 3. Next, in Sect. 4, we introduce our new co-browsing Cubic Gantt Chart visualization tool that allows multiple users to watch the charts at the same time. We also explain the functions of recording past users' activities in the visualization processes and replaying them to current users. In Sect. 5, we discuss the benefits of the proposed visualization tool in data analytics. Finally, we conclude the paper and discuss potential avenues for future work.

2   Related Works

2.1   Co-browsing Visualization Tools

The purpose of co-browsing is to share the view of webpage with multiple people and exchange ideas. This research field emerged in the 1990s [2, 3]. Early research focused on methods of sharing web pages with others in real time. There were various architectures to meet collaborative browsing needs in the 2000s [7, 8]. In the later 2000s to 2010s, this collaborative function was extended to share dynamic web pages [6, 9, 10], meaning web pages that support the synchronization of users' interactions.

There are also studies in which multiple users view the visualization tool simultaneously. Collaborative visualization improves upon the traditional visualization because each participant contributes to the common goal of analyzing the data [5]. In the 1990s, the first VR visualization systems CAVE and ImmersaDesk were proposed [11, 12]. The former is used for collaborative visualization. However, special hardware is necessary to use it. Another example is VizGrid, which applies grid computing [13]. VizGrid also needs a specific environment to generate 3D objects in the display to collaborate with other people.

We employ the web-based VR system to meet the demands of collaboration. Use of this collaboration function is possible with a web browser and without any specific hardware.

2.2   Recording and Replaying Visualization Processes

Robinson et al. defined "Re-visualization" and explained the difficulty to re-use, or re-visit what they had discovered [14]. Henry et al. proposed a method to save logs and replay actions in InfoVis systems with reduced modification of JAVA code [15]. This motivation was to evaluate the visualization tools. Other examples are also used for evaluating usability or improving the systems [16, 17, 18]. They also introduced a toolkit for Re-visualization [14]. This toolkit enables users to save visualization sessions. Users could view the display if they play or drag the scroll bar.

We combined the co-browsing and recording/replaying functions in our visualization tool for collaborative immersive visual data analytics.

3   Cubic Gantt Chart

The first Cubic Gantt Chart was proposed for visualizing e-learning activity data, such as how long and how often students look at each page [1]. Cubic Gantt Charts are based on traditional Gantt Charts, which were originally proposed by Henry L. Gantt in the 1910s and are suitable for basic management tasks. Rectangles are used to represent each task and its progress. The horizontal axis shows numeric values like time, while the vertical axis represents discrete attributes such as the type of tasks. Figure 1 is an example of a traditional Gantt Chart.

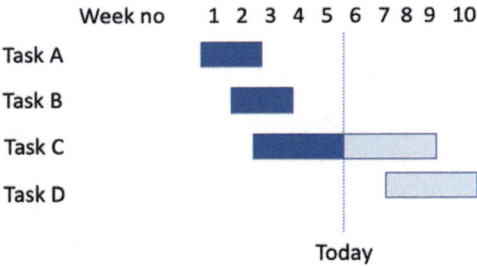

Fig. 1. Example of a traditional Gantt Chart

Cubic Gantt Charts exist in 3D space and have an additional axis to show the relationship between the three attributes. The small cubes in the tool are called "voxels" and each one represents one data point. Voxel length shows the numeric value if the related axis has continuous value, on the other hand, voxel position means the discrete value where the axis also has discrete value. Below are some examples using e-learning activity data.

In Fig. 2, the X axis represents time, which is a continuous value. In this case, time indicates the duration of one lecture. Y and Z axes show students and pages, which are discrete values, respectively. In this case, "students" refers to those who attended the class, and "pages" are the indices of PowerPoint slides. The example on the right shows the data of a single student. The length of the light green colored voxel represents spending a bit longer than time spent on other pages. The repetition of this voxel indicates that he/she went back to review a previous page after reading the pages the neon-green voxels represent.

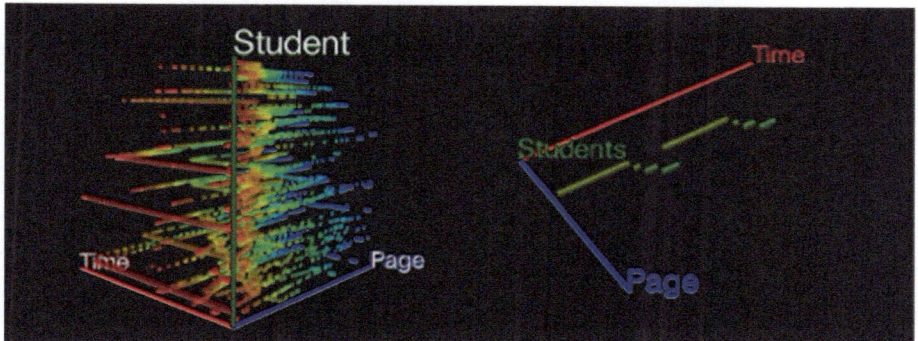

Fig. 2. Screen images of Cubic Gantt Charts [1], whole e-learning activity data (left) and an extract of one student's activity data (right).

4   Implementation

4.1   Overview of the System Architecture

We have developed our new visualization tool as a web-based system so that anyone can use the tool with any web browser on any hardware platform. Thanks to developing VR environments, VR functions can also be used in-browser. We implemented Co-Browsing and replay logs into our previous web-based visualization tool, Cubic Gantt Charts. The tool adopted the client-server mechanism and employed the design of ECSY, short for Entity Component System. The communication between the client and the server is done via WebSocket. ECSY is suitable to control all components which should be synchronized among users' devices work as a client after receiving the data from the server. We adopted Meta Quest2 VR headsets as immersive VR viewer devices of the visualization tool while we used an android smartphone to confirm web-based VR functionality across platforms.

Figure 3 shows ECSY architecture composed of Entity, Component and System. Each entity is comprised of several components. Each component has attributes such as position, rotation, scale, object shape and so on. There are several systems that updates their corresponding components. For example, when the user manipulates the tool via "Cubic Gantt Chart Entity" before the drawing function is executed, this visualization system calls the "movableSystem" that updates each entity position, the "rotatableSystem" that updates each entity rotation and the "zoomableSystem" that updates each entity size.

The role of "ButtonSystem" is calling the corresponding function when the user pushes a button. For example, when the Co-Browse button is pushed, the system communicates with the server and receives either the "master" or "slave" role.

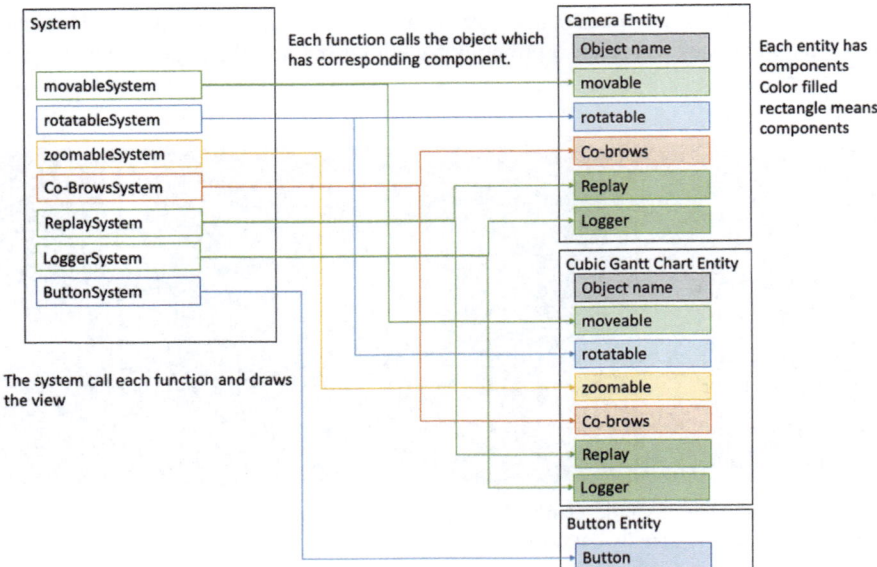

Fig. 3. Entity Component SYstem architecture

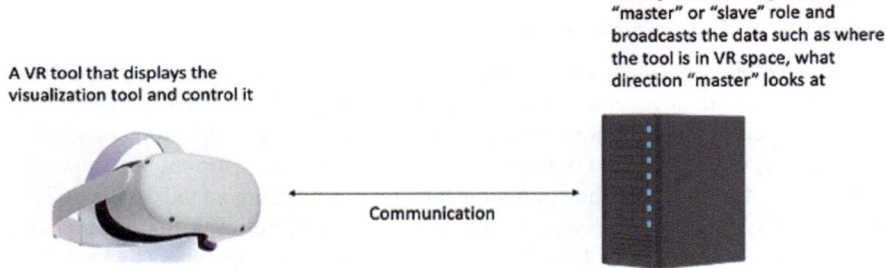

Fig. 4. System architecture between a client and the server.

Users can use this VR visualization tool by accessing the server via any device with a web-browser. This web-based tool communicates via the server with WebSocket (see Fig. 4). A user shares the screen with others by pressing the "Co-Browsing" button in the VR world. One user to whom the server assigns a "master" role can share the screen and control the visualization tool by manipulations such as translating, rotating and zooming. On the other hand, "slave" users can observe how the "master" manipulates the visualization tool. The system calls "Co-BrowsSystem" to send the entities position, rotation and size to the server if the device is assigned "master". The entities are updated from the "master" device data if the devices are assigned "slave". The function of "Co-BrowsSystem" is to broadcast the entity data on the "master" device and update all entity data on the "slave" device.

The user can also record his/her viewing and interaction log. Pressing the "Logger" button prompts the web-browser to send and store viewing and manipulation data in the server via socket communication. More specifically, the data includes where users are looking on the chart and how they control it. Pressing the "Replay" button, users can review past actions. After the user pushes the "Logger" button, the system calls "LoggerSystem" that always sends all entity position, rotation, and size information to the server.

4.2 Co-browsing Visualization Tool

Users who try to share the screen with others first make requests to the server to get the "master" role. Then they can share how they watch and interact with the visualization tool if there are no users who have already been given the "master" role. Other users are assigned "slave" roles after the server has assigned the "master" role. The "slave" users can watch in real time as the "master" views and controls the tool.

The "master" web-browser sends data, such as where the tool is in VR space, what direction the "master" looks at and what he/she was done in the tool via WebSocket. The server then broadcasts the data to all the other devices. The other web-browsers call functions to translate, rotate, and zoom in/out the tool following prompts based on the data sent from the server.

The system requests the server to give a role "master" or "slave" when the user pushes the "Co-Browsing" button, which means "ButtonSystem" is called. The server gives the "master" role to the device which requests first. After giving the "master" role to the device, the server gives only "slave" roles to the devices which request sharing the view. The system on the "master" device calls the "movableSystem", "rotatableSystem", and "zoomableSystem" functions which manipulate the visualization tool first. The system updates each entity position x, y, and z, rotation x, y, and z, and scale x, y and z values and then draws the view. Then "Co-BrowsSystem" is called to send the information of the entity attributes to the server. This data type is an array of object data. Figure 5 shows the data flow. The server broadcasts the data to all of the "slave" devices. The system on the "slave" device also calls the "Co-BrowsSystem" to update each entity position, rotation and scale according to the data sent from the server and then draws the new view. This function enables to keep the consistency among the users' view.

4.3 Record and Replay Logs

Users can record the logs of what they looked at and how they controlled the tool. Once they push the "Logger" button, the web-browser sends the data such as position, rotation and scale of the chart. The server receives the data and dumps the logs from each user. Use of the "Replay" button allows the replay of log data. Other users can get the same experience if they chose co-browsing mode.

Figure 6 shows the sequence of recoding and replaying logs to update the view of the visualization tool. The system becomes to call "LoggerSystem" to send each entity's position, rotation, and scale to the server after the user pushes the "logger" button. The server dumps each entity's position, rotation and scale data when the server receives the data. This data type is also an array of object data. This log data is linked by

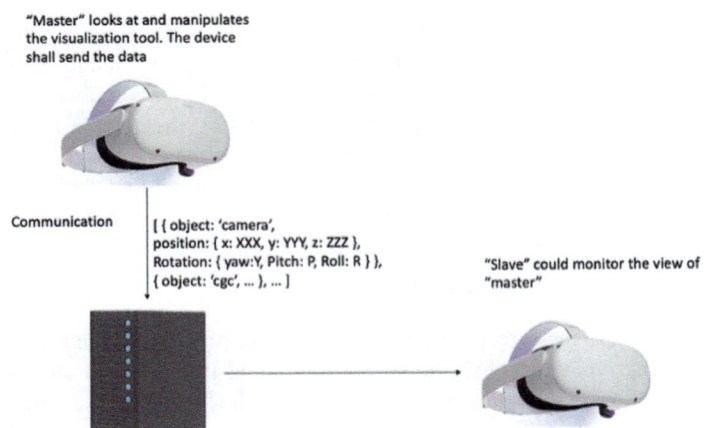

Fig. 5. Communication between two "master" and "slave" VR devices through the server.

each user's ID. The server records each user's previous actions. The server transmits the corresponding user's log data when the user pushes the "replay" button. The system calls the "ReplaySystem" to manipulate each entity's position, rotation and scale according to the log sent from the server and draws the new view.

The device assigned the "master" role also transmits each entity's attributes which will be updated by the log data to all of the "slave" devices. This communication flow is the same as explained in Sect. 4.2.

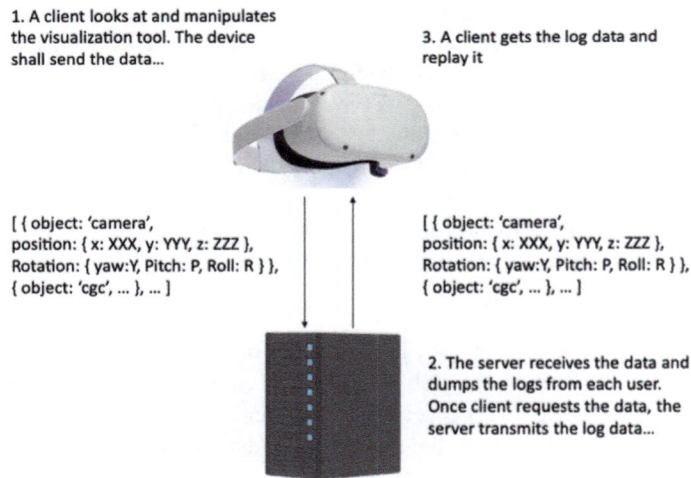

Fig. 6. Data flow between VR goggle and the server for recording and replaying activity data.

5   Discussion

5.1   Web-Based Visualization Tool

Barrett et al. proposed supporting cross platform collaboration as one of the grand challenges in collaborative analytics [4]. As a response to this challenge, we developed this VR visualization tool with JavaScript to enhance accessibility beyond previous visualization systems. Using JavaScript as the basis for this tool allows users to access it with smartphones, even though sections above explained how to watch this visualization tool with and described communication between VR headsets and the server in the previous section. An optional 2D display is available if users would not like to watch it in VR due to motion sickness, personal preference, or other reasons. We have not studied the difference between and impact of 2D version and VR versions. We suggest studies into these differences as one of our future works (Fig. 7).

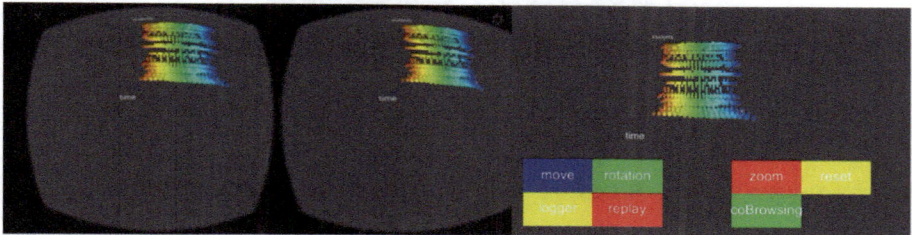

Fig. 7. An example of the view of the screen in VR mode on an android smartphone (left) and an example of the view of the screen with VR headset (right).

5.2   Impact of Co-browsing Function

Users can watch Cubic Gantt Charts to analyze the relationship between two of three attributes as we have already explained. We used the proposed visualization tool individually first and afterwards enabled the co-browsing function. Individual mode allows users to see limited fields of view which are not shared between users, making it easier for individuals to focus on specific sections of the visualized chart. However, the limited view makes it difficult to discuss data points, especially those seen from different viewing angles. On the other hand, in co-browsing mode, users could more easily track areas being discussed by other users, especially those which are seen from different viewing angles. In this research, we have not developed voice exchange functions for each device. We have used this system in the same room, but it is difficult to share opinions if users are in different locations. Developing a communication method between devices is necessary in the future works.

Even though projection of the chart is the same between users, it is difficult to find exact data points being discussed. Some VR headsets, such as Vive Pro Eye, have eye-tracking functions. The emphasis effect where the user looks into the visualization tool with this eye tracking function helps other users to understand where other users are looking. Devices with these functions are more expensive than devices which offer only VR functions, and future study is needed to implement eye-tracking into this visualization tool as one of the functions (Fig. 8).

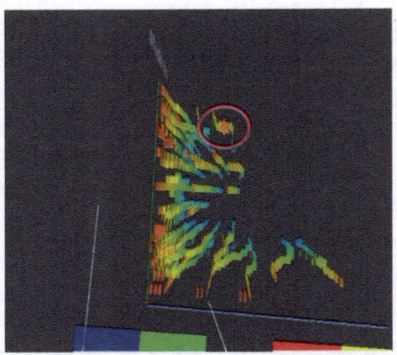

Fig. 8. One example of the emphasis effect and how it helps focus attention on precise areas being viewed

5.3 Usefulness of Recording Logs and Replaying Their Previous Actions

While using the visualization tool individually, users have different opinions from their own visual perspective. They cannot reproduce the previous view when they come up with their opinions after they enabled co-browsing function because the VR world is much more flexible when compared to 2D screens. Saving logs and replaying previous actions helps users quickly and accurately replicate actions taken on the tool, such as moving back and forth between viewing angles. This also enables users to share what they look at and how to manipulate if they enable the co-browsing function.

We will extend this function not only logging the interactions but also saving points of interest like the ReVise toolkit [14]. Another helpful feature may be to write down a memo in VR space, like we do on paper in the real world. We believe users could share notes easily in VR space with this system architecture.

There is also the potential to study the interaction logs [19]. Data from these logs could identify areas where users tend to look, which may be useful to analyze.

6 Conclusion

In this paper, we proposed new features allowing multiple users to watch Cubic Gantt Charts in VR space simultaneously. This function enables multiple users to analyze data and discuss its characteristics. This paper also introduced functions to record users' activity data and to replay the past data to current users. These two features are useful for users to share their experience with other users, and the use of VR is more accessible and convenient than our previous visualization tool. The proposed web-based VR system could work on any web-browser platform without any specific software.

Despite these proposals, there are still several points which should be further improved upon, as discussed in Sect. 5. In particular, voice exchange functions are necessary when users are in different locations to discuss directly via VR headset, rather than relying on external software applications. In addition, we should evaluate and potentially implement changes to methods of manipulating the tools in the VR world.

Acknowledgements. This research was partially supported by JSPS KAKENHI Grant Number JP22H03705.

References

1. Nakamura, S., Kaneko, K., Okada, Y., Yin, C., Ogata, H.: Cubic Gantt chart as visualization tool for learning activity data. In: 23rd International Conference on Computers in Education, ICCE 2015, pp. 649–658. Asia-Pacific Society for Computers in Education (2015)
2. Greenberg, S., Roseman, M.: GroupWeb: a WWW browser as real time groupware. In: Conference Companion on Human Factors in Computing Systems, pp. 271–272 (1996)
3. Bruce, K.B., Cardelli, L., Pierce, B.C.: Comparing object encodings. In: Abadi, M., Ito, T. (eds.) Theoretical Aspects of Computer Software. LNCS, vol. 1281, pp. 415–438. Springer, Heidelberg (1997). https://doi.org/10.1007/BFb0014561
4. Ens, B., et al.: Grand challenges in immersive analytics. In: Proceedings of the 2021 CHI Conference on Human Factors in Computing Systems, pp. 1–17 (2021)
5. Raje, R.R., Boyles, M., Fang, S.: CEV: collaborative environment for visualization using Java RMI. Concurr.: Pract. Experience **10**(11–13), 1079–1085 (1998)
6. Lowet, D., Goergen, D.: Co-browsing dynamic web pages. In: Proceedings of the 18th International Conference on World Wide Web, pp. 941–950 (2009)
7. Gerosa, L., Giordani, A., Ronchetti, M.: Symmetric synchronous collaborative navigation (2004)
8. de Jesús Hoyos-Rivera, G., Lima-Gomes, R., Courtiat, J.-P.: A flexible architecture for collaborative browsing. In: Proceedings of the Eleventh IEEE International Workshops on Enabling Technologies: Infrastructure for Collaborative Enterprises, pp. 164–169. IEEE (2002)
9. Niederhausen, M., Pietschmann, S., Ruch, T., Meißner, K.: Web-based support by thin-client co-browsing. In: Badr, Y., Chbeir, R., Abraham, A., Hassanien, A.E. (eds.) Emergent Web Intelligence: Advanced Semantic Technologies. Advanced Information and Knowledge Processing, pp. 395–428. Springer, London (2010). https://doi.org/10.1007/978-1-84996-077-9_15
10. Viegas, F.B., Wattenberg, M., Van Ham, F., Kriss, J., McKeon, M.: ManyEyes: a site for visualization at internet scale. IEEE Trans. Vis. Comput. Graph. **13**(6), 1121–1128 (2007)
11. Cruz-Neira, C., Sandin, D.J., DeFanti, T.A.: Surround-screen projection-based virtual reality: the design and implementation of the CAVE. In: Proceedings of the 20th Annual Conference on Computer Graphics and Interactive Techniques, pp. 135–142 (1993)
12. Czernuszenko, M., Pape, D., Sandin, D., DeFanti, T., Dawe, G.L., Brown, M.D.: The ImmersaDesk and Infinity Wall projection-based virtual reality displays. ACM SIGGRAPH Comput. Graph. **31**(2), 46–49 (1997)
13. Matsukura, R., Koyamada, K., Tan, Y., Karube, Y., Moriya, M.: VizGrid: collaborative visualization grid environment for natural interaction between remote researchers. FUJITSU Sci. Tech. J. **40**(2), 205–216 (2004)
14. Robinson, A.C., Weaver, C.: Re-visualization: interactive visualization of the process of visual analysis. In: Workshop on Visualization, Analytics & Spatial Decision Support at the GIScience Conference (2006)

15. Henry, N., Elmqvist, N., Fekete, J.-D.: A methodological note on setting-up logging and replay mechanisms in InfoVis systems. In: BELIV 2008 (2008)
16. Gerken, J., Bak, P., Jetter, C., Klinkhammer, D., Reiterer, H.: How to use interaction logs effectively for usability evaluation. In: BELIV (2008)
17. Vuillemot, R., Boy, J., Tabard, A., Perin, C., Fekete, J.-D.: Challenges in logging interactive visualizations and visualizing interaction logs. In: Proceedings of Workshop on Logging Interactive Visualizations and Visualizing Interaction Logs (2016)
18. Vuillemot, R., Boy, J., Tabard, A., Perin, C., Fekete, J.-D.: LIVVIL: logging interactive visualizations and visualizing interaction logs. In: Workshop IEEE VIS 2016, p. 33 (2016)
19. Sacha, D., Stoffel, A., Stoffel, F., Kwon, B.C., Ellis, G., Keim, D.A.: Knowledge generation model for visual analytics. IEEE Trans. Vis. Comput. Graph. **20**(12), 1604–1613 (2014)
20. Meta quest 2 picture. https://shop.ee.co.uk/accessories/pay-monthly/atp-meta-quest-2-128gb/details. Accessed 14 Apr

Hand Gesture Input Interface as Native Function of *IntelligentBox* Using Leap Motion Controller

Takumi Takeshita[1] and Yoshihiro Okada[1,2](✉)

[1] Graduate School of Information Science and Electrical Engineering, Kyushu University, Fukuoka, Japan
okada.yoshihiro.520@m.kyushu-u.ac.jp
[2] ICER of Kyushu University Library, Kyushu University, Fukuoka, Japan

Abstract. In this paper, the authors treat a native interface of *IntelligentBox* for two hands' gesture input. *IntelligentBox* is a 3D graphics software development system that provides 3D primitive components. Users can develop 3D applications as composite components composed from such 3D primitive components through direct manipulations of a mouse device on a computer screen. In the situation like COVID-19 pandemic, contactless input interfaces are significant. The authors have already proposed one dedicated component of *IntelligentBox* works as the input interface of two hands' gesture for its 3D applications. In this paper, the authors propose a new native interface for two hands' gesture input of *IntelligentBox* that allows users to create 3D applications without using any contact device like a mouse device.

Keywords: IntelligentBox · Development Systems · 3D-CG · VR · Gesture Input

1 Introduction

This paper treats contactless input interfaces because such interfaces are significant in the situation like COVID-19 pandemic. Especially, we propose a native interface for two hands' gesture input of *IntelligentBox* in this paper. *IntelligentBox* [1] is a 3D graphics software development system proposed in 1995. Although *IntelligentBox* is a very old system, our laboratory uses it as the platform for researches on 3D applications and we have developed many 3D applications actually using it so far. Users can develop 3D applications by composing composite components from 3D primitive components of *IntelligentBox* through direct manipulations using a mouse device on a computer screen. This is one of the remarkable points of *IntelligentBox*. Its application fields includes VR (Virtual Reality) [2–5]. The authors have already proposed OpenPose based action input interface [6] and Leap Motion Controller based gesture input interface [7] for VR applications of *IntelligentBox*. However, such interfaces are not native functions of *IntelligentBox*. When developing 3D applications, still users need to use a mouse device. This time, we implemented the new interface for two hands' gesture input using Leap

L. Barolli (Ed.): CISIS 2023, LNDECT 176, pp. 395–406, 2023.
https://doi.org/10.1007/978-3-031-35734-3_40

Motion Controller as a native function of *IntelligentBox*. Due to this function, users will come to develop 3D applications without using any contact device like a mouse device.

The rest of this paper follows: Sect. 2 describes related work. We introduce *IntelligentBox* and describe its essential mechanisms, are also introduce Leap Motion Controller in this Section. Section 3 describes the new interface of *IntelligentBox* realized as its native function for the input of two hands' gesture. Then, we discuss some problems in Sect. 4. Finally, we conclude the paper in Sect. 5.

2 Related Work

In 1995, we proposed *IntelligentBox* as a 3D graphics software development system. Its research aim was to establish a software architecture that allows users to develop 3D applications easier than ever because such development was tedious work needs much time. Related works of *IntelligentBox* are toolkit systems and programming libraries for 3D applications. Among of them are Open Inventor Toolkit [8], Coin3D [9] and so on. Open Inventor Toolkit and Coin3D are OpenGL based object-oriented programming libraries. Some of them provide a dedicated authoring tool that allows developers to create 3D applications. Even using such authoring tools, the development of practical complicated 3D applications still needs much time because they request developers to make text-based programs. For the development of 3D games, there are the two most popular game engines, e.g., Unreal Engine [10] and Unity [11]. Although these provide very powerful functions, basically developers need to make text-based programs.

Our *IntelligentBox* and *WebIB* (Web version of *IntelligentBox*) are component ware that provide developers with various 3D software components called *boxes*. Each *box* is implemented as a visible and manually operable functional 3D object. *IntelligentBox* also has a data linkage mechanism called slot connection to be described in the next section. These features of *IntelligentBox* and *WebIB* make it easier for even end-users to develop 3D applications including web 3D contents without text-based programs. This point is the main difference of *IntelligentBox* and *WebIB* from the others. Furthermore, in this paper, we propose the new input interface of two hands' gesture for making *IntelligentBox* more useful especially in the case like COVID-19 pandemic. There are many researches on gesture input interfaces. For instance, Nishino group proposed a real time motion generation system by human gesture [12]. Arrya Anandika, et al. proposed Virtual Keyboard [13] using Leap Motion Controller as a hand gesture input device. Xie Qingchao, et al. proposed Astronaut Virtual Training as one of the applications of hand gesture inputs using Leap Motion Controller [14]. Also, Jayash Kumar, et al. proposed an algorithm for Gesture and Hindi air writing recognition using Leap Motion Controller [15]. However, these are not researches on toolkit systems for developing 3D applications by hand gesture inputs like our *IntelligentBox*.

2.1 Essential Mechanism of *IntelligentBox*

Figure 1 shows a data-linkage mechanism called slot-connection, one of the essential mechanisms of *IntelligentBox*. As described in Sect. 1, 3D components of *IntelligentBox* are called *boxes*. Each *box* has multiple slots those are internal state variables of the *box* related to its functionality. Its one slot can be connected to one of the slots of the other *box*. This connection is called a slot connection. The slot connection means that the two state values of two *boxes* become always the same. This can be realized by the following messages those are a set message, a gimme (give me) message and an update message. These messages are sent when there is a parent-child relationship between the two *boxes*.

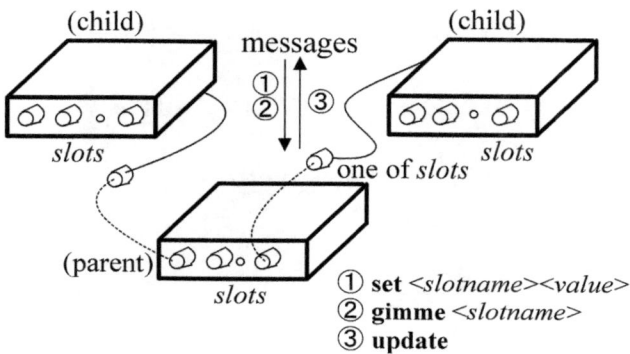

Fig. 1. Three standard messages between two *boxes*.

The message formats are below:

(1) set message: Parent *box* set <slotname> <value>, sent from child to parent.
(2) gimme message: Parent *box* gimme <slotname>, sent from child to parent
(3) update messages: Child *box* update, send from parent to all children

A <value> in a format (1) represents any value, and a <slotname> in formats (1) and (2) represents a user-selected slot of the parent *box* to whom these two messages are sent. Since a set message writes a child *box* slot value into its parent *box* slot, the child *box* works as an input component. Contrarily, since a gimme message reads a parent *box* slot value and sets it into its child *box* slot, the child *box* works as an output component. Update messages are sent from a parent *box* to all of its child *boxes* to tell them that the parent *box* slot value has changed. By these three messages, the two slots of a child *box* and its parent *box* are connected and their two functionalities are combined.

2.2　Leap Motion Controller

Figure 2 shows the photo of Leap Motion Controller, an input device that can capture hand and finger movements in real-time. The software development kit called Leap Motion SDK is available for C, C#, C++, Java, JavaScript, Python, and Objective-C (currently only Leap C and official C# are supported). The plug-ins for Unity and Unreal Engine are also available that converts the hand information into a cursor on a screen for manipulating desktop applications by the hand gesture [16]. In addition, when Leap Motion Controller is connected to a PC, a WebSocket server is always setup with port number 6437 on localhost. The server performs socket communication to output the hand information as a json data. Users can implement their own applications that supports the hand gesture input by accessing the server and getting the json data from it. Leap Motion SDK can acquire not only the coordinates of the palm of the hand and the tips of the fingers, but also the coordinates of their bones. In this research, we used this function to obtain the angle of the finger.

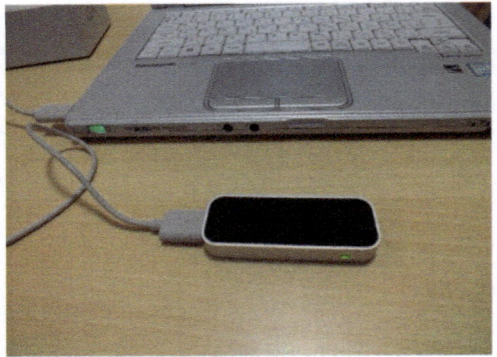

Fig. 2. Leap Motion Controller.

3   Hand Gesture Input Interface of *IntelligentBox* Using Leap Motion Controller

It was necessary to develop some programs for verifications of this function before implementing it. The programming language used for the development is C++.

3.1   Leap Motion Controller Used in *IntelligentBox*

As described in Sect. 2.2, there is a dedicated server program that handles Leap Motion Controller. By the client-server mechanism, the hand gesture data of Leap Motion Controller can be accessed to the server. The hand gesture data output from Leap Motion Controller server are provided to any applications, e.g., *IntelligentBox*, works as a client by accessing the server program as shown in Fig. 3.

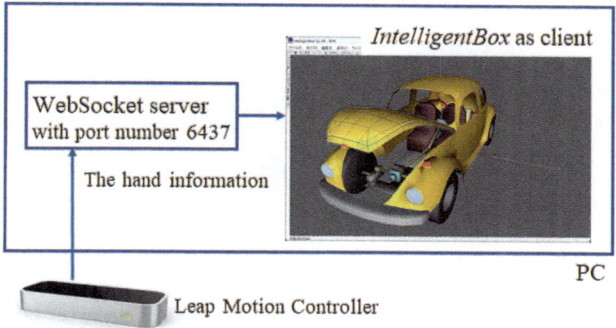

Fig. 3. Communication between the server program of Leap Motion Controller and *Intelligent-Box*.

3.2 Mouse Cursor Operations by Leap Motion Controller

3.2.1 Windows API

Using Windows API (Windows Application Programming Interface) by C and C++ programming language, it is possible to implement any GUI like *MessageBox* in the Windows OS environment. There are *Mouse_event* and *SetCursorPos* functions that can handle the input of a mouse device. *Mouse_event* function can handle the left and right button clicks and the rotation of the middle wheel. By specifying the action to be executed with *dwFlags* and other arguments such as the amount of movement of the mouse device and the amount of rotation of its middle wheel, these actions can be realized. The following is an example source code that perform a single left button click of a mouse device.

```
mouse_event(MOUSEEVENTF_LEFTDOWN, 0, 0, 0, 0);

mouse_event(MOUSEEVENTF_LEFTUP, 0, 0, 0, 0);
```

SetCursorPos is a function that can specify the position of the mouse cursor. By specifying the X and Y coordinates you want to move to X and Y respectively, you can move the mouse cursor to that position. Here, X and Y are the coordinates of the entire screen. If you want to specify the position within a current window, you need to use *GetWindowRect* function etc. to get the position and size of the window. The following is an example source code that moves the mouse cursor to the specified position (xPos, yPos) within the current window.

```
HWND leaphwnd = GetActiveWindow();

RECT windowrect;

GetWindowRect(leaphwnd, &windowrect);

int windowWidth = windowrect.right - windowrect.left;

int windowHeight = windowrect.bottom - windowrect.top;

xPos_abs = xPos + windowrect.left;

yPos_abs = yPos + windowrect.top;

SetCursorPos(xPos_abs, yPos_abs);
```

3.2.2 Tests for Mouse Cursor Operations

In operating the mouse cursor using the input of Leap Motion Controller, we tested which finger information corresponds to which operation.

First test is for the mouse cursor moving. Possible finger data for operating the mouse cursor include the position of the palm and the position of the tip of the index finger. If the position of the tip of the index finger is used for the position of the cursor, there is a possibility that problems such as hindering the subsequent left-click and right-click operations, or causing the cursor to blur due to the click action may occur. Therefore, in this research, X and Y coordinates of the palm of the right hand are employed as X and Y coordinates of the mouse cursor.

Second test is for left-clicking. Possible operations for the left-click include bending a finger and attaching any two types of fingers. In this research, the left-clicking is defined as the action of placing the right index finger tip and the right thumb together, which is easy to imagine from the action of clicking. Specifically, the distance is calculated using the coordinates of the tip of the index finger and thumb, and when the distance is shorter than a given value, it is assumed that the fingers are contacted.

Third test is for right-clicking. The right-click operation was performed by placing a finger in the same manner as the left-click. As the specification of Leap Motion Controller, close fingers such as the index finger and the middle finger may cause problems such as erroneously recognizing the middle finger as the index finger. Therefore, in order to prevent misrecognition as much as possible, in this research, a right click is generated when the right little finger and the right thumb are contacted.

3.2.3 Implementation of Mouse Cursor Operations

According to the tests, we implemented the program that converts the coordinates of the palm to the coordinates of the mouse cursor as follows:

```
int xPos = -PALM_X_MIN + min(max(rightHand.palmPosition().x,
  PALM_X_MIN), PALM_X_MAX);

int yPos = PALM_Y_MAX - min(max(rightHand.palmPosition().y,
  PALM_Y_MIN), PALM_Y_MAX);

HWND leaphwnd = GetActiveWindow();

RECT windowrect;

GetWindowRect(leaphwnd, &windowrect);

int windowWidth = windowrect.right - windowrect.left;

int windowHeight = windowrect.bottom - windowrect.top;

xPos = min(windowWidth - CURSOR_X_MAX, max(CURSOR_X_MIN, xPos *
  windowWidth / (PALM_X_MAX - PALM_X_MIN)));

yPos = min(windowHeight - CURSOR_Y_MAX, max(CURSOR_Y_MIN, yPos *
  windowHeight / (PALM_Y_MAX - PALM_Y_MIN)));

xPos_abs = xPos + windowrect.left;

yPos_abs = yPos + windowrect.top;

SetCursorPos(xPos_abs, yPos_abs);
```

First, get the coordinates of the four sides of the current window because the obtained palm coordinates are values between their maximum and minimum values. Next, calculate the width and height of the window using the coordinates of the four sides. Then, the cursor coordinates are calibrated to fit within the window by the following equation. After that, the coordinates are shifted by the position of the window and applied to the actual mouse cursor with the *SetCursorPos* function.

$$\frac{Palm\_X \times Width}{Palm\_X\_max - Palm\_X\_min} \tag{1}$$

$$\frac{Palm\_Y \times Width}{Palm\_Y\_max - Palm\_Y\_min} \tag{2}$$

Left and right clicks are implemented as follows:

```
double disLeft = sqrt(pow(rightBonePos[0][3].x -
  rightBonePos[1][3].x, 2) + pow(rightBonePos[0][3].y -
  rightBonePos[1][3].y, 2) + pow(rightBonePos[0][3].z -
  rightBonePos[1][3].z, 2));

double disRight = sqrt(pow(rightBonePos[0][3].x -
  rightBonePos[4][3].x, 2) + pow(rightBonePos[0][3].y -
  rightBonePos[4][3].y, 2) + pow(rightBonePos[0][3].z -
  rightBonePos[4][3].z, 2));

if (handtimer > 10) {// the time when recognizing the hand

  if (disLeft < 20) {

    if (!isLeftClick) { // left click

      mouse_event(MOUSEEVENTF_LEFTDOWN, 0, 0, 0, 0);

      isLeftClick = true;

    }

  }

  else if (isLeftClick) {

    mouse_event(MOUSEEVENTF_LEFTUP, 0, 0, 0, 0);

    isLeftClick = false;

  }

  if (disRight < 20) {

    if (!isRightClick) {// right click

      mouse_event(MOUSEEVENTF_RIGHTDOWN, 0, 0, 0, 0);

      isRightClick = true;

    }

  }

  else if (isRightClick) {

    mouse_event(MOUSEEVENTF_RIGHTUP, 0, 0, 0, 0);

    isRightClick = false;

  }

}
```

Using the coordinates of the tip of the thumb (*rightBonePos*[0][3]), the coordinates of the tip of the index finger (*rightBonePos*[1][3]), and the coordinates of the tip of the little finger (*rightBonePos*[4][3]), it is possible to find the distance between the index finger, thumb and little finger by the following expression. If the click judgment starts just after the hand is recognized, the judgment may become wrong when the posture of the fingers is not recognized sufficiently. To prevent this, the variable handtimer is set to delay the start of the judgment. After that, when the distance between the thumb and index finger or little finger is shorter than a given distance, the mouse_event function will be invoked as the left/right clicking, and when the distance exceeds the given distance, it will be treated as the mouse release.

$$\sqrt{(x_1 - x_2)^2 + (y_1 - y_2)^2 + (z_1 - z_2)^2} \tag{3}$$

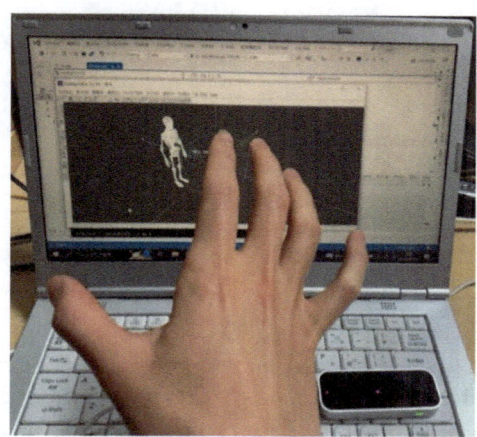

Fig. 4. Mouse cursor movement by hand movement.

3.2.4 Results of Mouse Cursor Operations

In order to confirm that the implemented program works as expected one, we operated the cursor with Leap Motion Controller using the following procedure.

- Connect Leap Motion Controller to PC using a USB cable.
- Start *IntelligentBox*.
- Select (Connect with Leap Motion Controller) from the (Leap) tab.

This will start the listener and make it ready to receive input from Leap Motion Controller. In this state, if you hold your right hand over Leap Motion Controller, on *IntelligentBox*. You can see that the mouse cursor moves as shown in Fig. 4. You can see that the hand is moved to the lower left of Leap Motion Controller, and the mouse cursor on *IntelligentBox* is also moved to the lower left of the window.

In addition, left-clicking can be performed by putting the index finger and thumb of the right hand together as shown in the left part of Fig. 5. In the right part of Fig. 5, you can see that the menu for right-clicking can be displayed by putting the thumb and little finger together.

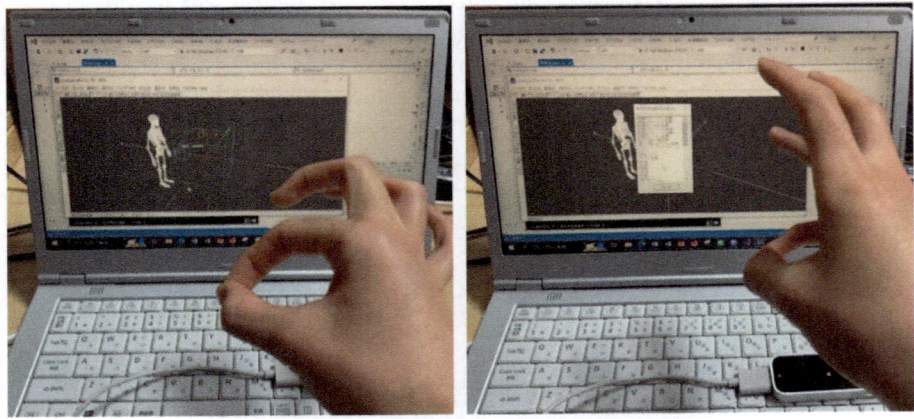

Fig. 5. Left-clicking (left) and right-clicking (right).

4   Discussion

As described in Sect. 3, it can be said that the attempt to introduce Leap Motion Controller into *IntelligentBox* and operate *IntelligentBox* with fingers was successful. However, in operating the mouse cursor, when you want to click a small button or one like it, the cursor may blur, making it difficult to align. Possible causes of this problem include the fact that the buttons are small in the first place, blurring occurs due to precision problems with Leap Motion Controller, and when you move your finger with a click action, the position of your palm also moves along with it. To solve this problem, we need to display a large virtual button on the screen of *IntelligentBox* and make it play the same role as a small button, and keep the sensor part clean to improve the accuracy of Leap Motion Controller. If we use the average of the coordinates of several frames and move the cursor based on this average value, the affection of this problem becomes little, or we should consider a different click operation method.

Also, regarding the click operation, it may not work well if the hand is too far from or too close to Leap Motion Controller. In this case, you cannot get the finger information well. In order to solve this, a method such as adjusting the maximum and minimum values of the palm coordinates obtained from Leap Motion Controller in the program should be considered.

5   Conclusion

In this research, as a method of operating *IntelligentBox* by non-contact gesture input, we developed a system that can operate mouse cursor by finger movement using Leap Motion Controller. Although there are still many issues such as the accuracy of movement and examination of operation methods, the results are promising in terms of enabling application development using non-contact fingers using Leap Motion Controller on *IntelligentBox* in the future.

As for future prospect, although it is currently being operated only with the right hand, the left hand can be used. The left hand will be able to operate the mouse wheel, click, etc. We will consider an operation method using both hands. Also, by introducing Leap Motion Controller as Virtual Mouse Pointers [2, 3], it is possible to develop 3D applications more intuitively using the fingers without any contact-type device. Since the range of operations is wider than the operation with only the mouse cursor, we believe that introducing Leap Motion Controller to Virtual Mouse Pointer will lead to more practical use of *IntelligentBox*. In the future, all *IntelligentBox* operations can be performed by Leap Motion Controller, so that all operations for creating applications can be performed with only uses' fingers without using any contact-type device such as a mouse device or data gloves. If this can be realized and *IntelligentBox* can be put into practical use, even school teachers and researchers with no programming knowledge will be able to develop 3D applications relatively easily.

Acknowledgements. This research was partially supported by JSPS KAKENHI Grant Number JP22H03705.

References

1. Okada, Y., Tanaka, Y.: IntelligentBox: a constructive visual software development system for interactive 3D graphic applications. In: Proceedings of Computer Animation 1995, pp. 114–125 (1995)
2. Okada, Y.: 3D visual component-based approach for immersive collaborative virtual environments. In: Proceedings of the 2003 ACM SIGMM Workshop on Experiential Telepresence, ETP 2003, pp. 84–90 (2003)
3. Okada, Y.: IntelligentBox as component based development system for body action 3D games. In: ACM SIGCHI International Conference on Advances in Computer Entertainment Technology (ACE 2005), pp. 454–457 (2005)
4. Okada, Y., Ogata, T., Matsuguma, H.: Component-based approach for prototyping of Tai Chi-based physical therapy game and its performance evaluations. ACM Comput. Entertain. **14**(1), 4:1–4:20 (2016)
5. Okada, Y., Kaneko, K., Fujibuchi, T.: IntelligentBox based training system for operation of radiation therapy devices. In: Barolli, L., Poniszewska-Maranda, A., Enokido, T. (eds.) CISIS 2020. AISC, vol. 1194, pp. 188–198. Springer, Cham (2021). https://doi.org/10.1007/978-3-030-50454-0_18
6. Yu, B., Shi, W., Okada, Y.: Action input interface of IntelligentBox using 360-degree VR camera and OpenPose for multi-persons' collaborative VR applications. In: Barolli, L., Yim, K., Enokido, T. (eds.) CISIS 2021. LNNS, vol. 278, pp. 747–757. Springer, Cham (2021). https://doi.org/10.1007/978-3-030-79725-6_75

7. Takeshita, T., Kaneko, K., Okada, Y.: Hand gesture input interface of IntelligentBox using leap motion controller and its application example. In: Barolli, L. (ed.) BWCCA 2021. LNNS, vol. 346, pp. 139–147. Springer, Cham (2021). https://doi.org/10.1007/978-3-030-90072-4_14

8. Open Inventor Toolkit. https://openinventor.com. Accessed Apr 2023

9. Coin3D. https://www.coin3d.org/. Accessed Apr 2023

10. Unreal Engine. https://www.unrealengine.com/. Accessed Apr 2023

11. Unity. https://unity.com/. Accessed Apr 2023

12. Haramaki, T., Goto, K., Tsutsumi, H., Yatsuda, A., Nishino, H.: A real-time robot motion generation system based on human gesture. In: Barolli, L., Leu, F.Y., Enokido, T., Chen, H.C. (eds.) BWCCA 2018. LNDECT, vol. 25, pp. 135–146. Springer, Cham (2019). https://doi.org/10.1007/978-3-030-02613-4_12

13. Anandika, A., Rusydi, M.I., Utami, P.P., Hadelina, R., Sasaki, M.: Hand gesture to control virtual keyboard using neural network. JITCE (J. Inf. Technol. Comput. Eng.) **7**(1), 40–48 (2023)

14. Xie, Q., Chao, J.: The application of leap motion in astronaut virtual training. IOP Conf. Ser.: Mater. Sci. Eng. **187**, 012015 (2017). https://doi.org/10.1088/1757-899X/187/1/012015

15. Sharma, J.K., Gupta, R., Sharma, S., Pathak, V., Sharma, M.: Highly accurate Trimesh and PointNet based algorithm for Gesture and Hindi air writing recognition (2023). https://doi.org/10.21203/rs.3.rs-2702018/v1

16. Ultraleap. https://www.ultraleap.com/product/leap-motion-controller/. Accessed Apr 2023

Author Index

Printed by Printforce, the Netherlands